Smart IoT System for Renewable Energy Resource

Smart IoT System for Renewable Energy Resource

Editors

Antonio Cano-Ortega
Francisco Sánchez-Sutil

MDPI • Basel • Beijing • Wuhan • Barcelona • Belgrade • Novi Sad • Cluj • Manchester

Editors

Antonio Cano-Ortega
University of Jaen
Jaen
Spain

Francisco Sánchez-Sutil
University of Jaen
Jaen
Spain

Editorial Office
MDPI
St. Alban-Anlage 66
4052 Basel, Switzerland

This is a reprint of articles from the Special Issue published online in the open access journal *Sensors* (ISSN 1424-8220) (available at: https://www.mdpi.com/journal/sensors/special_issues/iotenergy).

For citation purposes, cite each article independently as indicated on the article page online and as indicated below:

Lastname, A.A.; Lastname, B.B. Article Title. *Journal Name* **Year**, *Volume Number*, Page Range.

ISBN 978-3-0365-9182-7 (Hbk)
ISBN 978-3-0365-9183-4 (PDF)
doi.org/10.3390/books978-3-0365-9183-4

Contents

About the Editors

Antonio Cano-Ortega

Antonio Cano-Ortega received their M.S. degree in Industrial Engineering (Electrical) from the National Distance Education University UNED (Spain) in 2000 and Ph.D. Industrial Engineeringdegree from the UNED (Spain) in 2004. He has been an Associate Professor since 1996 in the Department of Electrical Engineering, University of Jaen. His research activities have been devoted to several topics, including power systems, smart grids and renewable energy. He is the author of more than 10 papers in journals included in the Journal Citation Report (JCR), and about 15 papers in the proceedings of International Conferences and several chapter books. He has been involved in research projects funded by Spanish Ministries.

Francisco Sánchez-Sutil

Francisco Sánchez-Sutil received the M.S. degree in Industrial Organization Engineering from the University of Jaen (Spain) in 2007 and Ph.D. degree from the University of Jaen (Spain) in 2016. He has been an Associate Professor since 1999 in the Department of Electrical Engineering, University of Jaen. His research activities have been devoted to several topics, including smart grids and renewable energy. He is the author of 27 papers in journals included in the Journal Citation Report (JCR) and about 6 papers in the proceedings of International Conferences. He has been involved in research projects funded by Spanish Ministries.

Preface

Distributed generation sources use renewable sources within the electricity systems to supply electricity to households, industry, etc. Distributed generation must be controlled and monitored for more efficient use.

Renewable sources allow clean energy to be supplied close to the points of consumption, thus reducing losses and improving energy efficiency.

IoT systems can be applied within smart grids to improve the operation of electricity systems. These systems consist of sensors, actuators, and communications equipment, allowing real time data to be obtained. IoT applications in smart grids can improve the operation of electricity grids and monitor the different magnitudes in real time.

The integration of renewable sources such as photovoltaic, wind, hydro and biomass in electricity systems reduces the greenhouse gas effect and emissions of polluting gases into the atmosphere, so their integration into smart grids is essential. In this sense, the development of applications for monitoring and control of renewable sources are fundamental tools and allow for better development of electricity systems.

The development of smart meters allows measurements of electrical variables to be taken and data to be sent via communications such as Wi-Fi, Bluetooth, Ethernet, etc., to the cloud, where they are subsequently processed and analysed. In this sense, both consumers and generation companies can analyse consumption at all times and create more efficient consumption patterns.

The development of electric vehicles and their integration into smart grids requires the development of IoT systems to monitor and control the charging of electric vehicles and their integration into the V2G grid or V2H home. This equipment has the mission to measure key parameters of the electrical system and to analyse the quality of the power supply.

Wireless networks are a fundamental part of IoT systems. It is necessary to take measurements in remote and difficult-to-access areas. The most commonly used wireless networks are Wi-Fi, Bluetooth, ZigBee, LoRa, SigFox, Nb-IoT, Bluetooth LE, etc.

Therefore, this book includes recent advances in the development and implementation of IoT devices applied to renewable energy for use in smart grids, so that readers can become familiar with new methodologies directly explained by experts in this scientific field.

Antonio Cano-Ortega and Francisco Sánchez-Sutil
Editors

sensors

MDPI

Article

Design and Implementation of a Cloud-IoT-Based Home Energy Management System

Felipe Condon [1], José M. Martínez [1], Ali M. Eltamaly [2,3], Young-Chon Kim [4,*] and Mohamed A. Ahmed [1,*]

1 Department of Electronic Engineering, Universidad Técnica Federico Santa María, Valparaíso 2390123, Chile
2 Electrical Engineering Department, Faculty of Engineering, Mansoura University, Mansoura 35516, Egypt
3 Sustainable Energy Technologies Center, King Saud University, Riyadh 11421, Saudi Arabia
4 Department of Computer Engineering, Jeonbuk National University, Jeonju 561-756, Republic of Korea
* Correspondence: yckim@jbnu.ac.kr (Y.-C.K.); mohamed.abdelhamid@usm.cl (M.A.A.)

Abstract: The advances in the Internet of Things (IoT) and cloud computing opened new opportunities for developing various smart grid applications and services. The rapidly increasing adoption of IoT devices has enabled the development of applications and solutions to manage energy consumption efficiently. This work presents the design and implementation of a home energy management system (HEMS), which allows collecting and storing energy consumption data from appliances and the main load of the home. Two scenarios are designed and implemented: a local HEMS isolated from the Internet and relies on its processing and storage duties using an edge device and a Cloud HEMS using AWS IoT Core to manage incoming data messages and provide data-driven services and applications. A testbed was carried out in a real house in the city of Valparaiso, Chile, over a one-year period, where four appliances were used to collect energy consumption using smart plugs, as well as collecting the main energy load of the house through a data logger acting as a smart meter. To the best of our knowledge, this is the first electrical energy dataset with a 10-second sampling rate from a real household in Valparaiso, Chile. Results show that both implementations perform the baseline tasks (collecting, storing, and controlling) for a HEMS. This work contributes by providing a detailed technical implementation of HEMS that enables researchers and engineers to develop and implement HEMS solutions to support different smart home applications.

Keywords: home energy management system; smart home; Internet of Things; cloud infrastructure

Citation: Condon, F.; Martínez, J.M.; Eltamaly, A.M.; Kim, Y.-C.; Ahmed, M.A. Design and Implementation of a Cloud-IoT-Based Home Energy Management System. *Sensors* **2023**, *23*, 176. https://doi.org/10.3390/s23010176

Academic Editors: Antonio Cano-Ortega and Francisco Sánchez-Sutil

Received: 25 November 2022
Revised: 18 December 2022
Accepted: 19 December 2022
Published: 24 December 2022

1. Introduction

Nowadays, the applications of the Internet of Things (IoT) are appearing in different domains, such as transportation [1], healthcare [2], agriculture [3], and power systems [4]. These IoT solutions aim to monitor and control various elements and devices in different scenarios that will ease tasks and provide useful applications for daily living [5]. In the electric power system, energy plays a central role in powering our homes and appliances. However, the metering process for estimating the consumption of a house is widely dependent on an electromechanical energy meter, which implies that utility companies have to employ personnel to perform the metering tasks monthly in order to bill their customers [6]. As for the appliance consumption within a house, residents may not be aware of individual power consumption for each appliance, therefore facing inefficient energy usage without even knowing. In this regard, Chile has set a goal to have around 6.5 million smart meters installed by 2025 [7]. This goal focuses mainly on using smart meters to lower energy demand by providing more detailed energy billing and interfaces of energy consumption through web interfaces and applications, as well as implementing new tariff systems.

IoT will play a key role in enabling several energy efficiency mechanisms, such as the Internet of Energy, smart grids, and smart homes. This is possible by using digital sensors and communication devices that enable a home energy management system (HEMS), which allows continuous consumption monitoring and appliance control, as well as supporting the communication between the utility and the power grid [8]. Data are collected using IoT devices and later transferred to the cloud-based system infrastructure from where it is stored and processed [9]. Data-driven applications, databases, and file storage systems are key features that can be designed and deployed in cloud-based infrastructure to support the IoT cloud-based requirement for several energy Internet applications.

The design and implementation of a proper architecture are the main challenges for enabling applications based on IoT-collected data on a global level. Several works have proposed HEMS-IoT architectures to solve these challenges. The common criterion used to define an architecture is the data processing and storage location. Three layers are found in the literature where data processing and storage can occur: edge, fog, and cloud.

In this paper, an overview of the HEMS-related work is presented, focusing on key elements, such as the design of a HEMS architecture enabled by IoT devices and the usage of local and cloud computing for data storage and processing. We propose a Cloud-IoT based home energy management system, which helps residents, landlords, researchers, and administrators manage the energy consumption within a house. The proposed HEMS implements a four-layer architecture, which is capable of collecting and storing energy consumption data. Consumption data are obtained from two kinds of devices: smart meters and smart plugs. The smart meter units are able to collect the energy consumption of the entire house, while smart plugs are capable of collecting energy consumption and controlling the power supplied to a single appliance. Two implementations were carried out following two different approaches. In one approach, a local HEMS was isolated from the Internet with a central processing unit. In the other approach, a cloud-based implementation used the cloud for data storage and processing. Both systems provide baseline features, such as collecting measurements from the devices, storing them in a database or performing load control actions, such as turning on/off a device manually or by scheduling. Results show the capability of both systems to perform the collection and storing features for energy consumption and to control the appliances by using smart plugs. The main contributions of this work can be summarized as follows:

- Design and implementation of a four-layer HEMS architecture that allows collecting and storing energy consumption data from appliances and the main load of the home.
- Two scenarios are designed and implemented: a local HEMS isolated from the Internet that relies on its processing and storage duties using an edge device and a Cloud HEMS using AWS IoT.
- To the best of the authors knowledge, this work is the first electrical energy dataset with a 10-second sampling rate from a real household in the city of Valparaiso, Chile, over a one-year period.
- Detailed technical implementation of HEMS will enable researchers and engineers to develop and implement HEMS solutions to support different smart home applications.

The rest of the paper is structured as follows: In Section 2, related work for HEMS is explained. Section 3 presents the proposed HEMS architecture. Section 4, presents two case studies: a local HEMS, and a Cloud HEMS. Section 5 elaborates on the technical details for the implementation of both systems. Section 6 discusses the results from the case studies and the future challenges for HEMS implementations. Finally, Section 7 concludes the paper.

2. Related Work

Several applications benefit from energy consumption data obtained from HEMS and appliance control. One example is peer-to-peer (P2P) energy trading which flips the traditional scenario, where electricity is transmitted from large-scale generators to consumers over long distances, while the cash flow goes in the opposite way. In contrast, P2P energy trading encourages multi-directional trading within a local geographical area [10]. With the increasing connection of distributed energy resources (DER), traditional energy consumers are becoming prosumers who can consume and generate energy [11].

Authors in [12] analyzed several IoT applications for smart grids, such as smart homes, smart metering, and energy management, among others. Challenges, issues, and future research regarding the use of IoT to enable Energy Internet (EI) applications were also discussed. A smart home incorporates various IoT-based smart technologies with the goal of providing security, convenience, comfort, energy efficiency, and entertainment which results in improving the quality of life within a residence. Ambient assisted living service, smart energy management technology service, and security are the predominant technology services associated with smart homes [13].

In [14], the authors presented an overview of IoT-enabled energy systems. Some of the outlined challenges include mapping every object into a unique virtual object which can be addressed with standard communication protocols. The authors also stated that given the variety of design decisions made by the system designers, there are different architectures to enable an IoT-based energy system, which implies that there is no unified architecture. In [15], the authors presented a smart load node (SLN) for enabling non-smart home appliances to operate efficiently in a smart grid paradigm. SLN is an innovative solution given that it does not require any modifications in the electrical wiring of a house, nor any modification on the appliances. SLN integrates within a HAN with other devices, such as smart meters and a load management unit (LMU), which enable various smart grid applications within a house, such as scheduling loads in a demand–response (DR) scheme. Authors in [16] presented a novel methodology including the concept of green building in order to reduce energy consumption. A key element stated by the authors is not only regarding the energy efficiency for appliances and at home but also to create awareness among residents on power conservation.

Authors in [8] presented a survey on HEMS which provides an aggregated and unified perspective on residential buildings. An overview of the literature on commonly managed household appliances was also presented. Home energy management systems (HEMS) aim to improve efficiency by providing control of smart home appliances, and this is feasible due to the use of the Internet of Things (IoT) [8]. HEMS relies on smart sensors, appliances, and advanced metering infrastructure (AMI) to achieve continuous monitoring. Authors in [16] presented a survey on the concepts, technical background, architecture, and infrastructure, among other challenges and issues regarding HEMS. The use of IoT will allow any smart device, also known as the "things" to interact with one or several sensors and other devices in the network, forming a wireless sensor network (WSN). This WSN can rely on a gateway for Internet connection, allowing the implementation of applications based on the collected data. Authors in [9] presented an overview, architecture, and implementation of IoT in energy systems. The HEMS follows some baseline features, such as monitoring of the main load of the property, individual loads of appliances, and control of appliances. Regarding the components that comprise a HEMS, such a system possesses sensing and measuring devices, smart appliances, a user interface, and a central platform.

Due to the services provided by public clouds, there has been an increasing interest in developing data-driven applications. Some of the data-driven applications that can be implemented in the context of a smart home are alarms on irregular load scenarios and scheduling the use of appliances in case of dynamic tariff systems. In [17], the authors presented a comparison between three cloud platforms: Amazon, Google, and Microsoft. MQTT messaging was used by IoT devices to send information to the cloud platforms, where a performance evaluation was carried out, not to benchmark the maximum message throughput, but rather to measure the service time of the provided message broker. Cost comparison and description of available tiers were also discussed.

Authors in [18] developed a demand response (DR) application on a HEMS in order to reduce utility operational costs and the consumer energy bill price. The proposed infrastructure by the authors is based on an edge–fog–cloud computing architecture, which allows for monitoring and control of residential loads. The testbed was carried out using Raspberry Pi as the HEMS and NodeMCU ESP8266 as smart plugs for energy-related measurements and controlling tasks. Results showed that the proposed system was able to schedule loads and reduce the energy bill when compared to the scenario without the DR algorithm. The proposed testbed scenario considered a dynamic tariff system. Another energy management system (EMS) was proposed by authors in [19], where a system was implemented at the IoT Microgrid Laboratory at Aalborg University. The IoT-based EMS showed the feasibility of using IoT devices to regulate consumption. Features, such as energy management using load priority, were presented in the results.

Authors in [20] presented a cloud-based platform that collects electricity consumption, indoor climate, and occupancy data in real-time using sensors. The energy monitoring platform was implemented in a smart villa. The architecture showed the devices' interaction over a star topology. The system utilizes ThinkEE, a cloud platform for connecting IoT devices. It also provided a web interface for data display and an energy management system for energy control. Authors in [21] described the building operation data, which includes electricity consumption and environmental measurements. The work provided information regarding the architecture of the system, which utilizes EMU, smart meters, and sensors for collecting the data. The data include one-minute interval measurements from 1 July 2018 to 31 December 2019 which are provided to support a variety of data-driven applications. In [22], the authors released I-BLEND, a 52-month electrical energy dataset at a one-minute sampling rate from commercial and residential buildings of an academic institute campus. The data collection of the system was carried out using a Raspberry Pi to collect measurements from smart meters, while also using the cloud for data storage and processing. The authors in [23] , introduced Plug-Mate, an IoT-based occupancy-driven plug load management system. Plug-Mate was able to deliver occupancy information, plug load type, and plug load usage preference. The solution was tested during a 5-month study in a university office with 10 participants. Results showed about 51.7% in the overall energy saving improvement among different plug loads and about 7.5% reduction in the building overall energy consumption.

Although previous research provided information on how to implement IoT-based HEMS and how to apply data-driven algorithms, such as demand response and load schedule, most of the solutions were implemented in a laboratory environment that does not represent the actual condition of a smart home. This work aims to fill the knowledge gap by providing a detailed description of the technical implementation of how to design and implement a HEMS that can be used for different applications. Two architectures are considered for local/cloud implementation using available IoT devices while deploying the system on a public cloud. Table 1 shows the comparison among previous research work.

Table 1. Comparison among previous research work.

Ref.	Year	Type	Description
[8]	2020	Survey	A survey on home energy management including main goals for operation and target strategies
[9]	2021	Survey	Comprehensive study of IoT business applications and smart energy systems
[10]	2018	Technical/Simulation	P2P energy trading was designed and simulated for energy trading among prosumers and consumers in a microgrid
[11]	2014	Technical/Simulation	Energy trading among prosumers in a microgrid to increase the utilization of renewable energy
[12]	2019	Survey	Comprehensive survey on IoT applications for smart grid and smart environments
[13]	2021	Review	Literature review on smart home adoption including motivations, barriers, and risks
[14]	2018	Review	Review on IoT-based energy system with respect to features, specifications, communication infrastructures, and privacy
[15]	2019	Technical/ Implementation	Design and implementation of a low-cost smart load node for monitoring and control non-smart residential load

Table 1. *Cont.*

Ref.	Year	Type	Description
[16]	2022	Review	Comprehensive review for home energy management system with respect to concepts, architecture infrastructure, and challenges
[17]	2020	Technical/Simulation	Performance analysis among three different Cloud-IoT platforms services for Amazon web service, Microsoft Azure, and Google Cloud
[18]	2021	Technical/Implementation	IoT-based infrastructure on edge–fog–cloud architecture to monitor and control residential loads to support demand response
[19]	2019	Technical/Implementation	IoT-based infrastructure for EMS. The system has been tested in a pilothouse named IoT Microgrid Living Lab, Denmark
[20]	2019	Technical/Implementation	Energy monitoring platform to collect real-time electricity consumption data in a smart villa, Doha Qatar
[21]	2020	Technical/Implementation	Detailed building operation data (electricity consumption and indoor environment) of seven-story building in Bangkok, Thailand
[22]	2019	Technical/Implementation	Electrical energy dataset (52 months) from commercial and residential building at one minute sampling rate, India
[23]	2022	Technical/Implementation	IoT-based plug load management system capable of providing occupancy and energy consumption information for smart building, Singapore
This work	2022	Technical/Implementation	Design and implementation of two HEMS architectures (local vs. cloud) in a real household environment located in Valparaiso, Chile

3. Cloud-IoT Home Energy Management System

Figure 1 shows the proposed Cloud-IoT HEMS architecture. It consists of four layers: perception layer, communication layer, middleware layer, and application layer.

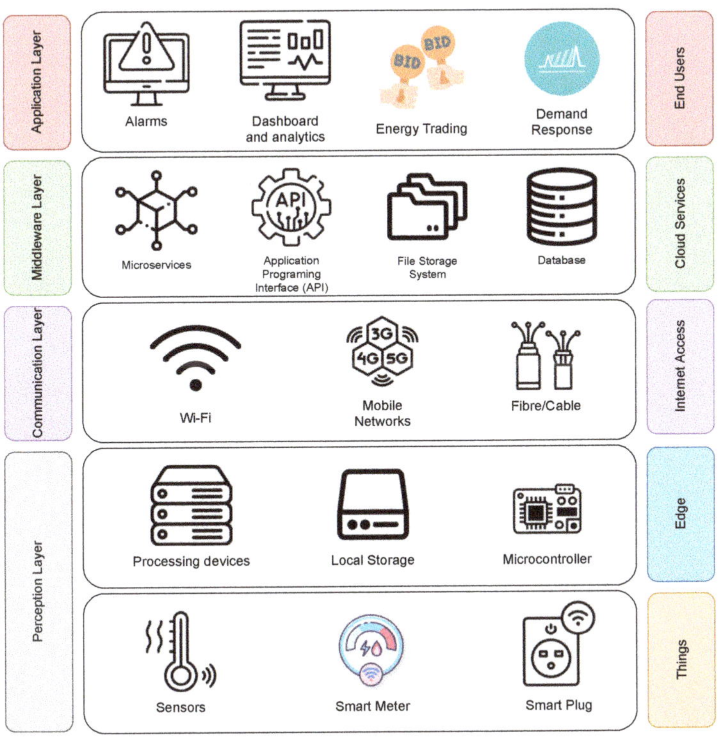

Figure 1. Cloud-IoT HEMS architecture.

3.1. Perception Layer

The perception layer enables data collection through sensors, as well as data storage and data processing tasks through edge devices. This layer considers several physical devices, such as sensors, smart meters, and smart plugs, which are often referred to as "things". Edge devices enable tasks, such as data storage, data processing, and performing actions, on things that are located in the edge sub-layer. The perception layer is a key element in the proposed architecture, working as the first step in the energy data collecting process and as the last step in the appliance controlling process.

- Things Layer: The things layer is composed of sensors and actuators. Sensors are used to collect information on important measurements of the smart home. Sensors, such as temperature, humidity, and light detection, among others, are used for comfort purposes. Regarding appliances' power consumption, two devices are introduced: smart meters and smart plugs. A smart meter is able to collect energy consumption information of the main loads of the smart home. It utilizes a current transformer (CT) and a voltage sensor to compute the power consumption. Smart plugs (SP) work as a middleware between the power line of the house and the appliance plug. The purpose of SPs is to collect energy consumption data from a specific appliance while also being able to control the energy delivered to the appliance using a relay.
- Edge Layer: The edge is the closest to the sensors and actuators. It provides the capability to collect data and perform command operations over "things". The edge layer considers a low latency but limited data storage and processing capabilities as available resources. From a HEMS perspective, the edge layer is found at the house level, where it is able to use this layer as a perception layer to collect the data of interest.

3.2. Communication Layer

This layer allows the connection of the perception layer with the middleware layer. Several communications technologies could be used to perform this task, among them WiFi, Zigbee, LoRa, Fiber, and mobile networks such as 4G and 5G. The election of the technology has various attributes, such as effective range (short range and long range), cost, coverage, and availability of a given communication system. An important role of the network layer is providing a home area network (HAN) for perception layer devices to join. The HAN allows communication between devices, also known as machine-to-machine (M2M) communication, which allows routing the data from the things to the edge or from the edge to the middleware layer.

3.3. Middleware Layer

The middleware layer is related to cloud-based services that rely on the received data to perform some tasks. The most common functions of a HEMS consider storing measurements in a database, performing data processing tasks through microservices, and providing an application programming interface (API) for managing data requests. These services can be implemented by using a public cloud, such as Amazon Web Services (AWS), Google Cloud Platform (GCP), and Microsoft Azure (Azure). Given the functionality, the middleware layer is similar to the edge layer but provides a scalable infrastructure to serve higher traffic and more demanding processing and storage tasks. Several technology stacks are available for developers to carry out the required implementation to achieve the abovementioned tasks. Some challenges and decisions that need to be addressed in order to provide a robust system consider database election, designing and deploying the required cloud architecture, programming language, and/or framework election. The middleware in this architecture serves as a bridge between the data perspective provided by the "things" and the intended smart home features or applications identified in the application layer.

3.4. Application Layer

A HEMS aims to provide data to several domains, such as Energy Internet, smart grids, and smart homes. Some data-driven applications for HEMS are, for example, demand response, P2P energy trading, and monitoring energy consumption for user awareness. These applications serve as the last layer in the proposed architecture which is the closest to the end user.

4. Home Energy Management System Design

Two case studies are proposed to compare different approaches for HEMS: a local HEMS and a cloud-based HEMS. On the one hand, a local HEMS system uses a central computing unit to handle data storage and processing tasks. On the other hand, a cloud-based system collects the data through a gateway and is able to enable several solutions and applications.

Both systems are designed to meet the following tasks, which enable HEMS to perform some features of a smart home:

- Monitoring of the energy consumption: HEMS should store data about the power consumption of the appliances monitored by the smart plugs and the main load of the smart home.
- Appliance control: HEMS should allow a resident to interact with the appliances connected to smart plugs to supply or deny energy.

In this work, the smart home includes some features that are present on both systems as a requirement. We recognize two categories that are common ground for the implementation of such systems:

- The things: Both case studies utilize the same end devices: smart plugs and smart meters. The task of the smart meter is to collect the total power consumption reading of the house. On the other hand, a smart plug is able to collect the power consumption and control the supplied power of a single appliance.
- Networking and communications: From a communication perspective, we considered that the smart home has WiFi capabilities, enabling devices to interact within a home area network (HAN). Devices such as smart plugs and smart meters have the capability to join the HAN using WiFi technology. An Internet Protocol (IP) address is assigned to each device that joins the network. The things have the capability to periodically send measurements of the energy consumption telemetry to the destination. Direct energy consumption requests can also be performed, following the messaging protocol that each device supports. The things send the collected data using MQTT, a popular Pub/Sub (publisher/subscriber) messaging protocol, where each device sends data over a unique topic.

Figure 2 shows a schematic diagram for HEMS design. On the left side, the system is composed of home appliances, a smart meter, smart plugs, and a computing unit serving as the local HEMS. This approach considers a solution that relies solely on monitoring and managing house appliance consumption in a local scenario. Such solution presents benefits from a privacy perspective, given that it is isolated from the Internet. The main drawback of this kind of approach is the limited computational and storage capacity of the HEMS, which is directly related to the provisioned on-premise hardware. From the Edge perspective, the devices such as smart plugs and smart meters are intended to collect energy measurements based on the energy consumption of certain appliances and the total energy consumption of the main home, respectively.

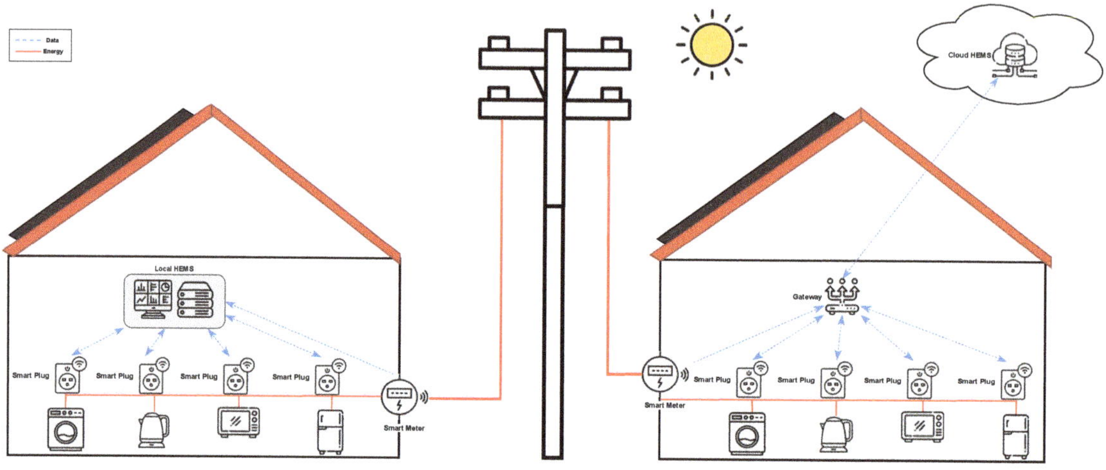

Figure 2. Schematic diagram for HEMS. **Left**: Local HEMS design, **Right**: Cloud HEMS design.

The design of the Cloud-IoT HEMS is presented on the right side of Figure 2, where the energy and data interactions are shown. Smart meters and smart plugs allow the collection of energy-related data. The main objective is to send the energy consumption data to the cloud. For this purpose, the things send information to a gateway, which allows bridging the data into the cloud.

These data are stored and processed by databases and cloud microservices. Several data-driven HEMS applications can be designed and implemented. One benefit of using public cloud services is the infinite scaling opportunity, which allows applications to scale as needed. Cloud providers manage the hardware and configurations required to enable any architecture to function appropriately, easing the development experience. This allows for cheaper costs when comparing cloud-based implementations to on-premise implementations for a scalable system.

5. Implementation

The testbed was carried out in a real house located in the city of Valparaiso, Chile. A wireless local area network (WLAN) was set up by a router that provides WiFi connectivity for all devices located in the house. The energy consumption data was acquired starting from July 2021 up to September 2022. Four appliances were used: a kettle, a washing machine, a refrigerator, and a microwave.

The selected smart plugs were the Sonoff POW R2, which were modified to fit inside a small electrical box, as seen in Figure 3a. This configuration provides one outlet for connecting the appliance and one plug for connecting the unit to the power outlet of the home. Each appliance was connected to a smart plug, as seen in Figure 3b. These devices were configured using the open-source firmware Tasmota. This firmware allows for an easier development experience enabling easier management of the configurations of smart plugs. The web interface was used to set up the WiFi connection with the HAN.

On the smart plugs, MQTT was set using the web user interface (UI) by specifying the required connection parameters, as seen in Figure 3c. Each smart plug uses a unique MQTT topic to publish and/or subscribe to messages. The Sonoff devices were configured to send one measurement every 10 s using the telemetry feature within Tasmota. Message data were transmitted using JavaScript Object Notation (JSON) structure. The topic structure for smart plugs is "tele/device-ID/sensor", which represents the telemetry event sent by a specific device informing its sensors data collection.

(**a**) (**b**) (**c**)

Figure 3. Sonoff POW R2 as smart plug, overview, usage, and configuration. (**a**) Smart plug unit; (**b**) Two smart plug units connected to a kettle (**left**) and a microwave (**right**); (**c**) MQTT configuration on Tasmota web interface.

For the smart meter, the eGauge EG4115 unit was selected, as seen in Figure 4a. This device uses current transformer (CT) sensors, as seen in Figure 4b, and voltage sensors to compute the energy consumption of the house with frequencies of one measurement per second. This device was connected via Ethernet to the router, joining the local area network (LAN) and obtaining an IP address. This device comes with XML API within its firmware, which enables any device in the network to request measurements of the unit.

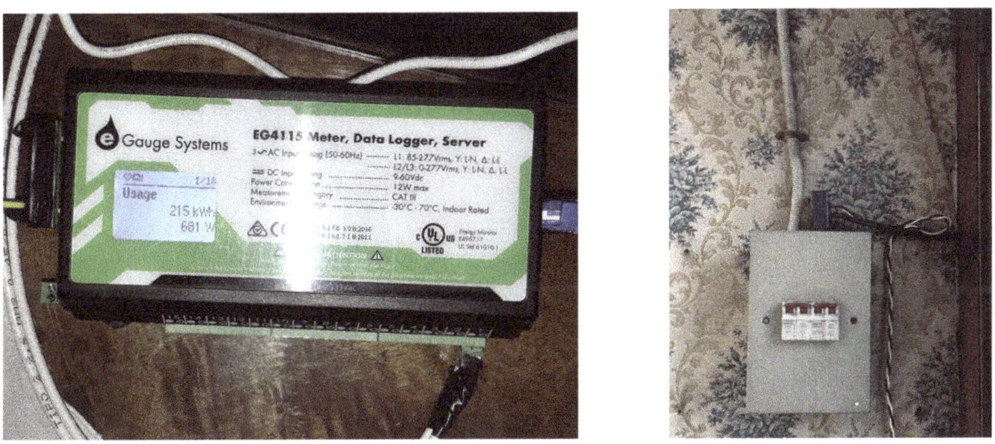

(**a**) (**b**)

Figure 4. eGauge EG4115 data logger used as smart meter to collect energy consumption data from the main load of the house.(**a**) eGauge data logger unit showing instant power consumption; (**b**) CT Sensor (blue device) used to collect the current of the main load of the house.

Messages sent by smart plugs and the smart meter are composed of several attributes that compose the payload of the message. Table 2 presents the data structured object provided on the event emission by the smart plug and the smart meter. These messages are structured in a data object based on a key-value pair structure. The attribute column in the table references the key of the data object, while the type column provides the data type of the value associated with the attribute. The description and measuring unit are provided for all attributes on the smart plug and the smart meter.

Table 2. The things, smart plug, and smart meter energy data collection attributes on messages.

Device	Attribute	Type	Description	Unit
Sonoff POW R2 Smart Plug	TotalStartTime	S	Starting timestamp for period computations	Date
	Total	N	Total Energy usage including Today	kWh
	Yesterday	N	Total Energy usage between 00:00 and 24:00 yesterday	kWh
	Today	N	Total Energy usage today from 00:00 until now	kWh
	Period	N	Energy usage between previous message and now	Wh
	Power	N	Current effective power load	W
	Apparent Power	N	Power load on the cable	W
	Reactive Power	N	Reactive load	W
	Factor	N	Power factor of the load	-
	Voltage	N	Current line voltage	V
	Current	N	Current line current	A
eGauge EG4115 Smart Meter	Date & Time	N	Timestamp of the current measurement	Unix Timestamp
	Usage	N	Current effective power load	kW
	Current	N	Current line current	A
	Voltage	N	Current Line Voltage	V
	Factor	N	Current Power factor	-

5.1. HEMS Local Implementation

The local implementation was carried out using a Raspberry Pi as the local HEMS, as shown in Figure 5. A Raspberry Pi was configured with Raspberry Pi OS-Lite, which is a Linux distribution developed to serve as the suggested operating system (OS).

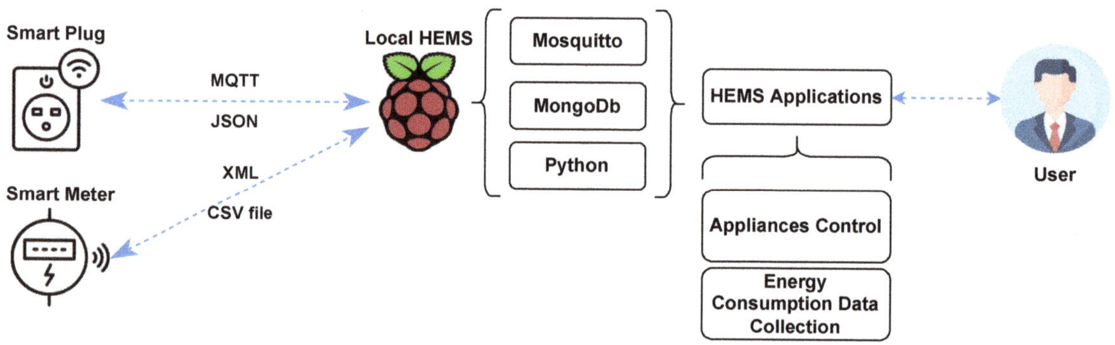

Figure 5. Schematic diagram for local HEMS implementation.

Several services were configured in the Raspberry Pi to allow this unit to establish MQTT communication, provide a database to store measurements, and perform lightweight processing tasks. Mosquitto is an open-source MQTT broker, which was used in the Raspberry Pi to participate in a publish–subscribe messaging scheme with the things. The MQTT broker was set up on port 1883.

MongoDB is a non-relational database that allows storing data in a JSON structure. A single MongoDB instance was set up on the Raspberry Pi on port 27017. Python is a high-level scripting language that was used to automatize the process of storing a document in the database for each measurement received. Three scripts were developed: the first one for collecting the measurements of the smart plugs, the second one to collect measurements of the smart meter, and the last one to handle appliance control commands. The first script uses the paho MQTT client, which connects to the MQTT broker and subscribes to each smart plug MQTT topic. PyMongo, a MongoDB client, is used to perform operations in the database. For each message received by the broker, the content of the message is saved into the database.

For collecting the smart meter measurements, the second Python script uses requests, an HTTP client for Python, to request the last four hours of energy consumption recorded by the eGauge unit. This request is fulfilled by a CSV file that contains the requested data. The file is saved in local storage within the Raspberry Pi. This script is scheduled to be performed every four hours using a cron job. Please note that the time of "4 h" could be adjusted based on application requirements. The last script provides the capability to send commands to a specific appliance topic for turning the relay of the smart plugs ON or OFF.

5.2. HEMS Cloud-Based Implementation

The Cloud HEMS system was implemented following the structure presented in Figure 6. A Raspberry Pi was configured as an MQTT broker using mosquitto. This broker was used as a host by the things to send the telemetry events using MQTT on individual topics. The MQTT broker was set up in a bridge mode, allowing it to route the inbound data to a new destination provided by the cloud provider.

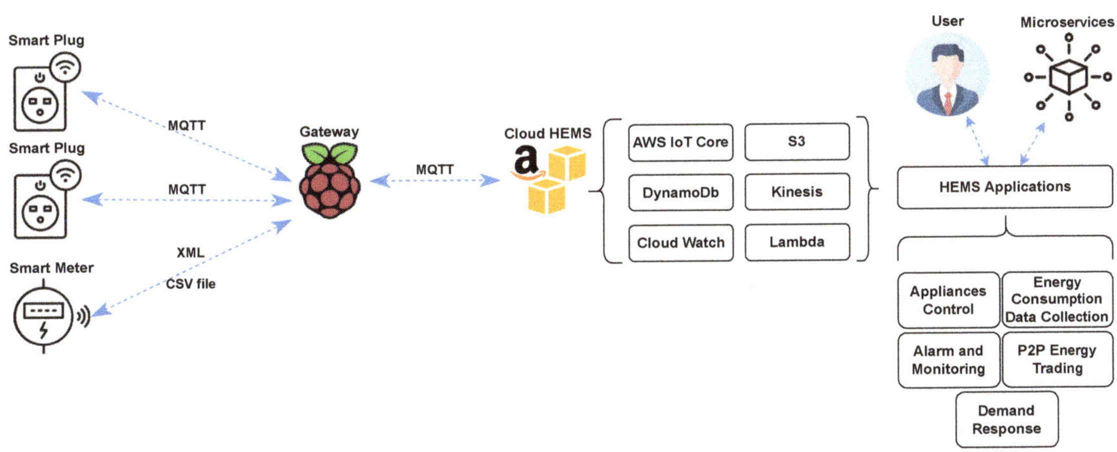

Figure 6. Schematic diagram for Cloud HEMS implementation.

AWS IoT Core was set up using the AWS console. The IoT Core provides an MQTT broker, which is able to manage MQTT connections from smart homes. IoT Core also allows setting up triggers for executing events when data is received. The implemented triggers and actions are presented in Figure 7.

Two rules are set in order to perform some actions. The first one considers that all the incoming messages that match the topic structure "tele/+/sensor" will be logged into AWS CloudWatch. The plus sign "+" is an MQTT single-level wildcard on AWS IoT Core rules that matches any value in that position; in this case, it represents the device ID from the smart plug unit. The rest of the actions are executed only when the second rule is met, which implies that the power consumption informed by the received message is greater than zero. This rule was set to optimize the storage and costs of the implemented system. When the rules are not matched by the incoming MQTT messages, these are discarded. Regarding the actions, the Cloud HEMS stores the data using DynamoDB, where smart plug data are stored in a JSON object using the schema and attributes presented in the data payload in Table 2. For storing the smart meter data, an S3 bucket is used to handle the CSV files provided by the unit over MQTT. Kinesis Firehouse is used to ingest the received MQTT message into a data stream enabling further processing, data events, and real-time-based applications.

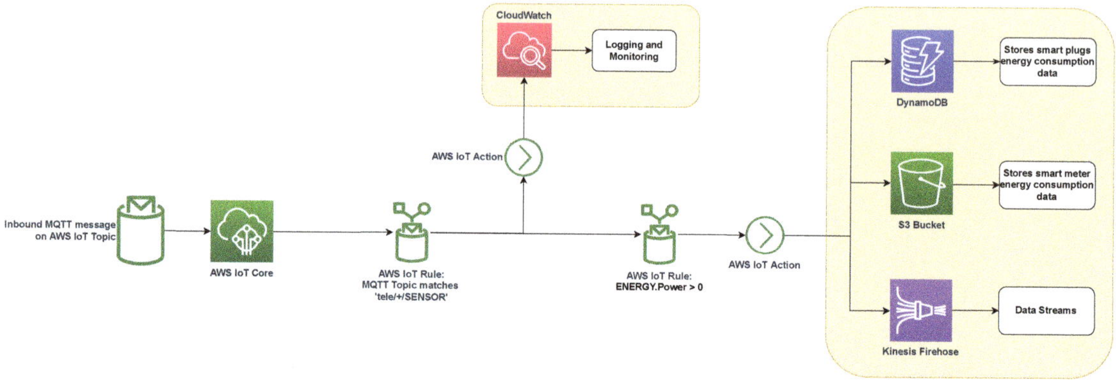

Figure 7. AWS IoT Core implemented rules and actions.

5.3. Case Study 1: Local HEMS Results

For the local HEMS, results show the capability of the system to store the energy consumption data of the appliances and from the main load of the house. Figure 8 shows a snapshot of the MongoDB collection that contains the measurements from the appliances collected by the smart plugs.


```
_id: ObjectId('6085b97000d06695bedf3ce3')
Time: "2021-04-25T14:48:16"
v ENERGY: Object
    TotalStartTime: "2021-03-29T20:20:27"
    Total: 3.623
    Yesterday: 0.002
    Today: 0.003
    Period: 3
    Power: 1662
    ApparentPower: 1673
    ReactivePower: 194
    Factor: 0.99
    Voltage: 225
    Current: 7.431
appliance: "kettle"
```


Figure 8. Database record of smart plug measurement of the kettle.

Each document contains the energy-related parameters for each telemetry event. The document contains one parameter that allows for appliance identification. By using the ID of the smart plug, it is labeled as the correct appliance with the same name attribute in the database document. This parameter is configured in the collection as an index, which allows for faster queries when used as a filter parameter. The records collected by the smart meter are shown in Figure A1, where they are stored as CSV files in a directory.

5.4. Case Study 2: Cloud HEMS Results

Cloud HEMS was deployed starting in July 2021, during which data were collected and stored in the cloud. The AWS S3 web panel allows checking the individual files stored by the system for the smart meter measurements, as seen in Figure A2. Each file is around 3MB, and it contains a four-hour window of measurements taken by the smart meter every second. These files follow the structured data provided by the eGauge unit.

Batch processes can be scheduled to perform analytics operations over the stored data. Figure 9 shows a pie chart for the energy consumption per appliance for the first week of August 2022. This information intends to serve the residents as feedback on their behavior regarding the energy consumption of each appliance.

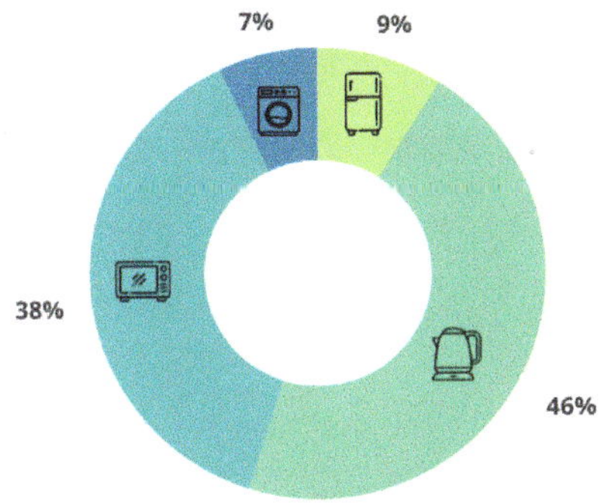

Figure 9. Energy consumption per appliance for resident awareness.

One feature that benefits from data analysis services provided by AWS in the cloud is shown in Figure A3, where an email notification alerts when the system has not detected any writing activity in the database. This enables the residents to be aware of appliance malfunction, energy blackouts, connectivity issues, or any other problem regarding the energy consumption in the home.

Figure 10 shows the data availability of the system regarding the collected data from the devices. The x-axis presents the available time period since the system started recording data from July 2021 up to the end of the study (grouped by months). The four appliances and the main load of the home are shown on the Y-axis. The values shown for each device and the month represent the percentage number of messages received from each device during a certain month over the total amount of messages that could be received considering the sample rate for each device. This graph shows that in some periods, such as August 2022, there were some constraints on receiving data from the house. This happened due to a system malfunction, blackouts, or scheduled maintenance from the power utility provider.

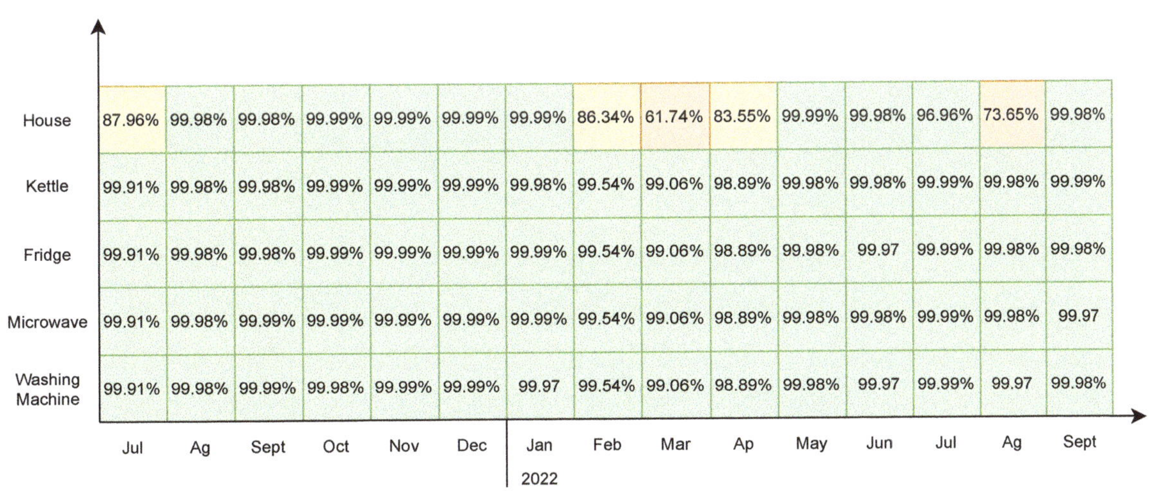

Figure 10. Analysis of system availability for the collected data.

Regarding data visualization, Figure 11 shows the main load of the house collected by the eGauge unit. The period shows the regular power consumption over the month of August with a gap of data. In Figure 12, a window of two hours shows the specific energy consumption for the kettle, refrigerator, microwave, washing machine, and the main load of the house on 2 August 2022. A slice of the data collected from our real house implementation with some instructions regarding the implementation of the HEMS can be found at our repository of the project https://github.com/pipegreyback/EViG-Server, (accessed on 1 October 2022).

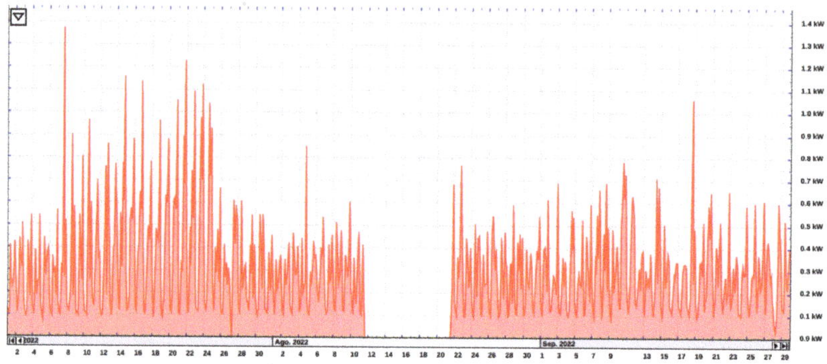

Figure 11. Total power consumption of the main load of the house using the eGauge unit during the period of July, August, and September 2022.

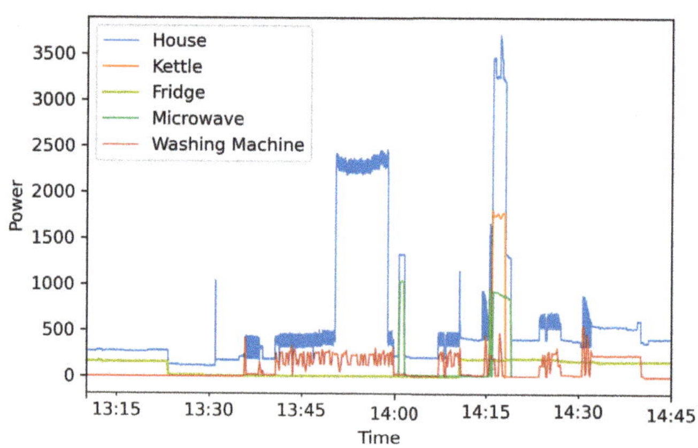

Figure 12. Power consumption data from the four appliances using smart plugs and the main load of the house.

6. Discussion

Data-driven applications provide an enormous benefit in various fields. In distribution power systems, the new applications of the Energy Internet, smart grids, and smart homes will allow, for example, disaggregated consumption per appliance using the total power consumption of the home. Some other applications allow load scheduling for appliances. These applications require reliable architecture and implementation that allow the collection, storage, and processing of such data. Such a challenge has been discussed widely, but there is no one standardized architecture that fits all solutions, thus the architects and developers need to identify the requirements of the system in order to design a reliable architecture that allows the system to function properly. Some of the considerations will enforce the data storage and processing to be deployed locally at the edge or remotely in the cloud. Nevertheless, a system such as the proposed HEMS could present different requirements when multiple houses implement the system. There are other constraints, such as a massive amount of data sent to the cloud from individual houses, which could result in high latency and degradation of the service. Thus, solutions that use a fog layer as data concentrators within a neighborhood, such as Neighborhood Area Network (NAN) should be considered when looking to deploy a HEMS in various houses within a neighborhood.

New tools such as AWS Cloud Development Kit (CDK) for speeding the provisioning process of services appear as an interesting alternative to provisioning infrastructure through web interfaces, such as the AWS console. AWS CDK follows the paradigm of Infrastructure as Code (IaaC) which enables the developers to write code in several programming languages such as Typescript, Python, and Go. The code represents the required services to be provisioned, with the respective configurations which are then synthesized and deployed into AWS through cloud formation.

6.1. Technology Adoption

The acquisition of dedicated sensors and actuators with Internet capabilities and IoT devices provide multiple benefits for different data-driven applications, but it comes with the challenge of technology adoption. In [24,25], the authors investigated the user perception on acceptance of monitoring energy consumption devices such as smart meters. In [24], the authors reviewed consumer beliefs regarding smart meters using behavioral decision research. Results showed that consumers are positively predisposed toward smart meters; however, the authors proposed recommendations for electric utilities in order to address misconceptions about smart meters' benefits and concerns over risks. Even though the use of smart energy management systems in the residential context provided energy savings, increasing user acceptance has been a challenge over field implementations [25]. The authors identified seven high-level categories based on a mixed-method approach providing a more holistic understanding of users' perception of smart energy management system adoption.

6.2. Implications of the Proposed Solution and Future Directions

The design of HEMS still entails a degree of uncertainty, given that developers and architects might propose different approaches on how to carry out such a system. This is due to multiple factors, such as the variety of IoT-enabled devices, networking and communication protocols, and data resolution requirements. The usage of IoT devices such as smart plugs, a common component of both implementations, provides the benefit of enabling existing households to benefit from smart grid applications without interventions of the property. Our implementation was able to perform a set of features available in HEMS for a real house in Valparaiso, Chile. However, another direction is the integration of other communication protocols such as Zigbee or LoRa. Such solutions will enable our system to support heterogeneous IoT systems, which provide flexibility while choosing IoT end devices such as smart meters and smart plugs.

7. Conclusions

In this paper, we designed and implemented two HEMS solutions that are able to store power consumption and control appliances by using edge (local) devices or the cloud. The proposed architecture consists of four layers: perception layer, communication network layer, middleware layer, and application layer. The local HEMS is isolated from the Internet and utilizes an edge device to serve as the main processing unit. The cloud-based HEMS utilizes a gateway to send the data to the cloud. Both implementations are driven by IoT devices to send data measurements or receive control signals. We reviewed, designed, and implemented the most common approaches on state-of-the art edge (local) HEMS and cloud-based HEMS. Both systems have some common features of HEMS; however, they differ in terms of privacy and scalability. In this regard, new challenges appear when multiple HEMSs need to be deployed in a community or a neighborhood area network. A hybrid approach could enable a more reliable and integral solution than using edge devices or the cloud as individual systems. Our ongoing work considers extending the developed HEMS to support different applications, such as energy disaggregation, anomaly detection, demand response, and peer-to-peer energy trading, further extending the system capabilities to enable real-time data processing applications through data streams.

Author Contributions: Conceptualization, M.A.A., J.M.M. and F.C.; methodology, M.A.A., J.M.M., F.C., A.M.E. and Y.-C.K.; software, F.C.; validation, M.A.A., J.M.M. and F.C.; formal analysis, M.A.A., J.M.M., F.C., A.M.E. and Y.-C.K.; investigation, M.A.A., J.M.M., F.C., A.M.E. and Y.-C.K.; writing—original draft preparation, F.C. and M.A.A.; writing—review and editing, M.A.A., J.M.M., F.C., A.M.E. and Y.-C.K.; funding acquisition, M.A.A. and Y.-C.K. All authors have read and agreed to the published version of the manuscript.

Funding: This work was supported in part by the Agencia Nacional de Investigación y Desarrollo (ANID) through the Proyecto Fondecyt de Iniciación en Investigación 2020 under Project ID11200178, DGIIP-UTFSM Chile, and in part by the National Research Foundation of Korea (NSF) Grant through the Korean Government under Grant (2021R1I1A305872911).

Institutional Review Board Statement: Not applicable.

Informed Consent Statement: Not applicable.

Data Availability Statement: Not applicable.

Acknowledgments: The authors would like to acknowledge Proyecto Fondecyt de Iniciacion en Investigacion 2020 under Project ID11200178, DGIIP-UTFSM Chile, and National Research Foundation of Korea (NSF) Grant through the Korean Government under Grant (2021R1I1A305872911).

Conflicts of Interest: The authors declare no conflict of interest.

Appendix A

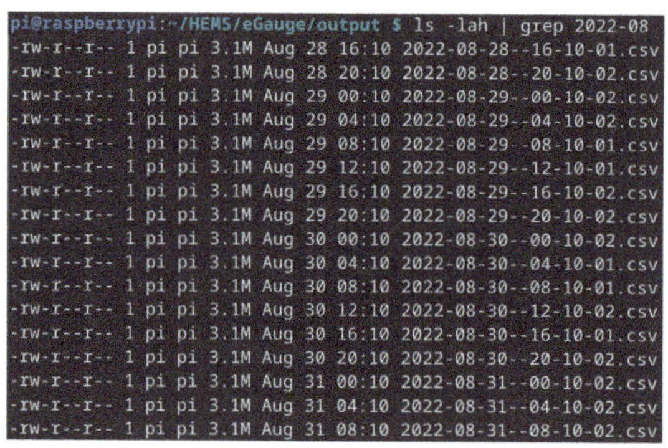

Figure A1. Smart Meter energy records from eGauge unit.

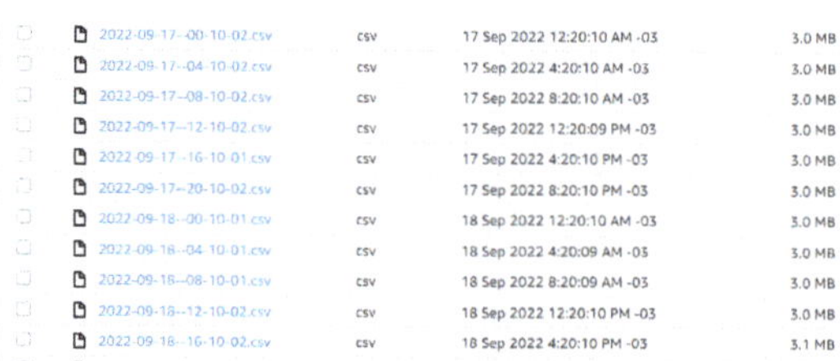

Figure A2. AWS S3 bucket with energy consumption records from smart meter.

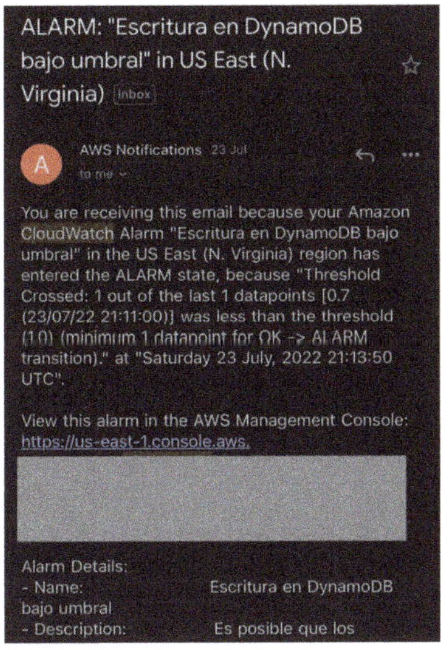

Figure A3. Email notification when a smart plug has not write into the database in a fixed period of time.

References

1. Luo, X.G.; Zhang, H.B.; Zhang, Z.L.; Yu, Y.; Li, K. A New Framework of Intelligent Public Transportation System Based on the Internet of Things. *IEEE Access* **2019**, *7*, 55290–55304. [CrossRef]
2. Bhuiyan, M.N.; Rahman, M.M.; Billah, M.M.; Saha, D. Internet of Things (IoT): A Review of Its Enabling Technologies in Healthcare Applications, Standards Protocols, Security, and Market Opportunities. *IEEE Internet Things J.* **2021**, *8*, 10474–10498. [CrossRef]
3. Friha, O.; Ferrag, M.A.; Shu, L.; Maglaras, L.; Wang, X. Internet of Things for the Future of Smart Agriculture: A Comprehensive Survey of Emerging Technologies. *IEEE/CAA J. Autom. Sin.* **2021**, *8*, 718–752. [CrossRef]
4. Ratasich, D.; Khalid, F.; Geissler, F.; Grosu, R.; Shafique, M.; Bartocci, E. A Roadmap Toward the Resilient Internet of Things for Cyber-Physical Systems. *IEEE Access* **2019**, *7*, 13260–13283. [CrossRef]
5. Bellini, P.; Nesi, P.; Pantaleo, G. IoT-Enabled Smart Cities: A Review of Concepts, Frameworks and Key Technologies. *Appl. Sci.* **2022**, *12*, 1607. [CrossRef]
6. Avancini, D.B.; Rodrigues, J.J.; Martins, S.G.; Rabêlo, R.A.; Al-Muhtadi, J.; Solic, P. Energy meters evolution in smart grids: A review. *J. Clean. Prod.* **2019**, *217*, 702–715. [CrossRef]
7. Biblioteca del Congreso Nacional de Chile. Experiencia Comparada Instalación de Medidores Inteligentes. 2019. Available online: https://obtienearchivo.bcn.cl/obtienearchivo?id=repositorio/10221/27088/1/BCN___Experiencia_comparada_medidores_inteligentes_25Marzo_edPM.pdf (accessed on 10 January 2022).
8. Leitão, J.; Gil, P.; Ribeiro, B.; Cardoso, A. A Survey on Home Energy Management. *IEEE Access* **2020**, *8*, 5699–5722. [CrossRef]
9. Ahmad, T.; Zhang, D. Using the Internet of things in smart energy systems and networks. *Sustain. Cities Soc.* **2021**, *68*, 102783. [CrossRef]
10. Zhang, C.; Wu, J.; Zhou, Y.; Cheng, M.; Long, C. Peer-to-Peer energy trading in a Microgrid. *Appl. Energy* **2018**, *220*, 1–12. [CrossRef]
11. Luo, Y.; Itaya, S.; Nakamura, S.; Davis, P. Autonomous cooperative energy trading between prosumers for microgrid systems. In Proceedings of the 39th Annual IEEE Conference on Local Computer Networks Workshops, Edmonton, AB, Canada, 8–11 September 2014; pp. 693–696.
12. Kabalci, Y.; Kabalci, E.; Padmanaban, S.; Holm-Nielsen, J.B.; Blaabjerg, F. Internet of Things Applications as Energy Internet in Smart Grids and Smart Environments. *Electronics* **2019**, *8*, 972. [CrossRef]
13. Li, W.; Yigitcanlar, T.; Erol, I.; Liu, A. Motivations, barriers and risks of smart home adoption: From systematic literature review to conceptual framework. *Energy Res. Soc. Sci.* **2021**, *80*, 102211. [CrossRef]
14. Shakerighadi, B.; Anvari-Moghaddam, A.; Vasquez, J.C.; Guerrero, J.M. Internet of Things for Modern Energy Systems: State-of-the-Art, Challenges, and Open Issues. *Energies* **2018**, *11*, 1252. [CrossRef]
15. Singh, S.; Roy, A.; Selvan, M. Smart Load Node for Nonsmart Load Under Smart Grid Paradigm: A New Home Energy Management System. *IEEE Consum. Electron. Mag.* **2019**, *8*, 22–27. [CrossRef]
16. Mahapatra, B.; Nayyar, A. Home energy management system (HEMS): Concept, architecture, infrastructure, challenges and energy management schemes. *Energy Syst.* **2022**, *13*, 643–669. [CrossRef]

17. Pierleoni, P.; Concetti, R.; Belli, A.; Palma, L. Amazon, Google and Microsoft Solutions for IoT: Architectures and a Performance Comparison. *IEEE Access* **2020**, *8*, 5455–5470. [CrossRef]
18. Veloso, A.F.d.S.; de Moura, M.C.L.; Mendes, D.L.d.S.; Junior, J.V.R.; Rabêlo, R.A.L.; Rodrigues, J.J.P.C. Towards Sustainability using an Edge-Fog-Cloud Architecture for Demand-Side Management. In Proceedings of the 2021 IEEE International Conference on Systems, Man, and Cybernetics (SMC), Melbourne, Australia, 17–20 October 2021; pp. 1731–1736. [CrossRef]
19. Palacios-Garcia, E.J.; Arbab-Zavar, B.; Vasquez, J.C.; Guerrero, J.M. Open IoT Infrastructures for In-Home Energy Management and Control. In Proceedings of the 2019 IEEE 9th International Conference on Consumer Electronics (ICCE-Berlin), Berlin, Germany, 8–11 September 2019; pp. 376–379. [CrossRef]
20. van Cutsem, O.; Bayram, I.S.; Maher, K.; Fosse, J.C. Demonstration of a Smart Villa Energy Monitoring Platform in Qatar. In Proceedings of the 2019 UK/ China Emerging Technologies (UCET), Glasgow, UK, 21–22 August 2019; pp. 1–4. [CrossRef]
21. Pipattanasomporn, M.; Chitalia, G.; Songsiri, J.; Aswakul, C.; Pora, W.; Suwankawin, S.; Audomvongseree, K.; Hoonchareon, N. CU-BEMS, smart building electricity consumption and indoor environmental sensor datasets. *Sci. Data* **2020**, *7*, 241. [CrossRef] [PubMed]
22. Rashid, H.; Singh, P.; Singh, A. I-BLEND, a campus-scale commercial and residential buildings electrical energy dataset. *Sci. Data* **2019**, *6*, 190015. [CrossRef] [PubMed]
23. Tekler, Z.D.; Low, R.; Yuen, C.; Blessing, L. Plug-Mate: An IoT-based occupancy-driven plug load management system in smart buildings. *Build. Environ.* **2022**, *223*, 109472. [CrossRef]
24. Krishnamurti, T.; Schwartz, D.; Davis, A.; Fischhoff, B.; de Bruin, W.B.; Lave, L.; Wang, J. Preparing for smart grid technologies: A behavioral decision research approach to understanding consumer expectations about smart meters. *Energy Policy* **2012**, *41*, 790–797. [CrossRef]
25. Tekler, Z.D.; Low, R.; Blessing, L. User perceptions on the adoption of smart energy management systems in the workplace: Design and policy implications. *Energy Res. Soc. Sci.* **2022**, *88*, 102505. [CrossRef]

sensors

MDPI

Article

Toward an Intelligent Campus: IoT Platform for Remote Monitoring and Control of Smart Buildings

Mohamed A. Ahmed [1,*], Sebastian A. Chavez [1], Ali M. Eltamaly [2,3], Hugo O. Garces [4], Alejandro J. Rojas [5] and Young-Chon Kim [6,*]

1 Department of Electronic Engineering, Universidad Técnica Federico Santa María, Valparaíso 2390123, Chile
2 Electrical Engineering Department, Faculty of Engineering, Mansoura University, Mansoura 35516, Egypt
3 Sustainable Energy Technologies Center, King Saud University, Riyadh 11421, Saudi Arabia
4 Departamento Ingeniería Informática, Universidad Católica de la Santísima Concepción, Concepción 4090541, Chile
5 Departamento Ingeniería Eléctrica, Universidad de Concepción, Victor Lamas 1290, Concepción 4070386, Chile
6 Department of Computer Engineering, Jeonbuk National University, Jeonju 561-756, Republic of Korea
* Correspondence: mohamed.abdelhamid@usm.cl (M.A.A.); yckim@jbnu.ac.kr (Y.-C.K.)

Abstract: With the growing need to obtain information about power consumption in buildings, it is necessary to investigate how to collect, store, and visualize such information using low-cost solutions. Currently, the available building management solutions are expensive and challenging to support small and medium-sized buildings. Unfortunately, not all buildings are intelligent, making it difficult to obtain such data from energy measurement devices and appliances or access such information. The internet of things (IoT) opens new opportunities to support real-time monitoring and control to achieve future smart buildings. This work proposes an IoT platform for remote monitoring and control of smart buildings, which consists of four-layer architecture: power layer, data acquisition layer, communication network layer, and application layer. The proposed platform allows data collection for energy consumption, data storage, and visualization. Various sensor nodes and measurement devices are considered to collect information on energy use from different building spaces. The proposed solution has been designed, implemented, and tested on a university campus considering three scenarios: an office, a classroom, and a laboratory. This work provides a guideline for future implementation of intelligent buildings using low-cost open-source solutions to enable building automation, minimize power consumption costs, and guarantee end-user comfort.

Keywords: intelligent campus; smart building; internet of things platform; remote monitoring and control

Citation: Ahmed, M.A.; Chavez, S.A.; Eltamaly, A.M.; Garces, H.O.; Rojas, A.J.; Kim, Y.-C. Toward an Intelligent Campus: IoT Platform for Remote Monitoring and Control of Smart Buildings. *Sensors* **2022**, *22*, 9045. https://doi.org/10.3390/s22239045

Academic Editors: Antonio Cano-Ortega and Francisco Sánchez-Sutil

Received: 22 October 2022
Accepted: 18 November 2022
Published: 22 November 2022

1. Introduction

The Chilean new energy efficiency law No. 21,305 was published on 13 February 2021, establishing Chile's national energy efficiency plan. Among the main aspects of energy efficiency that the new law targets are the national energy efficiency plan, energy management of large consumers, energy labeling/rating for buildings, and efficiency standards for vehicles [1]. Considering that buildings are responsible for about 60% of the total global electricity consumption [2], information on energy usage is fundamental for the development of different energy management system (EMS) solutions [3–23]. In particular, such solutions are critical for those users in charge of buildings administration. Knowing the different equipment power consumptions is vital for controlling the expenses associated with building operations.

The applications of IoT in smart homes [9–12,14] and smart buildings [2–8,13,15–23] have been discussed in many publications covering different domains, including surveys [6,7,10–12,15,18], architectures [5], frameworks [8,14], platforms [9], and algorithms [16].

However, most of the research is based on assumptions or simulations which ignore the practical issues from the real implementation. Examples of real implementation can be found in [3,4,9,12,19]. Furthermore, the condition of the electricity system differs between countries and regions, which then requires identifying the specific requirements and needs for each design. This work aims to fill the gap related to technical implementation in a realistic environment. The main objective of this work is to design and implement an IoT platform that integrates information from various smart sensor nodes and measuring devices connected to buildings for obtaining information related to power consumption in order to support energy management solutions. All collected data will be processed to deliver the relevant information to the end user, who can then perform remote actions accordingly. The main activities required to achieve the proposed objective are:

- Define the system requirements for intelligent buildings and prioritize these requirements for developing the solution, including data collection from different sensor nodes and measurement devices.
- Build the IoT platform for smart buildings, which includes four main layers: power layer, data acquisition layer, communication network layer, and application layer.
- Implement the back-end and front-end systems. The proposed solution involves the development of a network of sensors and measurement devices and the integration of the processing unit and the databases. The front-end implementation phase consists of developing a user interface for interaction, visualization, and data analysis.
- Implementation of the testbed and validation, where the designed prototype will be installed and tested in the context of a real application (Department of Electronic Engineering, Universidad Técnica Federico Santa María (UTFSM), Valparaíso, Chile), including an office, a laboratory, and a classroom to assess the functionality and usability of the proposed solution.

This paper is structured as follows: Section 2 presents a review of related work for smart buildings. Section 3 introduces the hierarchical energy management architecture for the intelligent campus. Section 4 presents the proposed IoT architecture for smart buildings. Section 5 discusses and analyses the proposed solution and the results. Section 6 presents the conclusion and future work.

2. Related Work

Smart buildings and building energy management systems (BEMS) are active research areas with different application domains such as demand response programs, optimizing building power consumption, integrating renewable energy systems, etc. In [2], the authors highlighted that IoT provides a new opportunity to integrate intelligence into building management systems. Such IoT solutions are cost-effective, enabling monitoring and managing the energy consumption of the buildings. The work summarized the application of IoT in buildings, including lighting, heating, ventilation, air conditioning (HVAC), flexible loads, human detection and diagnostics, and prognostics. A case study was discussed on how to use low-cost IoT devices to provide building management with key insights into human activity and occupancy detection.

In [3], the authors presented the design and implementation of an IoT gateway for a cloud-based building energy management system. The work focused on the software architecture and the software design of the gateway device, which acts as a master device, polling devices on the network and pushing the received data to the cloud. The gateway device was designed to support legacy protocols such as Modbus, BACnet, and HTTP RESTful interface devices. The developed software was evaluated with respect to RAM consumption under various stress tests and bandwidth utilization.

In [4], the authors proposed a fog-based IoT platform for smart building, which consists of five layers: end devices, network connectivity, fog processing, cloud processing, and security and privacy layer. The end devices include sensor nodes (temperature and humidity sensor, light sensor, PIR sensor, and accelerometer sensor) and actuator nodes (feedback action). The work focused on indoor ambience monitoring and occupancy monitoring. A prototype has been deployed and tested in a testbed room for door/window state detection, room occupancy detection, and room lighting sense and control. In [5], the authors proposed an IoT architecture for hybrid wind/PV/diesel/battery on a university campus. The proposed architecture consists of four layers: power layer, data acquisition layer, communication network layer, and application layer. The work is considered a case study on a university campus. However, the work focused on network modeling and simulation of the communication network layer for the hybrid energy system with respect to network topology, link capacity, and latency.

In [6], the authors presented a review of the concept of IoT and its potential application in smart buildings where the major components of the IoT system consist of devices/sensors, networks, cloud, analytics, and actuators/user interfaces. The conventional architecture for the smart building consists of three layers: the perception layer, network layer and application layer. The work discussed the challenges and recommendations for future research, including (1) security and privacy issues, (2) data acquisition, processing and storage issues, (3) feasibility and practicality issues, and (4) collaboration between IoT developers and the building industry. In [7], the authors presented a survey for different types of applications in smart buildings, including security control, energy management, monitoring and control of HVAC, water management, lighting system, fire detection, and health system for elders.

In [8], the authors proposed IoT based thermal model learning framework for a smart building based on low-cost IoT devices (smart thermostats). The data collected from the IoT platform installed inside the building has been used for validating the learning framework. In [9], the authors presented a low-cost solution for non-smart residential load appliances using smart load nodes. The integration of this solution does not require any change in the electric infrastructure of the house, as well as no modifications to the load appliances. The system considered wireless communication using WiFi in HAN, where the main measurements include voltage, current, power, and power factor.

In [10], the authors presented a comprehensive survey on the intersection of IoT and smart grid systems (IoT-aided smart grid systems), which includes architectures, applications, and prototypes. The work also presented different challenges and future research directions. In [11], the authors reviewed the architectures and functions of IoT-enabled smart energy grid systems. Special focus was given to IoT technologies such as sensing, communication, and computing. The work also reviewed security vulnerability, attack models, and existing threat summarizing mitigation techniques for such security vulnerabilities. In [12], the authors reviewed recent activities related to IoT-based energy systems. The work highlighted the potential areas to improve at different layers and reviewed communication technologies and standards related to energy systems. Some examples were discussed, including smart homes, smart power grids, and smart cities.

In [13], the authors presented a hierarchical IoT-based microgrid for energy-aware buildings. The proposed framework consists of the physical layer, information layer, control layer, and dispatch layer. The IoT microgrid laboratory at Aalborg university was introduced to explain how to implement the proposed scheme in a building. In [18], the authors presented a comprehensive review of thermal comfort in hospitals, identifying the current status of research and future research directions. The main research themes were influencing factors, field surveys, measures to improve and energy saving.

In [19], the authors proposed an IoT-based occupancy-driven plug load management system with the objective of reducing the energy consumption of plug loads and plug load automation. Six different strategies for plug load control, such as manual control, predefined schedule, occupancy-driven control, and hybrid control, were evaluated during a field study of 5-months within a university office space. In [22], the authors introduced the application of deep learning and IoT to control the operation of air conditioners to reduce energy consumption in a smart building. In order to count the number of persons in a certain area, the work considered the YOLOv3 algorithm. In [23], the authors provided architectural elements of connected indoor lighting systems within a building. In particular, application programming interfaces (APIs) were presented to support data access and lighting system control.

Table 1 summarizes the discussed related work highlighting the presence or absence of the different smart building IoT layers. The work proposed here aims to develop a hardware/software solution to enable gathering the information from smart sensors and energy monitoring devices and display it to the end users so they can take the necessary actions for the proper functioning of the building.

Table 1. Summary of related work for different IoT layers in related work for smart buildings.

Ref.	Type/ Case Study	Power Layer	DAQ Layer	Network Layer	Application Layer	Contribution
[2]	Technical/ Australia	YES	NO	NO	YES	Building energy management with key insight into human activity and occupancy detection.
[3]	Technical/ USA	NO	YES	YES	NO	Building energy management with a focus on the software architecture and software design for the gateway device to support various legacy protocols.
[4]	Technical/ Egypt	NO	NO	YES	YES	Fog-based IoT platform consists of five layers: end devices, network connectivity, fog processing, cloud processing, and security and privacy layer. The work focused on indoor ambience monitoring and occupancy monitoring.
[5]	Simulation/ Saudi Arabia	NO	NO	YES	NO	The work focused on network modeling and simulation of the communication layer for a hybrid energy system with respect to network topology, link capacity and latency.
[6]	Review	YES	YES	YES	NO	Overview of IoT technology for smart buildings. The components of the IoT system are devices/sensors, networks, cloud, analytics, and actuators/user interfaces.
[7]	Review	YES	NO	NO	YES	Survey on different types of applications in the smart building, including security control, energy management, monitoring and control of HVAC, water management, lighting system, fire detection and health system of elders.
[8]	Technical/ USA	YES	NO	NO	YES	IoT-based thermal model learning framework for smart buildings based on low-cost IoT devices (smart thermostats).
[9]	Technical/ India	YES	YES	YES	NO	A low-cost solution for non-smart residential load appliances using smart load nodes (SLN).
[10]	Review	YES	YES	YES	YES	A comprehensive survey on the intersection of IoT and smart grid systems (IoT-aided smart grid systems)
[11]	Review	YES	YES	YES	YES	Review the architectures and functions of IoT-enabled smart energy grid systems
[12]	Review	YES	YES	YES	YES	Review for recent activities related to IoT-based energy systems. Examples were discussed, including smart homes, smart power grids, and smart cities
[13]	Technical/ Denmark	YES	YES	YES	YES	A hierarchical IoT-based microgrid for energy-aware buildings
[18]	Review	YES	NO	NO	NO	A comprehensive review on thermal comfort in hospitals, identifying the current status of research and future research directions.
[19]	Technical/ Singapore	YES	YES	YES	YES	IoT-based occupancy-driven smart plug load management system that reduces plug load energy consumption. The occupancy information of users is collected using an indoor localization system.
[22]	Technical/ Finland	NO	NO	NO	YES	Application of deep learning and IoT to control the operation of air conditioners to reduce energy consumption in a smart building.
[23]	Technical/ Netherlands	YES	NO	NO	YES	Architectural elements of connected indoor lighting systems within a building. In particular, APIs were presented to support data access and lighting system control
Present Work	Technical/ Chile	YES	YES	YES	YES	Developed a hardware and software platform for remote monitoring and control of smart buildings. A real testbed has been designed, implemented, and tested at Universidad Técnica Federico Santa María, Valparaiso, Chile.

3. IoT-Based Architecture for Smart Buildings

3.1. Hierarchical Energy Management Architecture for Intelligent Campus

A university campus generally consists of a group of buildings connected to the main power grid. The energy management approach can be classified into different levels: campus energy management system (CEMS), building energy management system (BEMS), and office/laboratory/classroom level. The CEMS operates as an energy manager for the campus by collecting energy consumption data from each building through intelligent BEMS (iBEMS), as shown in Figure 1.

- Campus energy management system (CEMS): The CEMS monitors the power generation (renewable energy), energy storage, and consumption of university buildings and interacts with each BEMS to optimize energy usage. In addition, the CEMS receives the energy consumption data of each building, stores it in a database, and estimates the consumption and future generation based on historical data.
- Building energy management system (BEMS): In general, university buildings consist of a group of offices, laboratories, and classrooms. The BEMS collects the energy data and other weather information collected by different smart meters and sensors located in the building and interacts with the building loads.

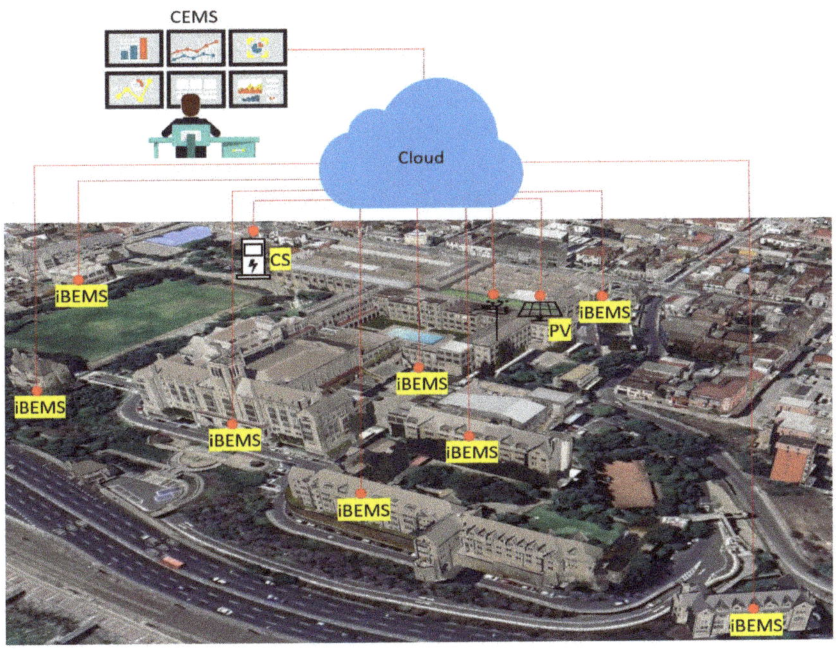

Figure 1. The schematic diagram for the intelligent campus. CEMS: campus energy management system; iBEMS: intelligent building energy management system; CS: charging station; PV: photovoltaic system.

The CEMS provides large-scale data acquisition, communication, and data processing for energy management in buildings which requires cooperation among each BEMS in order to meet the operator requirements for minimizing power consumption and costs. However, as the number of energy management units increases, many challenges are related to cost, latency and reliability. This work target one building (Building B) of the UTFSM Campus, Valparaiso, Chile.

The university campus (Casa Central, Valparaíso, Chile) was inaugurated in 1931. Nowadays, there is a lack of monitoring and control for energy consumption in university buildings. There is still a high cost for building automation systems for small and medium-sized buildings, which prohibits purchasing such solutions. Furthermore, there is still a limitation for compatibility with different vendors, devices, and communication technologies in case of relying on using a particular platform. The proposed IoT platform aims to develop a custom solution using cost-effective off-the-shelf sensors/devices for real-time monitoring and control of building energy consumption through a web interface.

Three case studies are considered: an office, a laboratory, and a classroom. Three case studies are considered: an office, a laboratory, and a classroom.

3.2. *IoT-Based Architecture for Smart Buildings*

The main objective of this work is to design and implement an IoT platform for real-time monitoring and control of energy consumption in buildings to achieve an intelligent campus. There are different IoT architectures., such as three-layer (perception layer, network layer, and application layer), four-layer (perception layer, network layer, service layer, and application layer), and five-layer (perception layer, network layer, service layer, application layer, and business layer) [5,10–14]. Considering the various IoT-based architectures, Figure 2 shows the main components of a three-layers IoT-based architecture for smart buildings, which consist of the perception layer, network layer, and application layer. The perception layer usually emphasizes energy usage, occupant activities, and environmental condition. Different wired/wireless communication technologies could be used for data transmissions, such as WiFi, ZigBee, Bluetooth, and LoRa in the network layer. The application layer corresponds to business, application, and service management. The collected data at the application layer explains the actual building energy usage, energy management, occupant information, etc.

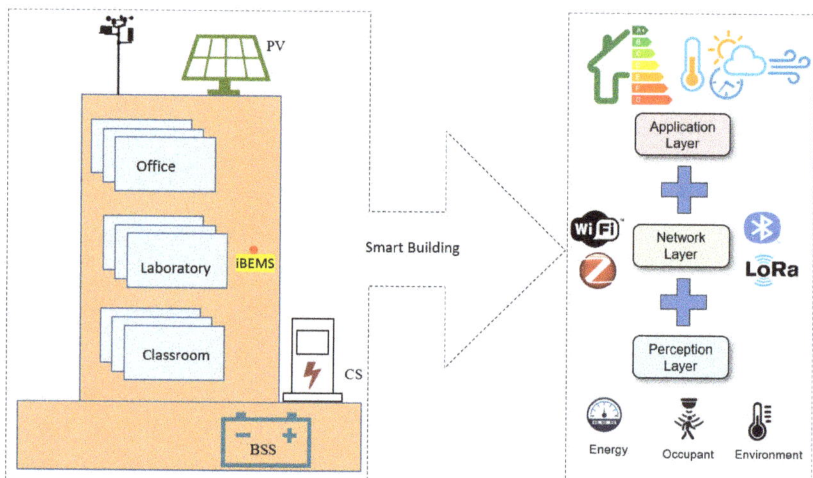

Figure 2. The schematic diagram for IoT-based architecture for smart buildings. iBEMS: intelligent building energy management system.

Figure 3 shows a detailed schematic diagram of the proposed IoT-based architecture for smart buildings. Our work aims to achieve building automation to minimize energy consumption costs and guarantee the comfort of the occupants. Our proposed architecture consists of four layers: power layer, data acquisition layer, communication network layer and application layer.

- Power Layer: the power layer consists of power generation, power storage, and loads that are connected to the power grid. In this study, the campus buildings are connected to the power distribution network (provided by Chilquinta). Other power generation sources may include solar panels, wind turbines, and batteries for power storage. The main power consumption in buildings may consist of HVAC, lights, and vehicle charging stations.
- Data Acquisition Layer: the data acquisition layer is responsible for capturing all the data coming from the power layer devices for making decisions. Examples of

sensor nodes are measuring devices from light, temperature, power consumption, and meteorological station.

- Communication Network Layer: Different communication technologies and protocols are defined from data acquisition devices in the communication network layer. The communication network layer receives the sensors' data and sends them to the application layer. Data might need to be sent over various networks, such as the local area network (LAN) and building area network (BAN). The most common communication technologies are ZigBee, Bluetooth, WiFi, and LoRa, using different communication protocols such as MQTT, CoAP, and Web Socket.
- Application Layer: the end-user can recognize the middleware services that allow data storage and interaction with building data in the cloud. All monitoring and status information received from the devices are stored and visualized. Real-time monitoring and control can then be achieved using different approaches such as energy management, safety, user comfort, and management of HVAC.

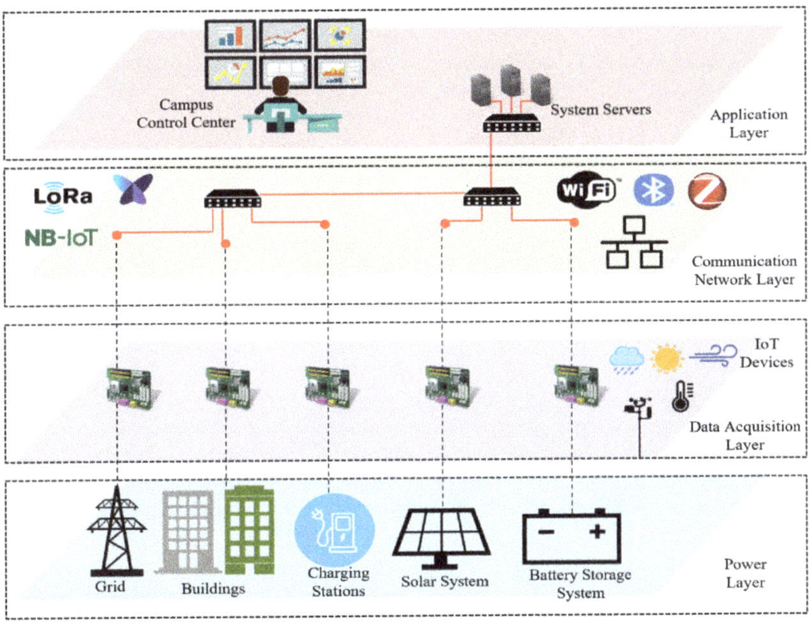

Figure 3. IoT-based architecture for the intelligent campus.

4. Smart Building Implementation

The proposed solution was implemented considering three locations with different needs. The details for each scenario and location are given below:

- Office Room: The first scenario is office B-349. This is a representation of professors' offices distributed in Jeonju pus. Most offices include computers with one or more monitors, printers, and plugs to charge mobile devices. All offices include fixed lighting activated by a switch on the wall. Because it is a relatively small space, the energy measurement of the entire room was not considered.
- Laboratory Room: The second scenario is laboratory B-110. In this laboratory, there are at least four permanent workstations where a computer can be connected and external monitors for each of these positions. In addition, there is a shared space to carry out different activities. In this case, the implementation of smart plugs is proposed to monitor computers and other equipment connected to the power network. On the

other hand, given the energy requirements of the space, it is suggested to install an energy meter in the electrical panel to monitor the total energy in the room.

- Classroom: The third scenario is classroom B-213. This classroom has luminaires that can be controlled with smart switches, one projector that can be controlled with smart plugs, and several sockets that allow students and teachers to connect their personal devices. An energy meter can be used on the electric board. In addition, an environmental measuring device can be installed to monitor the air quality during the classes.

Figure 4 shows the locations of different scenarios, while Table 2 shows the list of appliances and sensors considered at each location.

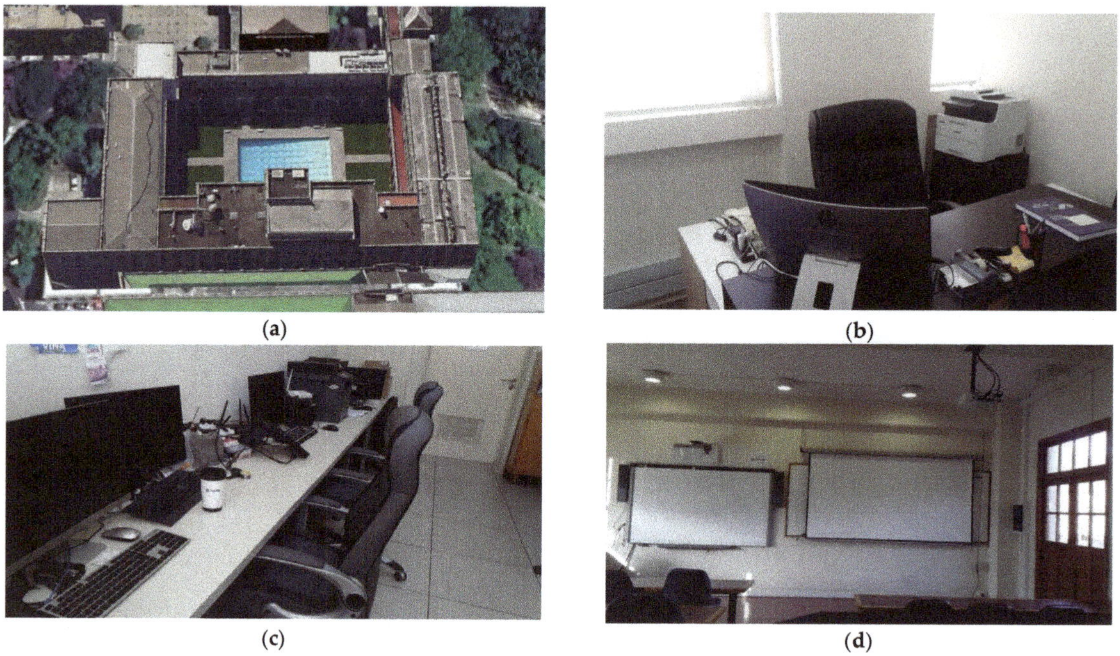

Figure 4. Locations of different implantation scenarios: (**a**) Engineering building B; (**b**) Office B-349; (**c**) Laboratory B-110; and (**d**) Classroom B-213.

Table 2. List of appliances.

Location	Details
Office Room (OR)	1 Computer, 1 Monitor, 1 Printer Lighting
Classroom (CR)	1 Projector, Many sockets Main electric board Lighting
Laboratory (LAB)	4 computers, 1 Printer, Many sockets, Lighting PV panels

4.1. Selected Alternatives Solutions for Devices, Technologies, and Services

The implementation is carried out considering the proposed 4 layers discussed in the previous section. The description of equipment and technologies used in each layer are given in Table 3 and Figure 5. Although the system is proposed and designed for the three different scenarios in different locations, we organized and assembled a testbed shown in Figure 6. The testbed will allow us to collect data from various appliances individually and/or together for later using it to validate different machine learning algorithms.

Table 3. List of sensors and measuring devices used for the testbed.

Layer	Details
Data Acquisition Layer	Smart Plug, Air quality sensor, Smart meter for the total power consumption, Data acquisition module for photovoltaic system, Data collection module for a weather station
Communication Network Layer	Cloud Service (Raspberry Pi, Node-RED) Network Layer (WiFi, LoRa, MQTT)
Application Layer	Digital Ocean, Node-RED, MySQL

Figure 5. Selected devices for data acquisition layer: (**a**) Raspberry Pi 4B; (**b**) PZEM-004t-100 A; (**c**) Sonoff Pow R2; and (**d**) Air quality LAQ4.

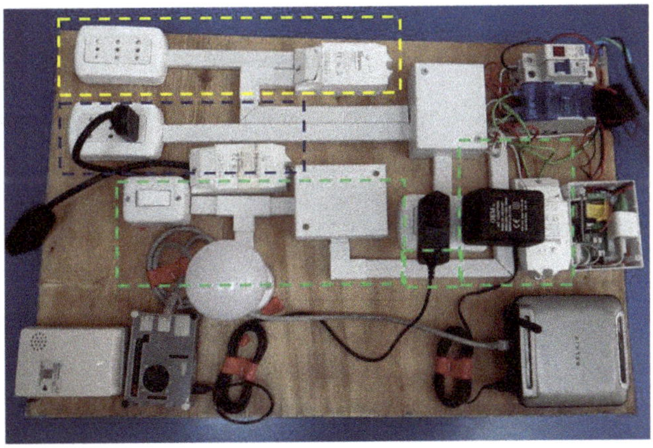

Figure 6. SThe schematic diagram for the testbed.

The testbed is powered by an external source connected directly to a small electrical panel. On the board, there is a pilot light to verify the electric power at the input, then a switch for protection in case of current overconsumption, and a tetrapolar bar to facilitate the connection of equipment on the testbed. The testbed also includes two circuits: the lighting circuit and the sockets circuit. Furthermore, other components include a small router that provides a local WiFi network for the equipment, a LoRa gateway, and a Raspberry Pi that implements the proposed architecture. Detailed characteristics, setup and comparison among different types of sensors, measuring devices, and network elements are given in Appendices A and B.

4.2. Power Layer

4.2.1. Monitoring Power Consumption of Appliances

For monitoring the power consumption of appliances, smart plugs of the "Sonoff POW R2" type are used [24]. All smart plugs were updated with ESPurna firmware and configured for wireless data transmission using WiFi [25]. Regarding the electric connection, the smart plugs need to be connected with electric cables and installed between the electric power supply and the appliances.

Different configurations for using smart plugs type "Sonoff POW R2" are considered for the testbed:

- Control the complete circuit: In this case, the whole lighting circuit was passing through the smart plug, which allows measuring and control of the lights (highlighted with green color in Figure 6)
- Control a socket: In this case, the connection was configured, which allows obtaining data from all equipment connected to the socket (highlighted with yellow color in Figure 6)
- Control a single device: This configuration allows a single device to be connected (highlighted with blue color in Figure 6)

4.2.2. Monitoring Total Power Consumption

The total power consumption measurement is carried out using the pzem-004t-100a module. The module measures the input voltage of the electrical panel under study and a current transformer sensor for the current. The module is connected to the ESP32 platform for subsequent data sending.

4.2.3. Monitoring Photovoltaic System

The main parameters considered for monitoring the photovoltaic panel include measuring the current with a current transform (CT) sensor. The sensor delivers a current proportional to the measurement current and a voltage divider for voltage measurement. Both measurements are captured with ADC pins of the Arduino nano development board. Data is sent to the NodeMCU development board Amica, which has WiFi communication and send the data by MQTT along with the date and time they were captured using a real-time clock (RTC).

4.3. Data Acquistion Layer

4.3.1. Monitoring Indoor Environmental Condition

The air quality sensor type "Dragino LAQ4" was used to monitor indoor environmental conditions [26]. The main parameters measured are total volatile organic compound, CO2 equivalent, temperature, and relative humidity of the air. To obtain such indoor environmental data, the configuration of LoRa Gateway is required. In this work, we use the Dragino Gateway LG308 [27].

4.3.2. Monitoring Weather Station

The meteorological information was measured using Davis Advantage Pro 2 Plus weather station [28]. The weather station obtains the data from the sensors physically connected to the station, then sends data wirelessly to the Vantage Pro 2 console. If connected to the datalogger, we can connect the console with a USB cable to the computer. Please note that the weather station is a closed system that does not allow the external manipulation of the data obtained.

4.4. Communication Network Layer

4.4.1. Network Layer

The communication protocols for data transmission are MQTT and LoRa, which have been considered for implementing the communication network layer. For the smart plugs, the communication using MQTT is activated and configured by choosing the MQTT section from the side menu displayed in the web interface of the smart plugs with ESPurna firmware installed. In the case of general measurement devices and photovoltaic panels, the integration of the MQTT protocol is carried out within the code which we were programmed with. The configurations for all devices connecting via MQTT are similar; that is, they connect to the broker running on the Raspberry Pi with the IP 192:168:2:2 through port 1883 and are differentiated by the topic in which they publish. The topics were defined as a descriptive manner of the measurement and the locations where the data are taken and displayed, as shown in Table 4.

Table 4. Description of topics connected to MQTT.

Name	Task	Location	Topic
Sonoff POW R2	Smart Plug	Office B-349	officeB349/enches01/ ESPURNAA9F0E4
Sonoff POW R2	Illumination	Office B-349	officeB349/ilminacionGeneral ESPURNA9CFBF8
Sonoff POW R2	Computer	Office B-349	officeB349/connectedDevice/ Computer/ESPURNA9CFBF8
DC Measurement	PV Panel	Outside	photovoltaicSystem/panel01
AC Measurement	General Electric Panel	Class B-213	classroomB213/generalPanel

In the case of the air quality measurement device (LAQ4), a procedure should be carried out for the configurations. An account is created on the things network server at (https://www.thethingsnetwork.org (accessed on 10 January 2022)) [29]; then, when registering, you must enter the start section (which leads to the address https://console.cloud.thethings.network/ (accessed on 10 January 2022)) where the Cluster "Nam1" located in the state of Carolina in United States was chosen, and once selected you enter the Gateways tab where you press the + add gateway button, filling in all the requested data, in particular, the *Gateway ID* which is the unique number associated with each gateway LoRa. For the case of this work, the names used are *GatewayOfBuilding*, and the gateway ID is gatewaylorabuilding01. Once the registration of the LoRa gateway is completed, one can return to the gateways tab and select the registered gateway to see information about this connection with the network server.

Once the gateway connection has been verified, we go to the applications tab and press the button + add application. Only 3 parameters must be filled, namely "Application ID", "Application Name" and "Description" that, for this case is filled with the data of smart-"buildingslora-sensors-usmcc" in Application ID and "lora sensors in usmcc" in Application Name. Once the application has been created and when entering it, press the button + add end device and enter the data for the desired sensor, which in this case is the LAQ4. Then, data can be entered corresponding to the manufacturer and model of the device. Next, the sensor's own parameters, called Registration Key, are entered. In addition, in this case, the value of "airqualitysensorusmcc" was added to the parameter End Device ID, and Register End Device was pressed. Once registration is complete, go back to the applications tab, and in the end devices, one can enter the newly created one. Finally, being inside the added device, it is possible to enter the side menu section Integrations and then MQTT, where the subscription data to the broker of the things network, data that will be used in the application layer to visualize the data.

4.4.2. Cloud Service

There are different service providers for the cloud layer (middleware). The major public service providers include Amazon, Microsoft, and Google [30,31]. Being leaders in the market, it is observed that the services are similar in almost all the services they provide. Due to the difficulty in calculating the costs associated with the service (the charge is made per hour of use and depends on the capabilities contracted), particularly when the project starts and the requirements can change; therefore, it was decided to use the service of virtual machines of Digital Ocean [32].

For this layer, a Raspberry Pi with the Raspberry Pi OS operating system is used, which is loaded into a micro-SD memory by downloading the installation software from (https://www.raspberrypi.com/software/ (accessed on 10 January 2022)), choosing the procedure to install and selecting the memory micro SD to use in the aforementioned Raspberry Pi. For later configuration, the system is accessed using a display with HDMI input and a mouse to activate the option to allow the connection by the SSH protocol. After doing this, it simply connects to the local network via the Ethernet cable. Then, on a computer within the same network, the connection is made via SSH (considering that the IP of the Raspberry Pi is 192.168.2.2).

Once the computer is connected to the Raspberry Pi via SSH, Mosquitto is installed as the selected MQTT broker [33]. Then, Node-RED is installed, with which the data is managed locally, following the official page's recommendation (nodered.org). To access the node-red programming palette, it is enough to be within the same local network of the Raspberry Pi and go to http://192.168.2.2:1880. Then, once inside the programming palette, the division is made into 3 flows that correspond to smart plugs (see Figure 7a), the metering device alternating current (see Figure 7b) and the direct current measuring device (see Figure 7c). In the 3 flows, MQTT data is received from each device and forwarded using the same protocol (MQTT) to the server hosted at Digital Ocean.

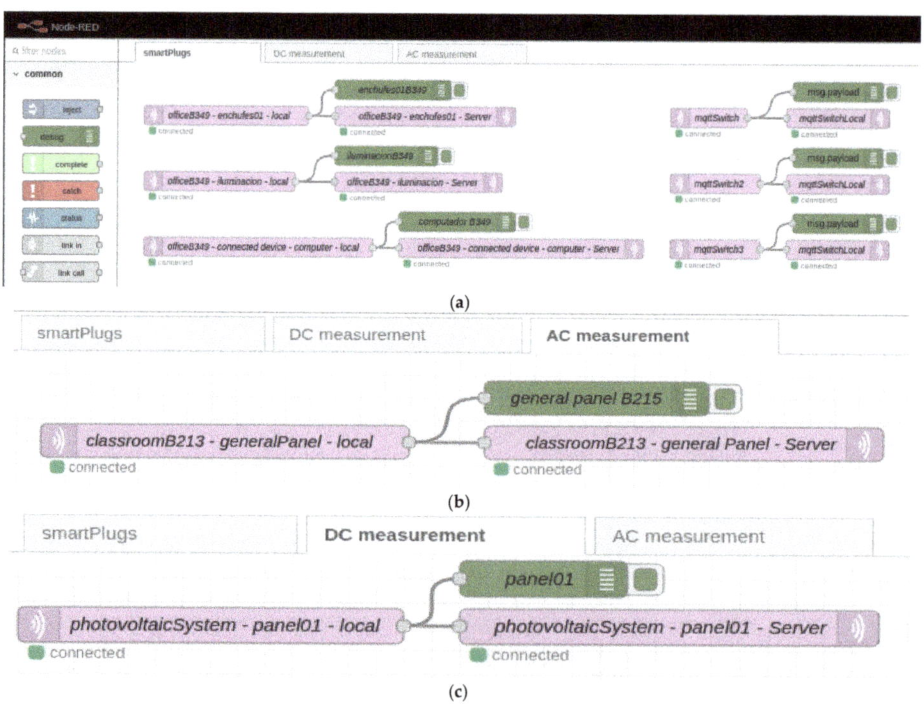

Figure 7. Node-RED configuration for forwarding data received from different devices (**a**) smart plugs; (**b**) general energy measurement; (**c**) direct current measurement.

4.5. Application Layer

The application layer is entirely supported on a virtual machine contracted with the services of Digital Ocean. The system supports a machine with 2 Gb of RAM and 50 Gb of hard disk. To acquire these services, one must enter the provider's page https://www.digitalocean.com/go/developer-brand (accessed on 10 January 2022) AND must be registered (considering that a valid credit card must be included during the registration process). After logging in to the page, a new project is generated, which, in this case, is called IoTPlatform4ManageEnergy and then the create button is pressed to create a new Droplet which is the way that Digital Ocean calls virtual machines, and this is created with 2 Gb of RAM, 50 Gb of hard disk and the Ubuntu 20.04 operating system is installed (LTS) x64. Once the Droplet has been created, it can be accessed with the fixed IP provided by the provider (165.232.139.50) via SSH. The Mosquitto broker was installed on this server to define communication via MQTT. In addition, Node-RED is installed to manipulate the data, and MySQL is installed to allow data storage. Figures 8 and 9 show the configuration for Node-RED for control and data visualization, respectively.

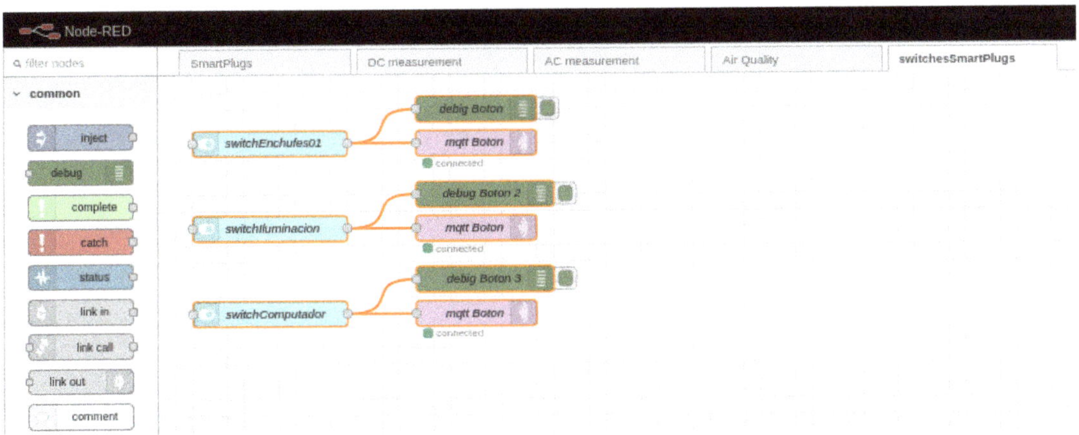

Figure 8. Configuration for Node-RED for the control of smart plugs.

The schematic diagram for the complete system is shown in Figure 10. The experiments were carried out on 11 November 2022. Different electric appliances in the laboratory were connected to validate the operation of the platform (Plug01: Hair Drayer-SiEGEN-Model SG-3049, Illumination: Led bulb-9W, Desktop Computer: TV Monitor-LG-24TL520S-PS). The dashboard, shown in Figures 11–13, shows an example for those observed from a computer; since Node-red is responsive, the visualization can adapt according to the display device, such as a smartphone or a tablet. Furthermore, with the implementation of the MYSQL database interaction block, the data is saved in the cloud.

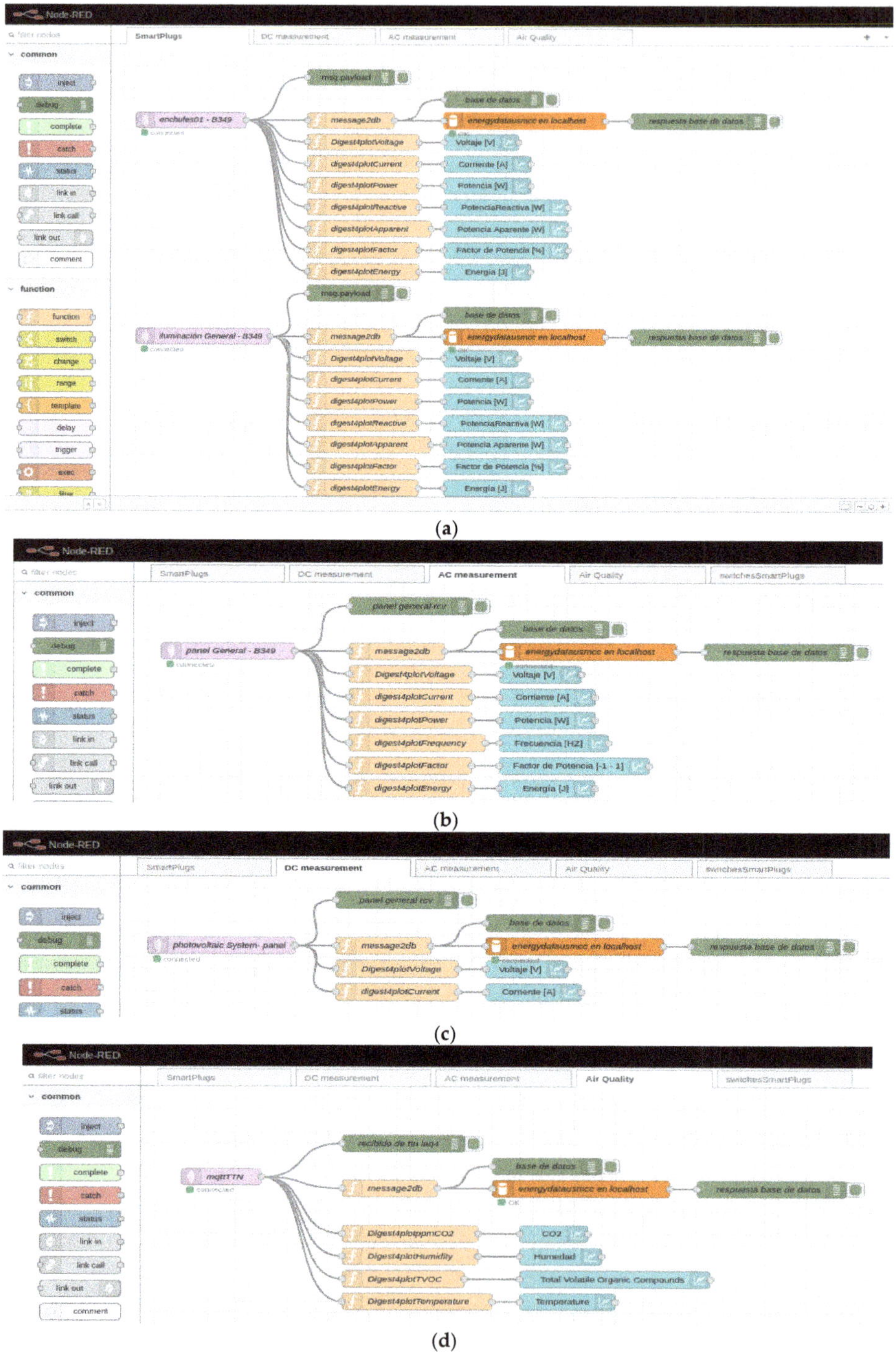

Figure 9. Configuration for Node-RED for data visualization: (**a**) Smart plugs; (**b**) Electric panel; (**c**) PV system; and (**d**) Air quality.

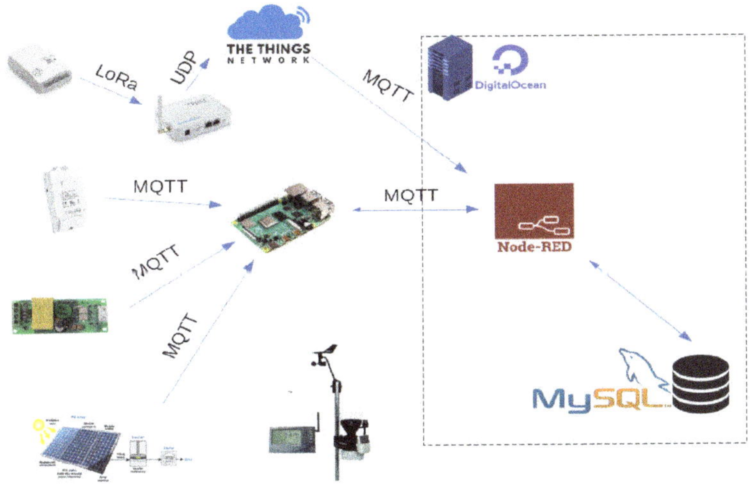

Figure 10. Schematic diagram for the complete system.

Figure 11. Real-time monitoring data from an office room.

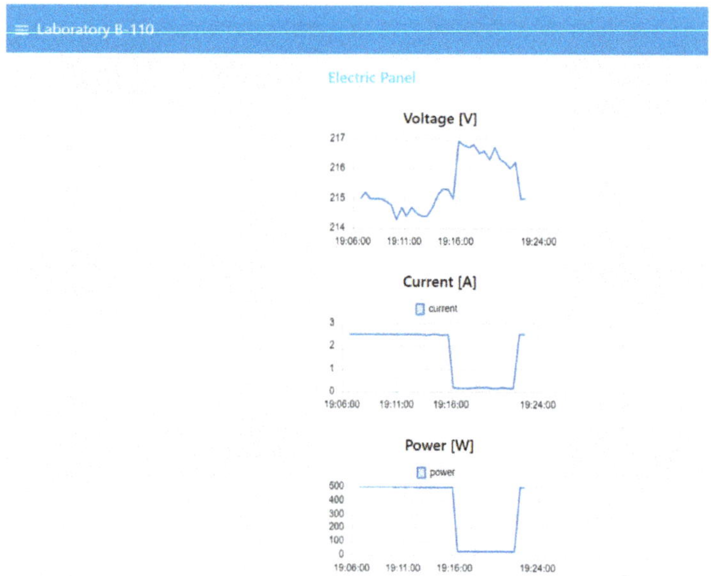

Figure 12. Real-time monitoring data from a laboratory room.

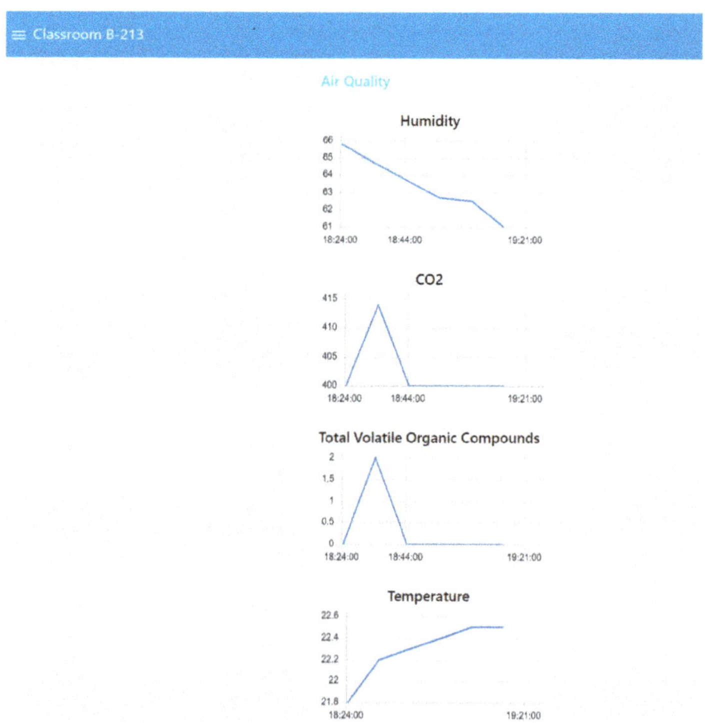

Figure 13. Real-time monitoring data from a classroom.

5. Discussion

This section provides a detailed analysis of the proposed IoT platform for building implementation and the future extension for large-scale performance on the university campus.

5.1. Analysis

The advances in IoT technologies will play a vital role in the development of different smart building solutions, such as smart lighting to minimize light load, smart HVAC to improve indoor comfort, smart plug loads to monitor and control usage, and smart energy management systems, toward achieving an intelligent campus. Among the key technologies which have been investigated in this work are smart plug loads to monitor and control various types of appliances (located in offices, labs, and classrooms), distributed energy resources including a PV system, outdoor data using a meteorological station, and indoor air quality monitoring. Detailed technical hardware/software configuration and implementations have been discussed, as well as the network connectivity among different subsystems. As real-time data monitoring is the first step toward transforming conventional buildings into smart buildings, the proposed solution will enable the building operator to view, analyze, and predict different appliance profiles and the occupancy of the buildings using a dashboard to visualize such real-time/historical data and alerts.

With respect to the cost, the proposed solution uses off-the-shelf components. The main electronic components used were Sonoff POW R2, PZEN-004T-100A, Raspberry Pi 4B, and Dragino LAQ4, which cost about CLP 20.900, CLP 26.812, CLP 78.390, and CLP 51.899, respectively. The total hardware cost of the proposed solution can be adapted for the university campus to support real-time data monitoring for power consumption from different buildings and environmental data with the objective of improving energy performance and building operation.

Given the experience gained during the development of the proposed IoT platform, the following are the guidelines that need to be considered for developing such a solution (in Chile or another region) that fits the user's requirements.

- Definition of requirements and services: This allows knowing the main problems and defining the objectives to be achieved during the full development of the proposed solution following the requirements:
- Define a network architecture for smart sensors and meters. Structure a network architecture that connects nodes, gateways, and servers seamlessly and efficiently to measure real-time electricity consumption and transmit the information obtained for the end-user.
- Develop a cloud storage server. This allows data storage and access to the information stored from anywhere through an API.
- Develop a platform for visualization. This platform enables the visualization of real-time data from each device/appliance connected to the platform.

5.2. Technology Adoption

The way toward an intelligent campus requires the acceptance of new technologies and the opportunities they provide. User perceptions on the acceptance and adoption of smart energy management in the workplace is an essential aspect that needs to be investigated [20,21,34]. In this regard, the work in Ref. [20] identified different types of behavior change interventions that are successful in saving energy in the workplace, such as education, persuasion, incentivization, and training. Another study in Ref. [21] proposed seven design implications that guide the future design of smart energy management systems in the workplace, including internal and external influence, user appeal, user control, ease of use, reliability, personalized and contextualized information, and data privacy. Among the open research topics are conducting a large-scale study in multiple countries (different geographical contexts and cultures) to identify overlaps in user perceptions. In addition, expanding the scope for other workplaces such as hotels, retail stores, and industrial sites [21].

5.3. Future Direction

There are different directions to extend the current work. From the power layer point of view, we aim to extend the current prototype to support an electric vehicle charging station, a battery storage system, smart lighting (dimmer lights and lights coupled with motion sensors), sensors for motion detection, and magnetic sensors for the doors. From the data acquisition layer point of view, our ongoing work aims to develop a low-cost meteorological station using off-the-shelf components to be able to access and control the data obtained. From the communication network layer point of view, heterogeneous communication technologies could be integrated to support different short-/long-range communications across the campus.

As a short-term goal, we aim to demonstrate the feasibility of the proposed IoT platform in one complete building to collect power consumption data during the whole semester with everyday activities and during the vacation period. Before implementing such a system, two different dashboards need to be developed: one for the end-user, who can control their plug loads, and the other for the building administrator. There is also a need to define which control strategy should be selected, which can be a manual or predefined schedule. In this regard, the authors in Ref. [19] performed different plug load automation strategies for a university office space which could be a starting point for the office scenario.

Other challenges are related to the internal electrical wiring for the building, as each floor may share various offices and laboratories from different departments. In addition, the absence of historical power consumption data for buildings/floors/zones. Therefore, historical power consumption data will need to be collected from the main electric panels using smart meters for individual rooms/zones. This step is essential to define spaces with high energy consumption. Based on energy consumption data from different floors/areas, the performance will be evaluated before and after applying different plug load management strategies. Special attention should be given to privacy and security, which should be considered at every step during the platform design. Emerging technologies such as blockchain, machine learning, and artificial intelligence are opening new opportunities to mitigate such security vulnerabilities.

6. Conclusions

This work proposed a cost-effective IoT solution for smart buildings to enable remote monitoring and control of power consumption at the appliance level. The proposed architecture consists of four layers: power layer, data acquisition layer, communication network layer, and application layer. The physical layer was characterized by different subsystems such as plug circuits, lighting, and the photovoltaic system. For the acquisition layer, measurement devices were used for electrical panels, smart plugs, direct current energy metering devices, air quality monitoring, and a meteorological station. The network layer was defined to gather the information captured from the physical layer and forward it to the remote server. A fog layer was implemented on a Raspberry Pi, and the data was handled with NodeRED. The communication technologies defined to obtain the information from the installed equipment were WiFi and LoRa, and the communication protocol to the server was through the MQTT. The Digital Ocean Droplet service was used as a server where the MQTT Broker was installed. For data management, NodeRED was installed for the general management of the data messages and the visualization by the end user. MySQL was installed, which allows storing the information in tables that were defined for each data acquisition device. This project was implemented using a testbed defined to characterize equipment and conditions in three different locations in Engineering Building, Universidad Técnica Federico Santa María, Valparaíso, Chile, including an office, a classroom, and a laboratory. This testbed allowed the design, implementation, and testing of the complete system in reduced space. At the end of the development of this work, a functional platform was obtained that brings together energy consumption data that will contribute to energy awareness and conservation.

Author Contributions: Conceptualization, M.A.A., A.M.E. and Y.-C.K.; methodology, M.A.A., S.A.C., A.M.E., H.O.G., A.J.R. and Y.-C.K.; software, S.A.C.; validation, M.A.A., S.A.C., A.M.E., H.O.G., A.J.R. and Y.-C.K.; writing—original draft preparation, M.A.A. and S.A.C.; writing—review and editing, M.A.A., S.A.C., A.M.E., H.O.G., A.J.R. and Y.-C.K.; funding acquisition, M.A.A., H.O.G., A.J.R. and Y.-C.K. All authors have read and agreed to the published version of the manuscript.

Funding: This work was supported in part by the Agencia Nacional de Investigación y Desarrollo (ANID) through the Proyecto Fondecyt de Iniciación en Investigación 2020 under Project ID11200178, Fondecyt Regular Grant project number 1220903 and Basal Center AC3E FB0008, and in part by the National Research Foundation of Korea (NSF) Grant through the Korean Government under Grant (2021R1I1A305872911).

Institutional Review Board Statement: Not applicable.

Informed Consent Statement: Not applicable.

Data Availability Statement: Not applicable.

Acknowledgments: The authors would like to acknowledge Proyecto Fondecyt de Iniciacion en Investigacion 2020 under Project ID11200178 and the National Research Foundation of Korea (NSF) Grant through the Korean Government under Grant (2021R1I1A305872911). Hugo O. Garces acknowledges the financial support of the Chilean Agency of Research and Development ANID, Chile, under the Fondecyt Regular Grant, project number 1220903. Alejandro J. Rojas acknowledges Basal Center AC3E FB0008.

Conflicts of Interest: The authors declare no conflict of interest.

Nomenclature

IoT	Internet of Things
EMS	Energy Management System
BEMS	Building Energy Management System
CEMS	Campus Energy Management System
HVAC	Heating, Ventilation, and Air Conditioning
HTTP	Hypertext Transfer Protocol
REST	Representational State Transfer
WiFi	Wireless Fidelity
HAN	Home Area Network
SLN	Smart Load Node
CEMS	Campus Energy Management System
CS	Charging Station
PV	Photovoltaic
LoRa	Long Range
BAN	Building Area Network
MQTT	Message Queuing Telemetry Transport
CoAP	Constrained Application Protocol
TTN	The Things Network
CT	Current Transformer
ADC	Analog to Digital Converter
RTC	Real Time Clock
PIR	Passive InfraRed

Appendix A

For all devices considered, the communication technology should be compatible with data acquisition, and great importance should be given to equipment and devices that are sold in the country (Chile). Table A1 shows the characteristics of the smart plugs. Due to their significant similarity, the only one that allows the power measurement of the device connected to it has been selected (Sonoff Pow R2). On the other hand, it is decided to use the same smart plug but connected in series to the lighting circuit.

Table A1. Characteristics of smart plug devices.

Model	Xiaomi Smart Plug *	Sonoff Pow R2 Smart Plug **	S26 WiFi R2 Smart Plug ***
Maximum current	16 A	16 A	16 A
Energy measurement	NO	YES	NO
Connectivity	WiFi 2.4 GHz	WiFi 2.4 GHz	WiFi 2.4 GHz
Price	19.990 CLP	20.990 CLP	14.990 CLP

* https://www.paris.cl/mi-smart-plug-wifi-MKA3SGOLRU.html (accessed on 22 October 2022); ** https://www.paris.cl/interruptor-wifi-diy-sonoff-pow-r2-MK4BFC5Y9J.html (accessed on 22 October 2022); *** https://www.paris.cl/sonoff-s26-r2/ (accessed on 22 October 2022).

For the energy measurement device, pzem-004t-100a equipment has been selected, as shown in Table A2. In the case of the temperature and humidity measurement device, significant similarities are observed in the characteristics of all the proposed alternatives. Dragino LAQ4 has been selected, which reads the most number of variables, as shown in Table A3. Table A4 shows the alternative gateway that allows obtaining the data from the LoRa sensors, where minor differences are observed between each device. Since the LG308 LORAWAN gateway is available in the laboratory, it has been selected for this work.

Table A2. Characteristics of energy metering devices.

Model	Sonoff POW R3 *	Vbestlife ZMAi-90 **	PZEM ***
Voltage input/output	100–240 V	90–250 V	80–260 V
Maximum current	25 A	60 A	100 A
Price	48.990 CLP	38.570 CLP	26.812 CLP

* https://www.paris.cl/interruptor-wifi-de-alta-potencia-sonoff-powr3-MKSS3WI6FZ.html (accessed on 22 October 2022); ** https://articulo.mercadolibre.cl/MLC-1069867809-medidor-de-energia-inteligente-wifi-de-consumo-unico (accessed on 22 October 2022); *** https://articulo.mercadolibre.cl/MLC-915082915-pzem-004t-voltaje-de-corriente-multimetro-modulo-80-260v-100 (accessed on 22 October 2022).

Table A3. Characteristics of devices with ambient sensors.

Model	LHT65 *	Sonoff SNZB-02 **	Sonoff TH16+AM2301 ***	Dragino LAQ4 ****
Parameters	Temperature Humidity	Temperature Humidity	Temperature Humidity	CO_2, Temperature Relative Humidity
Power	Battery 2400 mAh	CR2450—3V	Power connected	4000mAh Li-SOC12
Connectivity	LoRa	ZigBee	WiFi 2.4 GHz	LoRa
Price	37.890 CLP	9.890 CLP	14.990 CLP	51.899 CLP

* https://altronics.cl/sensor-lht65-lorawan?search=LHT65 (accessed on 22 October 2022); ** https://altronics.cl/sensor-temp-hum-zigbee-sonoff (accessed on 22 October 2022); *** https://sonoff.cl/pack-interruptor-diy-sonoff-th16-sensor-de-temperatura-y-humedad-am2301/ (accessed on 22 October 2022); **** https://altronics.cl/sensor-calidad-aire-laq4?search=LAQ4 (accessed on 22 October 2022).

Table A4. Characteristics of LoRa gateway.

Model	LG308 *	Laird RG191 **	LIG16-915 ***
Transceiver	SX1308/SX1276	SX1301/SX1257	SX1302
Frequency	915 MHz	915 MHz	915 MHz
WiFi Connectivity	WiFi 2.4 GHz	WiFi 2.4 GHz	WiFi 2.4 GHz
Price	249.900 CLP	499.990 CLP	154.690 CLP

* https://altronics.cl/lora-gateway-lg308 (accessed on 22 October 2022); ** https://mcielectronics.cl/shop/product/gateway-lora-915mhz-wi-fi-bluetooth-ethernet-laird-rg191-laird-25707/ (accessed on 22 October 2022); *** https://altronics.cl/lig16-gateway-lorawan (accessed on 22 October 2022).

The proposed alternatives for the local computing device with its main hardware characteristics are summarized in Table A5. Considering the highest percentage of the connectivity item and the cheap cost of the device, the Raspberry Pi 4B was chosen as the local computing device.

Table A5. Characteristics of computing devices.

Name	Core/RAM	Storage	Connectivity	Price
Raspberry Pi 4B *	4/4-8	SD	Gigabit Ethernet WiFi, BLE	78.390 CLP
Odroid-xu4 **	8/2	Flash board	Gigabit Ethernet	99.990 CLP
Jetson Nano Nvidia ***	4/4	microSD	Gigabit Ethernet	185.990 CLP

* https://altronics.cl/raspberry-pi-4-4gb (accessed on 22 October 2022); ** https://mcielectronics.cl/?s=Odroid-xu4&post_type=product&product_cat=0, (accessed on 22 October 2022); *** https://mcielectronics.cl/shop/product/kit-de-desarrollo-jetson-nano-nvidia-nvidia-27557/ (accessed on 22 October 2022).

Appendix B

Figure A1a shows the setup of the smart plug device, while Figure A1b shows the setup of the module pzem-004t-100a. Figure A1c shows the configuration of the solar panel measurement device. The CT sensor must be connected by wrapping the positive cable of the panel, and the terminals of voltage test (brown → ground and red → VCC) are connected in parallel to the current bus of the photovoltaic panel system. Figure A1d shows the configuration of the Dragino LAQ4 device. Figure A2 shows the data obtained using the weather station.

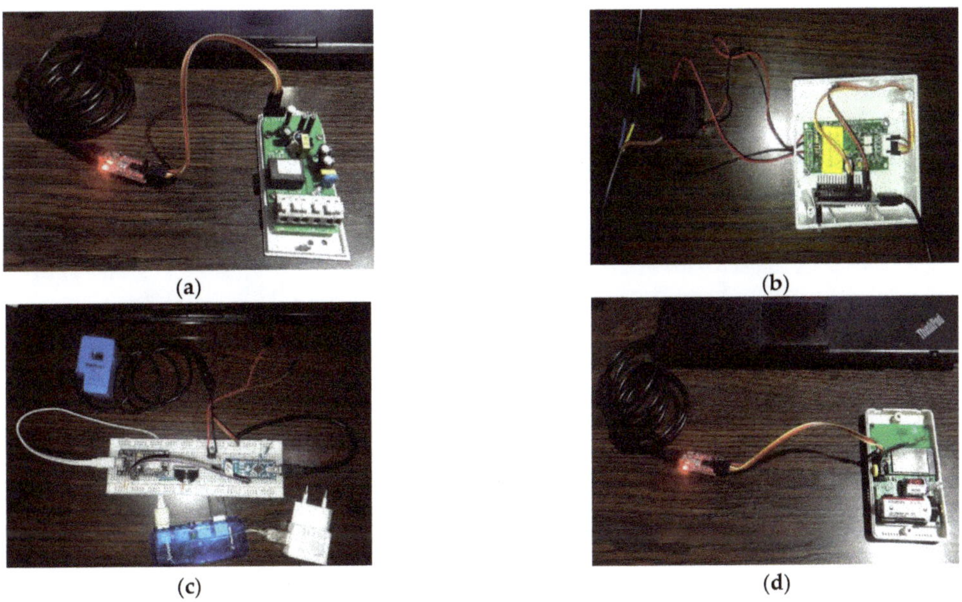

Figure A1. Setup of sensors and measuring devices: (**a**) smart plug device; (**b**) pzem-004t-100a module for total power consumption; (**c**) monitoring the PV system; and (**d**) Dragino LAQ4 for indoor environmental data.

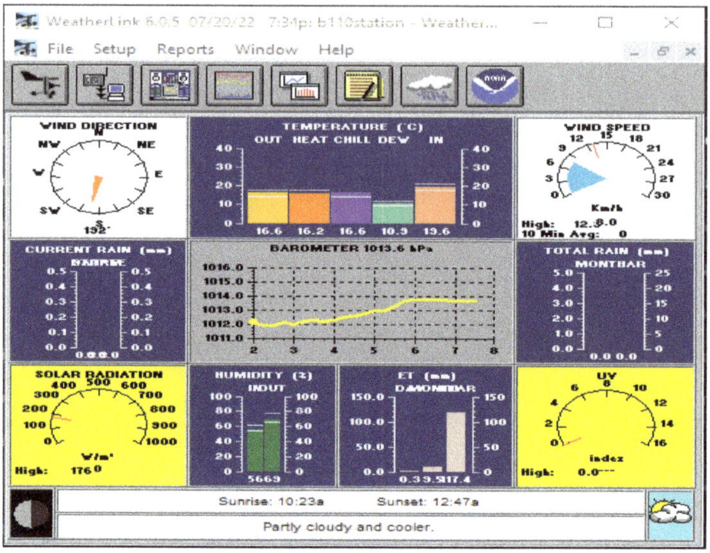

Figure A2. Weather station data obtained using Weather Link software.

References

1. Energy Efficiency Law and Plan, Ministry of Energy. Available online: https://energia.gob.cl/ley-y-plan-de-eficiencia-energetica (accessed on 10 June 2022).
2. Tushar, W.; Wijerathne, N.; Li, W.-T.; Yuen, C.; Poor, H.V.; Saha, T.K.; Wood, K.L. IoT for Green Building Management. *arXiv* **2018**, arXiv:1805.10635.
3. Nugur, A.; Pipattanasomporn, M.; Kuzlu, M.; Rahman, S. Design and Development of an IoT Gateway for Smart Building Applications. *IEEE Internet Things J.* **2019**, *6*, 9020–9029. [CrossRef]
4. Alsuhli, G.; Khattab, A. A Fog-based IoT Platform for Smart Buildings. In Proceedings of the 2019 International Conference on Innovative Trends in Computer Engineering (ITCE), Aswan, Egypt, 2–4 February 2019; pp. 174–179.
5. Eltamaly, A.M.; Alotaibi, M.A.; Alolah, A.I.; Ahmed, M.A. IoT-Based Hybrid Renewable Energy System for Smart Campus. *Sustainability* **2021**, *13*, 8555. [CrossRef]
6. Jia, M.; Komeily, A.; Wang, Y.; Srinivasan, R.S. Adopting Internet of Things for the development of smart buildings: A review of enabling technologies and applications. *Autom. Constr.* **2019**, *101*, 111–126. [CrossRef]
7. Vijayan, D.S.; Rose, A.L.; Arvindan, S.; Revathy, J.; Amuthadevi, C. Automation systems in smart buildings: A review. *J. Ambient Intell. Humaniz. Comput.* **2020**, 1–13. [CrossRef]
8. Zhang, X.; Pipattanasomporn, M.; Chen, T.; Rahman, S. An IoT-Based Thermal Model Learning Framework for Smart Buildings. *IEEE Internet Things J.* **2020**, *7*, 518–527. [CrossRef]
9. Singh, S.; Roy, A.; Selvan, M.P. Smart Load Node for Nonsmart Load Under Smart Grid Paradigm: A New Home Energy Management System. *IEEE Consum. Electron. Mag.* **2019**, *8*, 22–27. [CrossRef]
10. Saleem, Y.; Crespi, N.; Rehmani, M.H.; Copeland, R. Internet of Things-Aided Smart Grid: Technologies, Architectures, Applications, Prototypes, and Future Research Directions. *IEEE Access* **2019**, *7*, 62962–63003. [CrossRef]
11. Abir, S.M.A.A.; Anwar, A.; Choi, J.; Kayes, A.S.M. IoT-Enabled Smart Energy Grid: Applications and Challenges. *IEEE Access* **2021**, *9*, 50961–50981. [CrossRef]
12. Shakerighadi, B.; Anvari-Moghaddam, A.; Vasquez, J.; Guerrero, J. Internet of Things for Modern Energy Systems: State-of-the-Art, Challenges, and Open Issues. *Energies* **2018**, *11*, 1252. [CrossRef]
13. Wu, Y.; Wu, Y.; Guerrero, J.; Vasquez, J.; Palacios-García, E.; Guan, Y. IoT-enabled Microgrid for Intelligent Energy-aware Buildings: A Novel Hierarchical Self-consumption Scheme with Renewables. *Electronics* **2020**, *9*, 550. [CrossRef]
14. Choi, J.S. A Hierarchical Distributed Energy Management Agent Framework for Smart Homes, Grids, and Cities. *IEEE Commun. Mag.* **2019**, *57*, 113–119. [CrossRef]
15. Aliero, M.S.; Asif, M.; Ghani, I.; Pasha, M.F.; Jeong, S.R. Systematic Review Analysis on Smart Building: Challenges and Opportunities. *Sustainability* **2022**, *14*, 3009. [CrossRef]
16. Fanti, M.P.; Mangini, A.M.; Roccotelli, M.; Ukovich, W. A District Energy Management Based on Thermal Comfort Satisfaction and Real-Time Power Balancing. *IEEE Trans. Autom. Sci. Eng.* **2015**, *12*, 1271–1284. [CrossRef]
17. Plageras, A.P.; Psannis, K.E.; Stergiou, C.; Wang, H.; Gupta, B.B. Efficient IoT-based sensor BIG Data collection–processing and analysis in smart buildings. *Futur. Gener. Comput. Syst.* **2018**, *82*, 349–357. [CrossRef]

18. Yuan, F.; Yao, R.; Sadrizadeh, S.; Li, B.; Cao, G.; Zhang, S.; Zhou, S.; Liu, H.; Bogdan, A.; Croitoru, C.; et al. Thermal comfort in hospital buildings—A literature review. *J. Build. Eng.* **2022**, *45*, 103463. [CrossRef]
19. Tekler, Z.D.; Low, R.; Yuen, C.; Blessing, L. Plug-Mate: An IoT-based occupancy-driven plug load management system in smart buildings. *Build. Environ.* **2022**, *223*, 109472. [CrossRef]
20. Staddon, S.C.; Cycil, C.; Goulden, M.; Leygue, C.; Spence, A. Intervening to change behaviour and save energy in the workplace: A systematic review of available evidence. *Energy Res. Soc. Sci.* **2016**, *17*, 30–51. [CrossRef]
21. Tekler, Z.D.; Low, R.; Blessing, L. User perceptions on the adoption of smart energy management systems in the workplace: Design and policy implications. *Energy Res. Soc. Sci.* **2022**, *88*, 102505. [CrossRef]
22. Elsisi, M.; Tran, M.-Q.; Mahmoud, K.; Lehtonen, M.; Darwish, M.M.F. Deep Learning-Based Industry 4.0 and Internet of Things towards Effective Energy Management for Smart Buildings. *Sensors* **2021**, *21*, 1038. [CrossRef]
23. Pandharipande, A.; Zhao, M.; Frimout, E. Connected Indoor Lighting Based Applications in a Building IoT Ecosystem. *IEEE Internet Things Mag.* **2019**, *2*, 22–26. [CrossRef]
24. SONOFF Smart Plug Products. Available online: https://sonoff.tech/products/ (accessed on 10 January 2022).
25. ESPurna Firmware. Available online: https://github.com/xoseperez/espurna (accessed on 10 January 2022).
26. LoRaWAN Air Quality Sensor. Available online: https://www.dragino.com/products/lora-lorawan-end-node/item/174-laq4.html (accessed on 10 January 2022).
27. LG308 Indoor LoRaWAN Gateway. Available online: https://www.dragino.com/products/lora-lorawan-gateway/item/140-lg308.html (accessed on 10 January 2022).
28. Davis 6152 Wireless Vantage Pro2. Available online: https://www.davisinstruments.com/pages/vantage-pro2 (accessed on 10 January 2022).
29. The Things Network. Available online: https://www.thethingsnetwork.org/ (accessed on 1 May 2022).
30. Liu, Y.; Akram Hassan, K.; Karlsson, M.; Pang, Z.; Gong, S. A Data-Centric Internet of Things Framework Based on Azure Cloud. *IEEE Access* **2019**, *7*, 53839–53858. [CrossRef]
31. Pierleoni, P.; Concetti, R.; Belli, A.; Palma, L. Amazon, Google and Microsoft Solutions for IoT: Architectures and a Performance Comparison. *IEEE Access* **2020**, *8*, 5455–5470. [CrossRef]
32. Digital Ocean. Available online: https://www.digitalocean.com/, (accessed on 1 May 2022).
33. Eclipse Mosquitto. Available online: https://mosquitto.org/, (accessed on 1 May 2022).
34. Krishnamurti, T.; Schwartz, D.; Davis, A.; Fischhoff, B.; de Bruin, W.B.; Lave, L.; Wang, J. Preparing for smart grid technologies: A behavioral decision research approach to understanding consumer expectations about smart meters. *Energy Policy* **2012**, *41*, 790–797. [CrossRef]

sensors

MDPI

Article

Design and Development of an IoT Smart Meter with Load Control for Home Energy Management Systems

Omar Munoz, Adolfo Ruelas *, Pedro Rosales, Alexis Acuña, Alejandro Suastegui and Fernando Lara

Facultad de Ingeniería, Universidad Autónoma de Baja California, Blvd. Benito Juárez S/N, Mexicali 21280, Mexico
* Correspondence: ruelasa@uabc.edu.mx

Abstract: Electricity consumption is rising due to population growth, climate change, urbanization, and the increasing use of electronic devices. The trend of the Internet of Things has contributed to the creation of devices that promote the thrift and efficient use of electrical energy. Currently, most projects relating to this issue focus solely on monitoring energy consumption without providing relevant parameters or switching on/off electronic devices. Therefore, this paper presents in detail the design, construction, and validation of a smart meter with load control aimed at being part of a home energy management system. With its own electronic design, the proposal differs from others in many aspects. For example, it was developed using a simple IoT architecture with in-built WiFi technology to enable direct connection to the internet, while at the same time being big enough to be part of standardized electrical enclosures. Unlike other smart meters with load control, this one not only provides the amount of energy consumption, but rms current and voltage, active, reactive, and apparent power, reactive energy, and power factor—parameters that could be useful for future studies. In addition, this work presents evidence based on experimentation that the prototype in all its readings achieves an absolute percentage error of less than 1%. A real-life application of the device was also demonstrated in this document by measuring different appliances and switching them on/off manually and automatically using a web-deployed application.

Keywords: smart meter; power meter; internet of things; load control; energy meter; smart socket

Citation: Munoz, O.; Ruelas, A.; Rosales, P.; Acuña, A.; Suastegui, A.; Lara, F. Design and Development of an IoT Smart Meter with Load Control for Home Energy Management Systems. *Sensors* **2022**, 22, 7536. https://doi.org/10.3390/s22197536

Academic Editors: Antonio Cano-Ortega, Francisco Sánchez-Sutil and Antonino Laudani

Received: 24 August 2022
Accepted: 29 September 2022
Published: 5 October 2022

1. Introduction

Energy consumption all over the world is increasing mainly because of population growth, urbanization, and new technological trends that need a large amount of electricity to work, such as smartphones, electric cars, and the mining of cryptocurrencies. However, what is the problem with the high usage of power? We can examine this question from different perspectives, such as the production approach. According to [1], fossil fuels are the main source of energy worldwide, making up 62% of total consumption by 2021. The problems with this class of resources are that they are non-renewable, so the more they are used, the faster they disappear, and they also contribute significantly to global pollution and climate change.

Additionally, many power plants or equipment installed inside the distribution infrastructure are not ready to handle the new levels of energy consumption that are required by trend technologies. This results in inadequate power supply during peak hours for end-users or even complete power outages [2]. The above issues are common in microgrids, which are decentralized power systems composed of small, diverse sources of energy that operate independently or in parallel with the main grid.

The increase in power consumption comes with a series of challenges that can be addressed through better energy efficiency, which can be encouraged by the implementation of Smart Grids (SGs). This distribution method enables a balance between supply and consumption through an effective management based on the use of modern technologies of measuring and communication [2–5].

Traditional electric grids suffer from significant limitations because they do not have the capability of anticipating or responding to sudden failures that may occur within the structure. The nature of these power grids is their monodirectional communication with end-users. Therefore, the supplier company does not receive timely feedback about problems presented that might help it resolve them [6,7]. However, the bidirectionality of smart grids, which allow both electricity production and demand to be coordinated, may result in a better energy efficiency and in a higher level of customer comfort. For example, in a conventional system when the required energy is greater than the available, the supply company chooses to carry out total load shedding in areas of lesser commercial interest, whereas in smart grids, information is delivered practically instantly from the end user, so strategic areas or even appliances can be located to reduce consumption [2].

Following the information presented, how does a power grid become smart? To answer this question, we need to take into account the instrument that enables the main feature of SGs, that is, bidirectional communication. This characteristic is possible due to the Smart Meter (SM), which is considered the key component inside this distribution architecture. The smart meter is capable of measuring many electrical parameters, displaying locally or remotely the gathered information and sometimes controlling loads [2,3,5,6].

The efficient disconnection of loads due to insufficient energy generation is one of the most important problems in the field of smart grids; however, the consumer domain has been the least explored [8]. In isolated microgrids, for example, this type of situation is common, since the electricity produced is heavily reliant on renewable resources that are available and/or stored. In this situation, if the available power is divided by the number of dwellings inside the grid, the power will be variable. Therefore, this type of context is where load control management at the appliance level plays an extremely significant role. In recent years, in order to overcome the problem of the total electrical blackout, an important area of research has been attracting growing interest since it focuses on the design of Home Energy Management Systems (HEMS) that benefits both utility companies and end-users [9].

In HEMS, the main goal is to ensure the user's comfort while minimizing energy consumption so as to achieve a balance between the supplier and demand. In energy management on the domestic demand side, during the maximum usage window, there are multiple limitations to optimally schedule loads. According to [10], household devices can be categorized into two types: schedulable and non-schedulable. Moreover, schedulable appliances can be interruptible and non-interruptible [2]. For example, a water heater can be considered a programmable non-interruptible appliance, and a garden water pump can be a programmable interruptible appliance.

The Internet of things (IoT) is an ideal architecture for the creation of energy management systems that address the interaction with human beings, mainly because the devices with an IoT scheme are those that are in contact with an environment and provide feedback to the people through the internet [11]. Considering the above, and the importance of smart meters and the management of household appliances within HEMS in a smart grid system, this work presents the design and development of a smart meter within an IoT scheme aimed at monitoring and controlling loads on the domestic demand side according to their energy consumption, and thus consequently avoiding collapses or blackouts. The development of this device, which is called Smart Meter with Load Control (SMLC) in this document, took into account a few areas of opportunity found in related projects that are studied in the next section. As a result, our work differs from the others in the following ways:

- Simple IoT architecture (no gateway needed);
- Own design electronic implementation;
- Non-invasive electronic instrumentation;
- Integration of electrical parameters monitoring and load control;
- Scalability in standardized electrical installations;
- Presentation of calibration and validation of measurements.

The hardware application of the prototype presented in this document was carried out in different configurable priority environments, and the experiments took place in a laboratory space. However, the developed device can also be used to optimize the use of the energy generated by solar power plants, wind power plants, or some other green energy sources used in isolated systems such as in a microgrid.

The content of this paper has been organized as follows: After this introductory section, a review of the related works is presented in Section 2. The description and the process of development of the proposed device can be found in Section 3. Section 4 presents details of the experimental setup, an implementation in a real environment, the performance analysis, and results. Finally, the paper is concluded in Section 5.

2. Related Works

In the state of the art, different projects were found which address the management of loads at a domestic level from different approaches. For instance, Khan et al. in [12] conducted a systematic review of various home energy management schemes. Several topics were discussed, such as the advantages of HEMS, the coordination of Distributed Energy Resources (DER) (local generation) and/or appliances mixed with different tariff schemes that lead to an efficient electrical energy usage, and also the challenges of hardware that each architecture faces. In addition, Qureshi et al. investigated in [13] the existence of energy management systems for smart homes. According to the flaws that they found in the reviewed projects, they proposed an energy management scheme for smart homes based on the Internet of Things (IoT). Their design has a security mechanism to control end-to-end communication and the use of smart scheduling and time management for controllable and non-controllable household loads in order to monitor and reduce energy consumption.

Additionally, some researchers have studied the effects resulting from the demand control. For example, in [14], the National Renewable Energy Laboratory (NREL) of the United States conducted a study to identify the most effective way to reduce plug load energy use, using three different approaches. One of them and the most effective method was an automated energy management system which turns off equipment when it is unused for a certain period of time. In addition, 126 persons were tested with this technique and obtained a 21% energy reduction from the baseline.

Klein et al. in [15] simulated the operation of a multi-agent system. Their strategy is about taking real information from a building and combining it with parameters given by the occupants in order to manage and coordinate the different devices inside the building. A 12% reduction in energy consumption and a 5% improvement in occupant comfort are the impact they achieved. The proposal was never implemented in the real world. Similarly, a comprehensive automation system for buildings was discussed in [16], where they demonstrated, through a simulation, how the use of electrical energy is reduced by controlling objects like heating ventilation, air-conditioning, lighting, and plugs.

On the other hand, some other studies are focused on the development or implementation of algorithms in the demand-side energy management framework. That is the case of Ahmed et. al in [17]. For HEMS architecture, they created a new real-time load controller with a scheduling technique based on a Binary Tracking Search Algorithm (BBSA). The goal of this project is to achieve energy savings and limit household peak demand based on the scheduled operation of various appliances according to specific time, resident comfort restrictions, and priorities. Similarly, ref. [18] implemented a reinforcement learning algorithm to a home energy management system with the purpose of optimizing the household electric appliances power demand. It is important to highlight that, in the presented work, a smart meter is the source of data for the applied algorithm. According to the simulations, the approach this research took can save between 6.23% and 11.54% of electricity costs.

Rocha et al. published in [19] an artificial intelligence (AI) algorithm for energy management on the demand side in smart homes. With a new methodology, they combined three AI techniques to solve the planning of power demand in smart homes and reach a harmony between the cost of energy and user comfort. Using the techniques of Elitist

Non-dominated Sorting Genetic Algorithm II, Support Vector Regression and K-means clustering, demand management was implemented taking into account the fluctuations in the price of electricity over time and the priority of appliances. Furthermore, they were able to consider forecasts of a distributed generation for the next day and determine user comfort levels.

Other authors have also covered the topic of HEMS from the perspective of developing and implementing devices for switching on/off household appliances, or measuring the amount of energy consumed by those appliances. For instance, Kang et al. introduced a light-powered remote control system that consumes absolute zero power in standby mode. The goal of this scheme is to reduce the energy usage of appliances when they are in standby mode. In their design, a 15 mW laser diode is mounted on a commercial remote controller. A 2 cm × 2.5 cm photovoltaic array, an autonomous connection circuit (ACC), and a latch type power relay are mounted on a receiver unit it does not have any power supply. This receiver is a bridge between the appliance and the AC power line, so it can completely de-energize equipment when they receive the shutdown signal from the remote control. The receiver does not have any power supply, but when it receives light, it energizes a capacitor and connects the appliance again [20].

The NREL in [21] presented research focused on plug load control and behavioral change in office buildings. The study consisted of a deployment of advanced power strips (APS) in GSA offices along with two plug load reduction strategies: schedule timer and load sensing. Under the test conditions, APS implementation resulted in an average electricity savings of 21% for laptops, 35% for printers, 7% for monitors, 12% for under-cabinet lights, and 48% for shared equipment (office and kitchen combined). The APS characteristics were four receptacles for plug-in devices, a fuse that trips at 1800 watts (W) and a manual reset button, which allows the user to override the controls that were programmed into the device. The APS used does not have direct internet communication so it has to transfer data through Zigbee to a gateway, which must be within 50 m (164 feet) of the APS.

Park et al. proposed in [22] a Smart Energy Management Systems SEMS based on a smart power strip and motion sensors. The power strip uses ZigBee wireless communication and relays to control sockets, as well as current transformers and an integrated circuit to measure energy consumption in individual plugs. The SEMS can turn on/off loads in two ways, depending on the test room activity based on motion sensors (whether or not people are present) or according to a predetermined time of use. This SEMS does not use an IoT architecture but rather has a computer server that allows the user to set timers and view only the power and energy consumption.

In [8], Spanò et al. proposed an architecture for a smart meter based on the Internet of Things with the intention to be part of the smart grid infrastructure. The scheme presented is focused on the end-user in order to enable smart home applications such as smart plugs. The device is capable of turning on/off electronic devices and also providing some electrical parameters such as active, reactive and apparent power, power factor, and rms current and voltage. The smart plug is based on the energy measurement unit ADE7953, and it uses a shunt resistor as a current sensor, which requires an invasive application to function. This outlet does not have direct internet communication, but it transmits directly to a gateway using ZigBee technology, which is responsible for sending all the information to the cloud.

In the same way, Tsai et al. [23] worked on a residence energy control system based on a wireless smart socket and IoT. Their implementation has three major components including smart socket, home gateway, and energy controller. The smart socket was equipped with a digital power meter which supports between 50 V and 350 V, and current from 10 mA to 15 A. The smart socket itself does not have direct internet communication. It transmits the information via ZigBee to a gateway equipped with 64 MB SDRAM, ZigBee module, 100 Mbps Ethernet inter- face, and USB I/O interface. The energy controller was developed on a server with an Intel i5-2300 2.8 GHz processor, 16 GB RAM, 1 TB hard disk, and a Linux 3.8.13 operating system. With all that hardware, the system provides four control modes, including peak-time control, energy-limit control, automatic control, and user control. In addition, they showed how the proposed scheme could save up to 43.4% of energy for some appliances in one weekday, but there is no electrical parameters' validation.

Pawar and Vittal K in [2] worked on the design and development of an intelligent energy management system integrated into the IoT framework and addressed to a smart grid environment, which is based on a smart socket module. The electronic circuit they made is big compared to conventional plugs because it is built based on existing electronic modules, such as the Arduino Uno, a relay module for load control and an Xbee for wireless communication. It also has the LEM LV-25P voltage transducer which requires a transformer to be implemented and the LEM LA-55P current transducer. Their system only provides the electrical parameters of rms current and voltage, apparent power, power factor, and energy in watt-hours. As in [8], the smart socket module does not have direct internet communication, so it sends all the collected data by Xbee to a gateway, which can upload the information to a cloud database.

Similarly, in [24], an IoT smart socket for electricity control in a home environment was presented. The system uses two invasive current sensors, two relays to switch on/off up to two loads per device, an AC/DC converter to supply the whole circuit from the line power, and the Wemos D1 Mini development board with a WiFi module to control the complete system and enable internet communication. All the components and connections were enclosed inside a wall socket. However, the system did not include any voltage sensors, so in order to compute the power consumption, a voltage of 220 V rms was assumed. In the web application, the user can monitor current from the smart socket plugged, turning on/off the electricity switch manually and setting a timer for turning on/off the smart socket.

An Internet of Things smart energy meter for monitoring energy usage in a device-level was presented in [25]. Their concept consists of an outlet capable of obtaining rms current and voltage and active power and energy, but it does not have the feature of controlling the load of what is connected to it. Karthick et al. in [26] designed and built an IoT-based smart compact energy metering system to monitor and control energy usage and power quality with demand-side management for a commercial building. In their scheme, there are groups of primary and secondary loads to control and monitor their consumption, but there is not a measurement of energy in individual household devices. The system as a whole has a distributed architecture, which has a central measurement system based on the PZEM-004T (sensor with an invasive application) and different smart switches. Each component uses the ESP8266 to communicate with a Raspberry Pi, which is responsible for calculating some other electrical parameters and sending the information to the cloud.

To conclude with this section, ref. [27] conducted an investigation that provided valuable information for the design and implementation of smart energy management systems. The authors focused on providing a better understanding of user perception and motivations when adopting energy management systems for plug loads in the workplace. With a comprehensive analysis of what they obtained in the research, they proposed seven design implications that could improve the following areas in SEMS: external and internal influence, user appeal, user control, reliability, ease of use, personalized and contextualized information, and data privacy. The same authors, but in [28], worked with strategies to improve the implementation of plugs with load control. Tekler et al. state that real-world applications in this area remain relatively unexplored due to several issues related to deployment viability, energy-saving potentials, and system acceptance. For the above, they presented a novel IoT-based occupancy-driven plug load management system, called "Plug-Mate", designed to reduce plug load energy consumption and user burden through intelligent plug load automation. The researchers spread 30 smart plugs inside a university office space that recorded users' real-time plug load power consumption, which was transmitted to a gateway device via Z-wave communication protocol. They proposed and applied different levels of plug load automation, including manual, predefined schedules, and occupancy-driven, all of them implemented from an online user interface. With the above strategies, they achieved an average energy savings of 51.7% among different plug load types evaluated.

Discussion

As a result of the study of the existing works that address home energy management systems by developing and implementing devices with load control and energy monitoring, it was concluded that there are aspects that need to be improved. For example, those based on IoT architectures that require more than one device for internet connection make it more difficult for users to implement and adapt. In addition, some approaches do not use their own electronic circuits, limiting them to the characteristics provided by smart meter manufacturers or electronic module manufacturers. Additionally, customized circuits make it easier to develop devices that are scalable with standardized electrical systems.

Likewise, how current is measured is a topic to take into account. An important drawback of invasive sensors, such as shunts, is that they are unavoidably electrically connected to the current to be measured and the sense circuit, which means there is no isolation and that the whole circuit is less protected. The above is not the case of current transformers or some Hall effect sensors [29].

Even though the main idea in HEMS is the efficient consumption of energy, there is no reason to only present this electrical parameter. Smart meters with load control that provide more electrical parameters (such as active, reactive, apparent power, power factor, line frequency, etc.) can be used in a wide range of scenarios, including detecting appliance failures, or detecting loads automatically with machine learning algorithms.

Furthermore, how the user interacts with energy management systems is very important for adaptation. Some applications only focus on automatic control and do not provide direct manual control for the user. Some others, however, use manual control only and do not incorporate automatic functions to make their systems more efficient. In addition, many studies do not present how their devices were calibrated and a validation of their measurements, which adds an uncertainty regarding how correct the information they provide is. Moreover, a demonstration of how the proposed system functions is very important, so it is critical to illustrate how the system behaves with real appliances in real situations.

A comparison is presented in Table 1 among existing works related to the application of energy management systems by developing and implementing devices that control load and monitor energy consumption.

Table 1. Comparison among existing works that apply energy management systems by developing and implementing devices that control load and monitor energy consumption.

Authors	(1)	(2)	(3)	(4)	(5)	(6)	(7)	(8)	(9)	(10)	(11)	(12)
Kang et al. [20]	No	No	Yes	Yes	No	No	Yes	No	1	No	No	Yes
Metzger et al. [21]	Yes	No	No	Yes	Yes	No	No	yes	4	No	No	Yes
Park et al. [22]	No	No	Yes	Yes	Yes	No	No	Yes	3	No	No	Yes
Spanò et al. [8]	Yes	No	Yes	No	Yes	Yes	Yes	Yes	1	Yes	Yes	Yes
Tsai et al. [23]	Yes	No	Yes	-	Yes	Yes	Yes	Yes	1	No	No	Yes
Pawar et al. [2]	Yes	No	Yes	Yes	Yes	Yes	No	Yes	1	No	No	Yes
Phangbertha et al. [24]	Yes	Yes	No	No	Yes	No	Yes	Yes	2	Yes	No	Yes
Muralidhara et al. [25]	Yes	Yes	Yes	No	Yes	No	Yes	Yes	1	-	Yes	Yes
Karthick et al. [26]	Yes	No	Yes	yes	Yes	Yes	Yes	Yes	-	No	No	Yes
Tekler et al. [28]	Yes	No	No	-	Yes	No	Yes	Yes	1	No	No	No
Us	Yes	Yes	Yes	Yes	Yes	Yes	Yes	Yes	2	Yes	Yes	Yes

Note: (1) Access to Internet; (2) Simple IoT architecture (no gateway needed, on board internet connection); (3) Electronic board design and implementation; (4) Non-invasive current sensor; (5) Energy consumption; (6) Other electrical parameters (V rms, A rms, VA, VAR, pF, etc.); (7) Remote manual load control; (8) Automatic load control; (9) Number of sockets per device; (10) Scalability in standardized electrical installations; (11) Include calibration or validation of measurements; (12) Implementation in a real environment.

3. Description and Development of the Smart Meter with Load Control

In order to have a better understanding of the usefulness of the device proposed in this document, an example of application in an isolated microgrid context can be seen in Figure 1. As mentioned above, with this type of generation and distribution grid, the amount of electricity that homes can use is limited by the availability of renewable sources, and, when power is insufficient, it is necessary to limit it among users. Therefore, the use of a smart central meter placed before the power panel is essential to know how much power is consumed in each dwelling in real time. As a result of the information that is collected for each house and the avoidance of a total power outage in the grid, the smart meter with load control in an outlet format becomes an important tool, since it is capable of sending electrical data (rms voltage and current, active and reactive energy, power factor, and active, reactive and apparent power) through the internet about the electrical equipment connected to it, allowing better and substantiated decision-making. Furthermore, with this device, strategic appliances could be turned on or off remotely, and they could also be categorized in order to schedule and prioritize their usage.

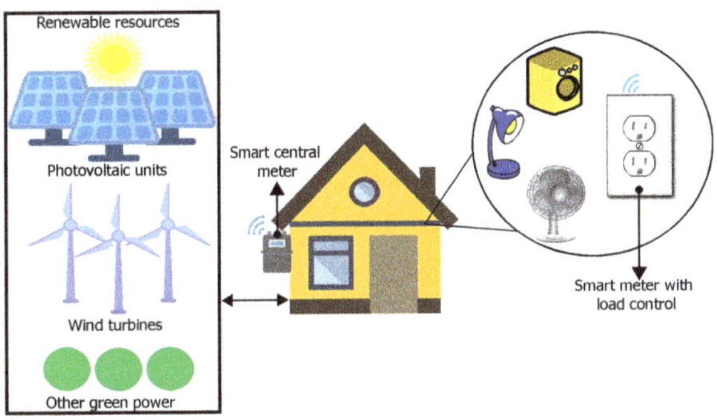

Figure 1. Environment of usage of the smart meter with load control.

3.1. Architecture

The smart meter and load controller is used here to replace the traditional electrical outlet, measure energy consumption at a device level, and allow the on/off switching of equipment connected to it. In addition, the complete functionality of the system is monitored wirelessly through a web application. In order to achieve the above features, the device's scheme is built around the ADE7758 as an energy measurement unit, the ESP32 microcontroller, the CST-1020 current transformer, a resistive attenuator for the voltage input, the SRA-05VDC-CL relays, and an integrated power supply HLK-PM01. The architecture used can be seen in the diagram of Figure 2.

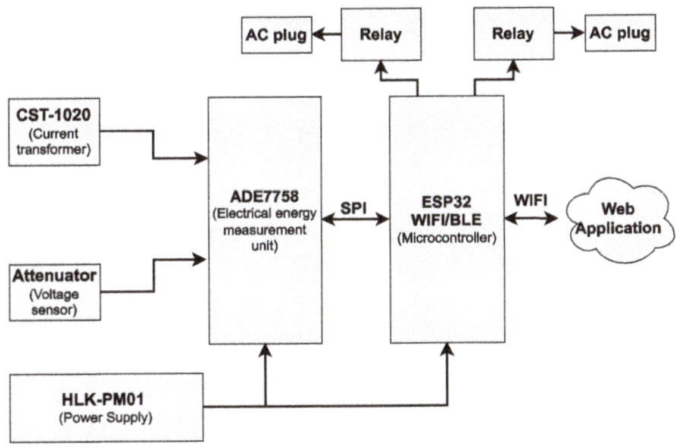

Figure 2. Architecture of the smart meter with load control.

3.1.1. Energy Measurement Unit: ADE7758

The ADE7758 is a high accuracy three-phase electrical energy measurement IC that supports the implementation of IEC 60687, IEC 61036, IEC 61268, IEC 62053-21, IEC 62053-22, and IEC 62053-23 standards. It has an SPI serial communication interface and two pulse outputs to interact with external equipment. The ADE7758 incorporates a second order Delta-Sigma type ADC, a digital integrator, reference circuits, a temperature sensor and also the implementation algorithms to determine the active, reactive and apparent power, active and reactive energy, and rms voltage and current calculations, all in a dynamic range of 1000:1. Many three-phase configurations can be used, either for delta or star services of three or four cables, but it can be also implemented for single-phase systems; such is the case in this project.

3.1.2. Microcontroller: ESP32

The ESP32 is a 2.4 GHz Wi-Fi and Bluetooth microcontroller created by Espressif Systems and manufactured by TSMC with 40 nm ultra-low power technology. The product is designed to be robust and reliable in a variety of applications and power scenarios, and to provide optimal RF performance and power consumption. ESP32 is designed for mobile applications, wearable electronics and projects based on the Internet of Things platform. It features all the state-of-the-art characteristics of low-power chips, including fine-grained clock gating, multiple power modes, and dynamic power scaling. In addition, the ESP32 includes a dual core CPU, a 520 KiB SRAME memory, and peripheral interfaces, such as I2C, SPI, I2S, UART, CAN BUS, etc.

3.1.3. Current Sense Input

The ADE7758 has six analog inputs divided into two sets for current and voltage measurement. The current group consists of three pairs of fully differential voltage inputs: IAP and IAN, IBP and IBN, and ICP and ICN, of which just the first two were used. These fully differential voltage input pairs have a maximum differential signal of ±0.5 V. Due to the above, as well as the size of the traditional electrical outlets and the fact that they typically handle up to 15 A, an insert mount transformer (CST-1020) was used as current sensor. It has a turns ratio of 1000:1 and is capable of handling 20 A. It is also able to operate at 50 Hz as well as 60 Hz. For the electronic instrumentation, shunt-type load resistors were placed at the output of the secondary winding of the transformer to generate a voltage signal that is directed to IAP and IAN. In addition, RC low pass filters with a corner frequency of 4.8 kHz were used on these analog inputs, Figure 3.

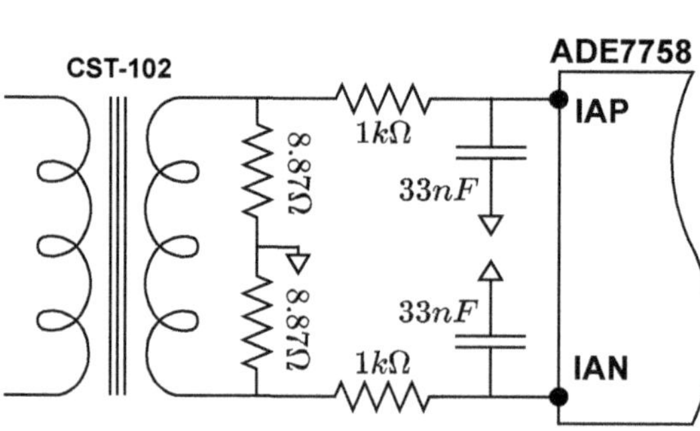

Figure 3. Electronic instrumentation of current input.

3.1.4. Voltage Sense Inputs

Figure 4 shows the phase voltage channel signal path on the SMLC circuit. The voltage group has three single-ended voltage inputs: VAP, VBP, and VCP. These single-ended voltage inputs have a maximum input voltage of ±0.5 V with respect to VN. Only VAP and VN were used here. The line voltage is attenuated using a simple resistor divider network before it is presented to the ADE7758. The attenuation network with a ratio of 1000:1 on the voltage channels is designed such that the corner frequency (3 dB frequency) of the network matches that of the RC (anti-aliasing) filters on the current channels inputs.

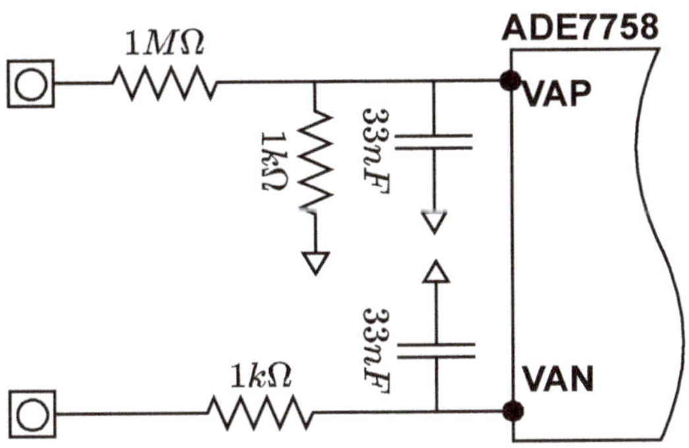

Figure 4. Electronic instrumentation of voltage input which allows a maximum input of 353 V rms.

3.1.5. Load Switcher

In conventional outlets, there are at least two sockets to power different instruments at the same time; this is why the SMLC is designed to switch on/off two plugs individually. This function is achieved by using two electromechanical relays, specifically the SRA-05-VDC-CL. According to its technical specifications, the coil's nominal voltage is 5VDC, its nominal current is 120 mA, and it can handle loads up to 20 A and switch currents up to 10 A. This relay was chosen because of the good relationship between size and performance.

3.2. Printed Circuit Board and Enclosure

All the components of the architecture were taken into account in designing a schematic and a two-layer PCB, which is shown already manufactured with a size of 3.3″ × 3″ and with all the elements soldered in Figure 5. The SMLC aims to replace the traditional domestic electrical outlet so the electronic circuit was placed inside a 4″ × 4″ metal electrical wall box. On the circuit board, neutral and phase were connected to the voltage inputs, the hot wire was also passed through the current transformer, and the relays were wired to a duplex socket, where each plug was labeled as "A" and "B". Figure 6 is the final prototype of the SMLC.

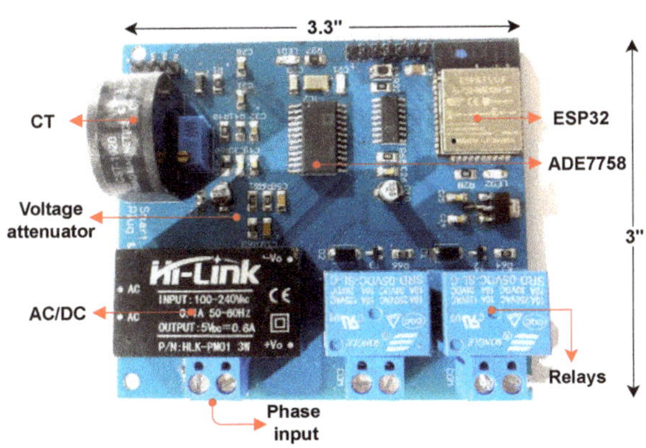

Figure 5. Printed circuit board and components of the SMLC.

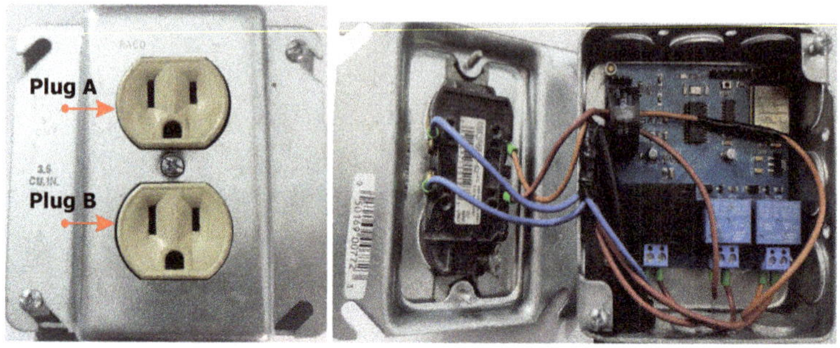

(a) (b)

Figure 6. (**a**) Plugs of the SMLC; (**b**) electrical wiring of the SMLC.

3.3. Calibration

The calibration of the SMLC is a procedure for configuring some registers of the ADE7758 through spi communication, for which the manufacturer supplies a method called line accumulation [30]. In this process, the target is to determine the offset for rms voltage and current, the gain for active, reactive, and apparent power, as well as the phase delay and the offset for active and apparent energy. Part of the calibration process is to measure electrical parameters under different electrical load conditions and compare them with a reference meter, for which the HIOKI PW3360-20 was used, whose characteristics can be found in [31].

3.3.1. Calibration of rms Voltage and Current

Adding an offset to the input signals helps to reduce the noise or previous offset that can appear while a measurement is in process. This can be accomplished by modifying the$xIRMSOS$ (0x36) and $xVRMSOS$ (0x33) registers of the ADE7758. To calculate the value to establish in these registers, the following steps were carried out:

1. Activation of zero crossing detection on the input phase by modifying the $LCYCMODE$ (0x17) register;
2. Modification of the register $MASK$ (0x18) to allow the interrupt pin to be activated with a zero crossing of phases;
3. Set up the calibration system to achieve a test rms current, nominal rms voltage, and minimum rms current and voltage;
4. Average of N samples from the lecture of the registers $xIRMS$ (0x0A) and $xVRMS$ (0x0D) after each interruption caused by the zero crossing detection;
5. Calculation of the offsets with Equations (1) and (2).

$$xIRMSOS = \frac{1}{16384} \times \frac{(I_{TEST}^2 \times IRMS_{IMIN}^2) - (I_{MIN}^2 \times IRMS_{ITEST}^2)}{I_{MIN}^2 - I_{TEST}^2}. \quad (1)$$

$$xVRMSOS = \frac{1}{64} \times \frac{(V_{NOM} \times VRMS_{VMIN}) - (V_{MIN} \times VRMS_{VNOM})}{V_{MIN} - V_{NOM}}. \quad (2)$$

6. Adjustment of the registers *xIRMSOS* (0×36) and *xVRMSOS* (0×33) with the values calculated.

For the calibration system, a test current (I_{TEST}) of 10 A rms and a minimum current (I_{MIN}) of 0.052 A rms were used. In Equation (1), $IRMS_{IMIN}$ corresponds to the value of the register $xIRMS$ (0x0A) when I_{MIN} is measured, as well as $IRMS_{ITEST}$ with I_{TEST}. The register gave an average of 5221 with a current of 0.052 A rms and 100 samples, whereas with 10 A rms 1,045,378. With the values of the readings, the calculation of $xIRMSOS$ resulted in 140. On the other hand, $xVRMSOS$ was calculated using a nominal voltage (V_{NOM}) of 123 V rms and a minimum voltage (V_{MIN}) of 20 V rms. In Equation (2), $VRMS_{VNOM}$ is the reading of the register $xVRMS$ (0x0D) when measuring V_{NOM}, and $VRMS_{VMIN}$ when measuring V_{MIN}. This register averaged 565,547 with 100 samples of 123 V rms and 94,088 with 20 V rms; therefore, the outcome of Equation (2) was −40.

3.3.2. Gain Power Calibration

This calibration is primarily used to adjust active, reactive, and apparent power measurements. The ADE7758 accomplishes this by utilizing its three registers: xWG, $xVARG$, and $xVAG$ (0x2A to 0x32), which can be used to increase or decrease the amplitude of the reading. In order to calculate the mentioned gains, the following steps were taken:

1. Clearing of the xWG, $xVARG$, and $xVAG$ registers;
2. Selection of phase A, B, or C for a line period measurement with register $MMODE$ (0x14);
3. Set up the ADE7758 for the line accumulation mode by writing to $LCYCMODE$ register;
4. Set the number of half-line cycles for line accumulation by modifying the register $LINECYC$ (0x1C);
5. Modification of the interrupt mask with the register $MASK$ (0x18) in order to enable the interrupt signaling the end of the line cycle accumulation;
6. Set up the calibration system. To obtain the gain for active and apparent power, it is necessary to work with a test current and a nominal voltage with a unity power factor;
7. Reset the interrupt status register by reading $RSTATUS$ (0x1A);
8. Read the energy registers $xWATTHR$ and $xVAHR$ after the interruption of line accumulation has occurred and store the values;
    (a) Calculate the values to be written to xWG register according to the following equation:

$$xWG = \left(\frac{WATTHR_{EXPECTED}}{WATTHR_{MEASURED}} - 1 \right) \times 2^{12}, \tag{3}$$

before obtaining xWG, an expected value in the register of active energy must be determined, which is represented by:

$$WATTHR_{EXPECTED} = \frac{4 \times 3200 \times I_{TEST} \times V_{NOM} \times cos(\theta) \times AccumTime \times APCFDEN}{1000 \times 3600}, \tag{4}$$

where θ represents the phase angle between the voltage and the current, and $AccumTime$ is the total energy accumulation time inside the ADE7758 according to the number of half-line cycles selected. Tacum can be determined as

$$AccumTime = \frac{No. of\, half\, cycles}{2 \times line\, frequency \times No.\, used\, phases}, \tag{5}$$

whereas $APCFDEN$ is

$$APCFDEN = INT\left(\frac{16000 \times \frac{V_{NOM}}{V_{MAX}} \times \frac{I_{TEST}}{I_{MAX}}}{\frac{3200 \times I_{TEST} \times V_{NOM}}{1000 \times 3600} \times cos(\theta)} \right). \tag{6}$$

    (b) Calculate the values to be written to the $xVAG$ register using the following equation:

$$xVAG = \left(\frac{VAHR_{EXPECTED}}{VAHR_{MEASURED}} - 1 \right) \times 2^{12}, \tag{7}$$

$VAHR_{EXPECTED}$ is the same as $WATTHR_{EXPECTED}$ as long as a unity power factor is being used for the calibration system.

9. Write the outcomes of Equations (3) and (7) into registers xWG and $xVAG$.
10. Set up the calibration system. To obtain the gain for reactive, it is necessary to work with a test current and a nominal voltage with a power factor of 0.5.
11. Repeat step 7.
12. Read the energy register $xVARHR$ after the interruption of line accumulation has occurred and store the values.
13. Calculate the values to be written to the $xVARG$ register using Equation (8):

$$xVARG = \left(\frac{VARHR_{EXPECTED}}{VARHR_{MEASURED}} - 1 \right) \times 2^{12}, \tag{8}$$

where $VARHR_{EXPECTED}$ is

$$VARHR_{EXPECTED} = \frac{4 \times 3200 \times I_{TEST} \times V_{NOM} \times sin(\theta) \times AccumTime \times VARCFDEN}{1000 \times 3600}, \tag{9}$$

and $VARCFDEN$ is calculated as

$$VARCFDEN = INT\left(\frac{16000 \times \frac{V_{NOM}}{V_{MAX}} \times \frac{I_{TEST}}{I_{MAX}}}{\frac{3200 \times I_{TEST} \times V_{NOM}}{1000 \times 3600} \times sin(\theta)} \right). \tag{10}$$

14. Write the outcome of Equation (8) into register $xVARG$.

The calibration system consisted of a nominal voltage of 123.9 V rms at 60 Hz, a test current of 10 A rms with unity power factor, which means there is no phase shift between the voltage and current signals ($\theta = 0$), and 128 half cycles for the line accumulation time. Once all the electrical physical parameters had been established, the energy consumption measurements were taken. The results were 12,862 and 12,824 for registers $xWATTHR$ and $xVAHR$, respectively, while, using Equation (4), the expected value was 11,950 for both registers. Utilizing Equations (3) and (7) and the data previously obtained, xWG and $xVAG$ were calculated, resulting in -279 and 290, respectively. According to [30], the gain adjustment for reactive power requires a power factor of 0, which was not achievable with the available loads; therefore, the lowest possible power factor was used: 0.1190, which means an angle of 83.0617° between the voltage and the current. In this calibration, the nominal voltage was 125.85 V rms, the test current was 5.06 A rms, and the line accumulation was 128 half cycles. This scenario caused the reactive energy register $xVARHR$ to return 6565, while the expected value was 6143 based on Equation (9). Using the previous data in Equation (8), the gain for active power to write in the register $xVARG$ was –264.

3.3.3. Phase Calibration

The ADE7758 includes a phase calibration register in each current channel $xPHCAL$ (0x3F to 0 x41) to compensate small phase errors caused mainly by current transformers, complex phase errors must be fixed by adjusting the values of the antialiasing filters from Figure 3. Phase calibration consists of adding a time delay that can be in a positive or negative direction. To calculate the degree of phase shift of the signal and the value to be written in the $xPHCAL$ register, the following steps were followed:

1. Repeat steps 1, 2, 3, 4, and 5 of the gain calibration to select the phase to calibrate, set the line accumulation mode, define the number half cycles in the line accumulation, and set the interrupt mask;
2. Set up the calibration system. Two active power measurements are required for this calibration, one with a nominal voltage and a test current with a unity power factor and another with a power factor of 0.5;
3. Reset the interrupt status register by reading $RSTATUS$ (0x1A);

4. Read the active energy register $xWATTHR$ after the interruption of line accumulation has occurred and store the values;
5. Repeat steps 2, 3, and 4 but using a power factor of 0.5 in the calibration system;
6. Calculate the phase error in degrees with the Equation (11):

$$phaseError(^\circ) = Arcsin\left(\frac{digitalError}{\sqrt{3}}\right), \tag{11}$$

to determine $phaseError(^\circ)$ it is necessary to obtain a digital phase error, which can be computed by:

$$digitalError = \frac{xWATTHR_{pf=0.5} - \frac{xWATTHR_{pf=1}}{2}}{\frac{xWATTHR_{pf=1}}{2}}. \tag{12}$$

7. Find the value to be written in the register $xPHCAL$ with the following equation:

$$xPHCAL = phaseError(^\circ) \times \frac{9.6\mu s}{Ph/Lsb/W} \times \frac{\frac{1}{line\ frequency \times 9.6\mu s}}{360}, \tag{13}$$

where

$$PH/Lsb/W = \begin{cases} 1.2\mu s & digitalError < 0 \\ 2.4\mu s & digitalError > 0 \end{cases}.$$

8. Modify the register $xPHCAL$ with the outcome of Equation (13).

For the phase calibration, as mentioned in step 2, two active energy measurements are required. The first one was carried out in a unity power factor with a nominal voltage of 123.6 V rms at 60 Hz and a test current of 7.21 A rms. With the above electrical conditions and 128 half cycles for the line accumulation time, the active energy register $xWATTHR$ showed a value of 8641. In the second measurement a power factor of 0.505, a nominal voltage of 123.4 V rms at 60 Hz, and a current of 7.22 A rms were used, resulting in a value of 4470 in the $xWATTHR$ register. A digital phase error of 0.0346 was obtained substituting the measured values in Equation (12), and this datum was used in Equation (11) to find that the phase had a shift of -1.1447°. For the phase error correction, the value to be written in the $xPHCAL$ register was obtained with Equation (13), resulting in a value of -44.

3.3.4. Power Offset Calibration

This calibration serves to meet exceptional performance within the dynamic measurement range of 1000:1, especially when power consumption levels are very low. The ADE7758 has offset registers for the active and reactive power, $xWATTOS$ (0x39 to 0x3B) and $xVAROS$ (0x3C to 0x3E), whereas the offset in the apparent power measurements is affected by adjusting the rms offset registers. This calibration must be performed with a test current as close as possible to the minimum current within the dynamic range, and a greater number of half cycles for the line accumulation is also required to avoid the effect of quantization errors. In order to calculate the power offsets, the following steps were taken:

1. Repeat steps 1, 2, 3, 4, and 5 of the gain calibration to select the phase to calibrate, set the line accumulation mode, define the number half cycles in the line accumulation, and set the interrupt mask;
2. Set up the calibration system with a nominal voltage and a test current achieving a unity power factor;
3. Reset the interrupt status register by reading $RSTATUS$ (0x1A);
4. Read the active energy register $xWATTHR$ after the interruption of line accumulation has occurred and store the values;

5. Calculate the value to be written in the register $xWATTOS$ with:

$$xWATTOS = \frac{offsetW \times 4}{AccumTime \times CLKIN} \times 2^{29}, \quad (14)$$

where $CLKIN$ is the oscillator frequency used for the ADE7758, and $OffsetW$ can be obtained as

$$offsetW = \frac{xWATTHR_{I_{MIN}} \times I_{TEST} - \left(xWATTHR_{I_{TEST}} \times \frac{No.of\,half\,cycles\,I_{MIN}}{No.of\,half\,cycles\,I_{TEST}}\right) \times I_{MIN}}{I_{MIN} - I_{TEST}}. \quad (15)$$

6. Modify the value of the register $xWATTOS$;
7. Modify the calibration system with a nominal voltage and a test current at a power factor of 0;
8. Reset the interrupt status register by reading $RSTATUS$ (0x1A);
9. Read the reactive energy register $xVARHR$ after the interruption of line accumulation has occurred and store the values;
10. Calculate the offset for the reactive energy with Equation (16).

$$xVAROS = \frac{offsetV \times 4}{AccumTime \times CLKIN} \times 2^{29}, \quad (16)$$

$offsetV$ is equal to

$$offsetV = \frac{xVARHR_{I_{MIN}} \times I_{TEST} - \left(xVARHR_{I_{TEST}} \times \frac{No.of\,half\,cycles\,I_{MIN}}{No.of\,half\,cycles\,I_{TEST}}\right) \times I_{MIN}}{I_{MIN} - I_{TEST}}. \quad (17)$$

11. Write the Outcome of Equation (16) in the Register $xVAROS$

In order to obtain the value of $xWATTOS$, the calibration system was set to 123.5 V rms as nominal voltage and 0.0525 A rms as a minimum test current with a unity power factor. For the line accumulation, 4096 half cycles were used, resulting in a reading in the active energy accumulation register of 1991, a value that corresponds to $xWATTHR_{I_{MIN}}$. For $xWATTHR_{I_{TEST}}$, a nominal voltage of 123.4 V rms, a test current of 7.2 A rms, and 128 half cycles for line accumulation were used, resulting in a value of 8826 in the $xWATTHR$ register. Once the $xWATTHR_{I_{MIN}}$ and $xWATTHR_{I_{TEST}}$ readings were obtained, both with different accumulation times and a 10 Mhz clock ($CLKIN$), the $offsetW$ was calculated with Equation (15) giving an outcome of 63, which was used in Equation (14) for $xWATTOS$, resulting in 397. In the same way, to calculate reactive power offset, a 123.5 V rms voltage and a minimum test current of 0.06 A rms were used, but now with a power factor as close to 0 as possible, 0.1260 in this case. The line accumulation was performed using 4096 half cycles, resulting in a value of 1831 in the reactive energy register, which was used as $xVARHR_{I_{MIN}}$. For $xVARHR_{I_{TEST}}$, a nominal voltage of 123.3 V rms and a test current of 7.22 A rms with a power factor of 0.1260 were used. These electrical conditions, along with 128 half cycles, resulted in a reading of 7655 for the reactive energy register. As a result of the above measurements, $offsetV$ had a value of 206 according to Equation (17), and, using that in Equation (16), $xVAROS$ had a value of 1396.

3.4. IoT Integration of the SMLC

One of the novelties of this work is that the SMLC is capable of providing all the electrical parameters previously mentioned in real time through the internet, and it can also receive orders to turn on/off any device that is connected to it. According to [32], there are six levels of IoT implementations and the integration of the SMLC corresponds to the last one because of the following characteristics: (1) it is designed to be an independent end node within a network of multiple SMLCs, the printed electronic circuit developed enables the wide scalability where each SMLC has an internet connection thanks to the ESP32 microcontroller. (2) The information that is sent by the SMLC is stored in a cloud database, specifically MongoDB Atlas. (3) Only electrical variables are calculated by the SMLC; everything else needed is computed in the server. (4) In order to visualize the data, we developed a web app using Node js and Express as a framework, and the PaaS (Platform as a service) Heroku for deploying. The web application can be seen in Figure 7.

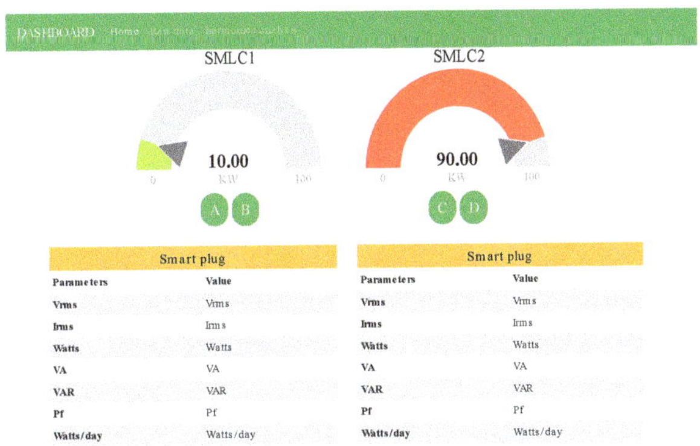

Figure 7. Web application to monitor two smart meters with load control.

It is important to point out that the communication between the SMLC and the server is carried out by the network protocol based on the publish/subscribe method MQTT over TCP/IP sockets, mainly because its publish operation is faster and consumes less energy. On the other hand, to avoid data going through the server and after that to the webapp, we implemented MQTT over websockets to receive data directly from the broker. This technique allows for achieving a latency under 500 ms, from when the measure is taken to it is shown in the cloud interface. An example of how data are transferred can be seen in Figure 8.

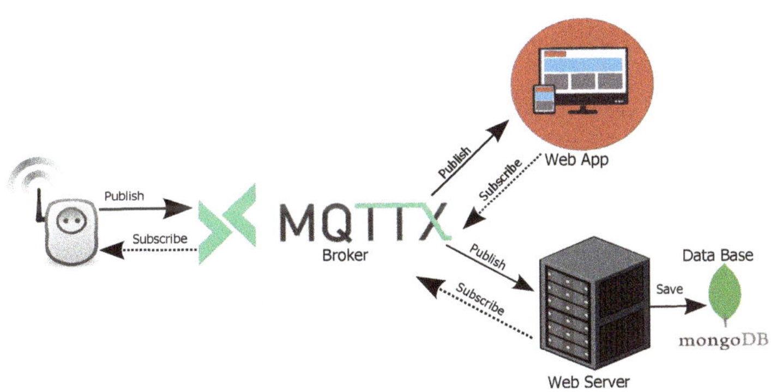

Figure 8. Data flow in the IoT scheme for the SMLC.

4. Experiments and Results

4.1. Data Validation

In the measurement tests, a modular electrical training system with switchable loads was used to control the physical scenarios of the experiments and thus create different electrical conditions. The modules applied were an AC variable power supply and banks of resistors and inductors. To compare the SMLC readings, the HIOKI PW3360-20 network analyzer was used as a reference measurement instrument. The setup for the validation can be seen in Figure 9.

Figure 9. Modular electrical training system used for calibration and validation of the SMLC readings.

In order to achieve better data validation, all the electrical parameters the SMLC can provide were tested twice, once in a unit power factor and once in a 0.5 power factor. The first experiment consisted of making different current values (from 10 A rms to 0.5 A rms) flow through the SMLC using the modules of the power supply (at 127 V rms) and the banks of resistors ($pf = 1$). Under the above conditions, the readings of rms current and active and apparent power were taken. In Table 2, all the outcomes of the SMLC and how they were compared with the HIOKI PW3360-20 measurements can be found. Calculating each error in the readings yields a mean absolute percentage error (MAPE) of 0.1340, and we also carried out a linear regression analysis, Figure 10, to determine the coefficient of determination R^2 that can be used to weigh how well the SMLC behaves against the reference.

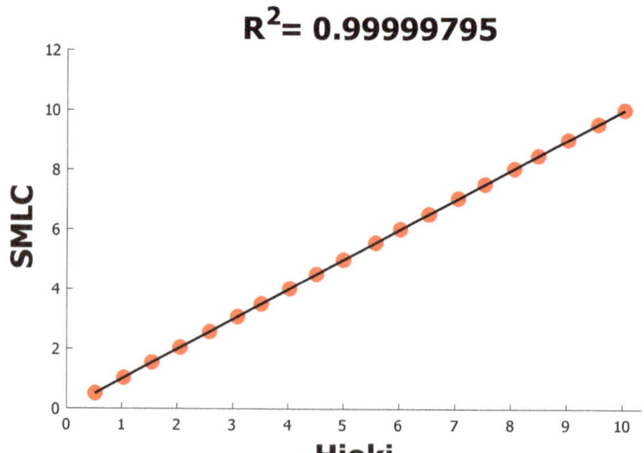

Figure 10. R^2 of SMLC against HIOKI PW3360-20 using rms current measurements.

Table 2. Comparison of rms current samples between HIOKI PW3360 and the SMLC.

HIOKI PW3360	SMLC	%Error
10.031	10.036	0.0498
9.559	9.566	0.0732
9.023	9.03	0.0776
8.489	8.494	0.0589
8.06	8.058	0.0248
7.531	7.535	0.0531
7.049	7.053	0.0567
6.532	6.529	0.0459
6.022	6.026	0.0664
5.571	5.568	0.0539
4.999	4.998	0.0200
4.516	4.512	0.0886
4.028	4.025	0.0745
3.516	3.513	0.0853
3.092	3.088	0.1294
2.582	2.58	0.0775
2.056	2.052	0.1946
1.553	1.55	0.1932
1.038	1.033	0.4817
0.516	0.512	0.7752
% MAPE		0.1340

As mentioned in the previous paragraph, in addition to the rms, current other electrical variables were tested, and the R^2 and the mean absolute percentage errors for all the different parameters are shown in Table 3. For instance, the MAPE for active power was 0.1639 and the R^2 0.99999460, whereas the apparent power results were identical to those for active power because, for this experiment, there was no phase shift between current and voltage, which means a unit power factor.

Table 3. Resume of all the mean absolute percentage errors and R^2 of different electrical variables under a unit power factor.

Electrical Parameter	%MAPE	R^2
Rms current	0.1340	0.99999795
Active power	0.1639	0.99999460
Apparent power	0.1639	0.99999460

After finishing the unit power factor trial, distinct power factors were evaluated too, and in order to do so, some banks of inductors were added to the setup to mix them with the resistor loads and thus lag the voltage against the current. Initially, the experiment had a power factor of approximately 0.95, and after that, it gradually decreased in steps of 0.05. The complete readings are in Table 4 along with the error resulting from the comparison with the HIOKI PW3360-20 values, the mean of which was 0.7976%. Moreover, the linear regression analysis is in Figure 11, with an R^2 of 0.99974430.

Table 4. Comparison of power factors samples between HIOKI PW3360 and the SMLC.

HIOKI PW3360	SMLC	%Error
0.9512	0.95	0.1262
0.9074	0.906	0.1543
0.8508	0.848	0.3291
0.8047	0.802	0.3355
0.7522	0.746	0.8242
0.7024	0.697	0.7688
0.6578	0.652	0.8817
0.6028	0.598	0.7963
0.5526	0.548	0.8324
0.5037	0.5	0.7346
0.4499	0.446	0.8669
0.4076	0.404	0.8832
0.3511	0.346	1.4526
0.2963	0.294	0.7762
0.2486	0.245	1.4481
0.2055	0.204	0.7299
0.1667	0.164	1.6197
% MAPE		0.7976

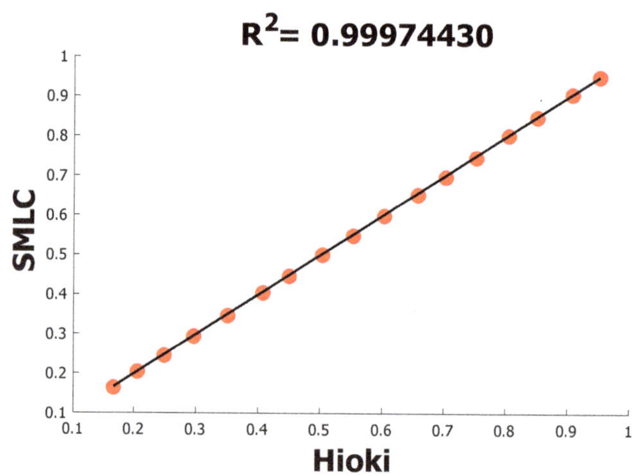

Figure 11. R^2 of SMLC against HIOKI PW3360-20 using power factors' measurements.

Similarly to the last experiment, every time a power factor reading was made, other electrical values were taken, in this case rms current and voltage and active, apparent, and reactive power. To summarize the results, Table 5 contains the MAPE for every parameter according to SMLC and HIOKI PW3360-20 measurements, as well as the corresponding R^2. In this table, it can be seen that now active and apparent powers have different values; this happens because the power factor is not a unit anymore, which gives a MAPE for active powers of 0.4623% and an R^2 of 0.99992047, whereas, for the apparent powers, the MAPE was 0.0900% and the R^2 0.99994137. It was possible to evaluate the reactive power using this test, which resulted in a MAPE of 0.2443% and R^2 of 0.99974305.

Table 5. Resume of all the mean absolute percentage errors and R^2 of different electrical variables under different power factors.

Electrical Parameter	%MAPE	R^2
Power factor	0.7976	0.99974430
Rms current	0.0633	0.99997885
Active power	0.4623	0.99992047
Apparent power	0.0900	0.99994137
Reactive power	0.2443	0.99974305

In order to test how well the SMLC calculates the energy consumption, the source power module (at 126.15 V rms) and the load banks were set, obtaining an rms current of 6.1 A rms and a 0.5 power factor. The SMLC provides this information every second, but to synchronize the kW/h samples with those of the HIOKI PW3360-20, they were compared in steps of one minute. The mean absolute percentage error of the SMLC active energy measurements versus the HIOKI PW3360-20 readings was 0.4242%, and its R^2 value was 0.99988799, whereas, for the reactive power, it was 0.1095% and 0.999999309, respectively. The last evidence is summarized in Table 6.

Table 6. Resume of the mean absolute percentage errors and $R^2$of active and reactive power using 126.15 V rms and 6.1 A rms.

Electrical Parameter	%MAPE	R^2
Active energy	0.4242	0.99988799
Reactive energy	0.1095	0.99999309

A final experiment was conducted to validate the rms voltage estimation by subjecting the SMLC voltage inputs to values from about 15 V rms to 125 V rms in steps of 5 V rms using the variable part of the power source module. The complete samples are listed in Table 7, along with their respective errors. Considering all the readings, the MAPE resulted in 0.3774% and, according to the linear regression analysis of Figure 12, R^2 in 0.99996019.

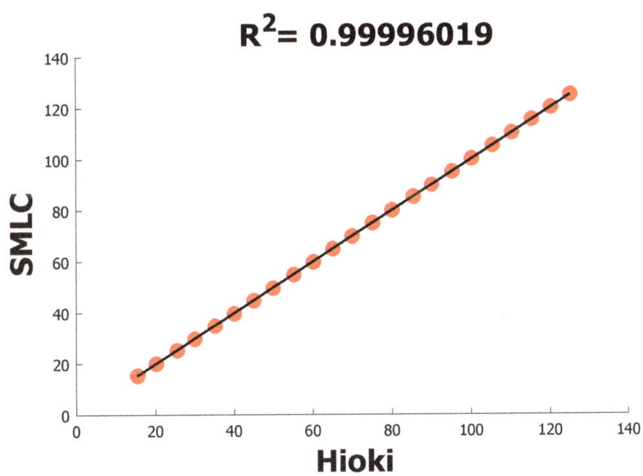

Figure 12. R^2 of SMLC against HIOKI PW3360-20 using rms voltage readings.

Table 7. Comparison of rms voltage samples between HIOKI PW3360 and the SMLC.

HIOKI PW3360	SMLC	%Error
15.44	15.4	0.2591
20.16	20.026	0.6647
25.44	25.19	0.9827
29.95	29.707	0.8114
35.12	34.86	0.7403
40	39.638	0.9050
45	44.672	0.7289
49.94	49.605	0.6708
55.15	54.82	0.5984
60.13	59.812	0.5289
65.11	64.833	0.4254
70.02	69.834	0.2656
75.14	74.946	0.2582
80.05	79.906	0.1799
85.42	85.271	0.1744
90.01	89.874	0.1511
95.24	95.105	0.1417
100.14	100.14	0.0000
105.41	105.349	0.0579
110.38	110.329	0.0462
115.42	115.461	0.0355
120.23	120.241	0.0091
125.18	125.123	0.0455
% MAPE		0.3774

4.2. Implementation

To demonstrate its functionality, the SMLC was applied in a real-life operative environment, where it measured electrical parameters and controlled what was connected to it. The setup of the application is shown in Figure 13, where it can be seen how a central smart meter was placed in the main power panel to compare the information provided by the SMLC, which had a coffee maker and an electric heater plugged in; according to the manufacturer, those devices have an active power of 540 W and 720 W, respectively. With the above arrangement, two tests were run, a manual trail where one user turned on and off the loads, and another trail where the loads switched automatically.

(a) (b) (c)

Figure 13. (**a**) Central smart meter next to the main breaker box; (**b**) current transformers of the central smart meter over the two phases of the house; (**c**) coffee maker and electric heater plugged in the SMLC.

Three states were monitored in the manual test: the first with both appliances off, the second with just the electric heater on, and the third with both devices on at the same time. Figure 14 shows how in the beginning there was no active power in the SMLC while in the whole house there was almost 1500 W. At 19:42:10, the button "A" (from Figure 7) was pressed, causing the electric heater to be turned on, resulting in a reading after a momentary peak of 715 W in the SMLC and a rise to 2184 W in the entire house. After almost one minute, at 19:43:02, the coffee maker was turned on using the button "B" in the web application. The SMLC's active power went from 715 W to 1245 W after a short transient, which means 530 W for the coffee maker; the central smart meter reported 2776 W. The coffee maker was switched off at 19:43:02, decreasing the active power in the SMLC to 637 W, while the whole house to 2092 W.

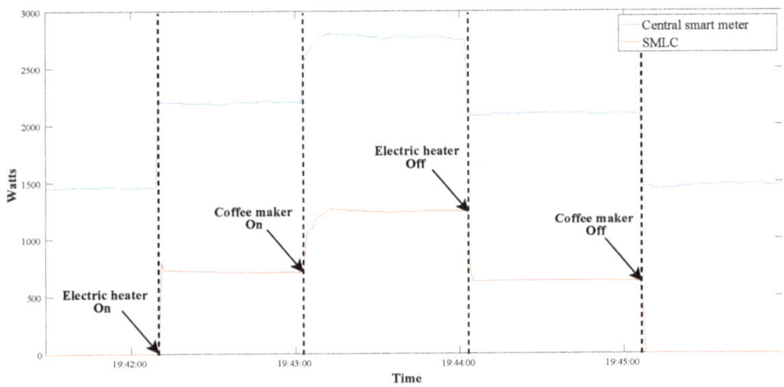

Figure 14. Behavior of the SMLC readings during the manual control of the test loads.

Another experiment was carried out using the same set up that appeared in Figure 13. This time the loads were switched on/off automatically according to an active power setpoint, which was 5100 W for the entire house. If the power used in the house exceeds 5100 W for 30 s, the SMLC would have to turn off the electric heater or the coffee maker. The first household appliance being switched would be the one connected to plug "B", and if the total power continues over 5100 W, the next one would be plug "A". Figure 15 presents the results of the test. At first, the house had a power of 3590 W, then the electric heater and the coffee maker were turned on and the power increased to 4826 W, still under the setpoint. It occurred at 00:43:04 that the power jumped to 5935 W for more than 30 s, which caused the SMLC to turn off the plug "B" (coffee maker), going to 5345 W. After turning off the coffee maker, the total power was still over 5100 W, so 30 s after the first shut-off, the SMLC opened the plug "A", lowering the power to 4808 W in the house.

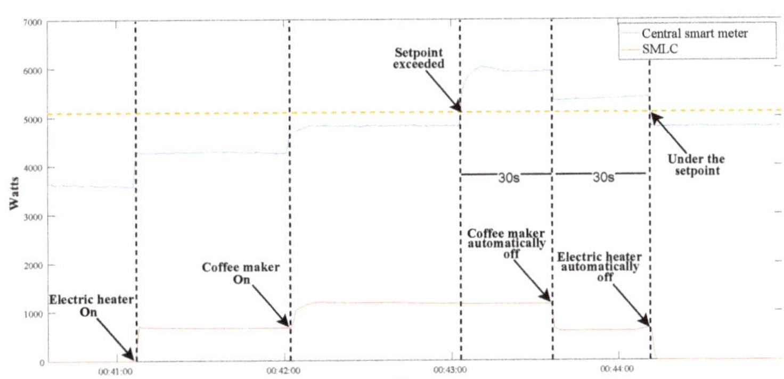

Figure 15. Behavior of the SMLC readings during the automatic control of the test loads.

5. Conclusions

The development of the smart meter with load control for a home energy management system was presented in this paper. The whole practice involved the design and implementation of the electronic instrumentation, the creation of a simple IoT scheme model, a calibration process, the measurement validation, and the demonstration of the system in a space environment.

In contrast to those works that require a gateway to send measurements to the cloud [2,8,21–23,26,28], the SMLC electronic proposal based on the ESP32 simplifies the IoT architecture of the entire system because it enables direct internet communication without an extra device. In addition, using integrated circuits specifically designed for calculating electric parameters facilitates future international certifications and assures accurate measurements. Furthermore, the use of a CT as well as the custom-made PCB allows scalability of the prototype in electrical installations, since the circuit board fits in standard $4'' \times 4''$ metal electrical wall boxes, and, thanks to the CT as a current sensor, a non-intrusive connection can be carried out.

Following the calibration steps for the ADE7758 provided in [30], the SMLC was able to provide readings with a mean absolute percentage error below 0.5% in all its electrical parameters tested with a unit power factor, particularly 0.1340%, 0.1639% and 0.1639% in rms current and active and apparent power. Similarly, but in measurements under no-unit power factor conditions, the MAPE was less than 1%, for example, 0.7976%, 0.0633%, 0.4623%, 0.090%, and 0.2433% in power factor, rms current, and active, apparent, and reactive power, respectively. Active energy consumption exhibited an error of 0.4242% and reactive energy 0.1095%. Readings of rms voltage also showed errors below 0.5%, specifically a MAPE of 0.3774%.

Furthermore, the fact that the SMLC not only provides the active power, but a variety of electrical parameters, including rms current and voltage, reactive and apparent power, and power factor, is an advantage over [20–22,24,25,28], as it can be used in future applications. For example, according to Angelis et al. [33], meters that offer the above kind of readings are necessary to implement automatic appliance recognition in HEMS.

Real-time monitoring of electrical energy consumption could not be enough for its efficient use and saving, but it is also essential to facilitate its control. Thanks to the electronic implementation of the SMLC, both functions were possible, and the SMLC includes two electromechanical relays, which demonstrated to be effective switching elements as can be seen in Section 4.2, where they turned on/off different household appliances. In addition, how users interact with energy management systems is crucial to their adaptation. The loads in the SMLC can be switched off/on manually and automatically, unlike [20] that only offer manual control, and [2,21,22] only automatic control.

Future research will focus on how the device proposed in this work could be a useful tool for the areas of smart grids and microgrids, primarily because it allows the opportunity to know exactly how energy is being used in individual appliances, as well as enabling remote control of them—aspects that can help to limit the energy consumption.

Author Contributions: Conceptualization, O.M., A.R. and P.R.; methodology, O.M. and A.R.; software, O.M. and A.R.; validation, O.M., A.R. and P.R.; formal analysis, O.M. and A.R.; investigation, O.M., A.R. and P.R.; writing—original draft preparation, O.M. and A.R.; writing—review, O.M., A.R., A.A., A.S. and F.L.; visualization, O.M., A.R., P.R., A.A., A.S. and F.L.; supervision, O.M., A.R., P.R., A.A., A.S. and F.L. All authors have read and agreed to the published version of the manuscript.

Funding: This research received no external funding.

Institutional Review Board Statement: Not applicable.

Informed Consent Statement: Not applicable.

Data Availability Statement: The data presented in this study are available on request from the corresponding author.

Conflicts of Interest: The authors declare no conflict of interest.

References

1. Dale, S. *BP Statistical Review of World Energy 2022 | 7*, 1st ed.; Technical Report; BP p.l.c.: London, UK, 2022.
2. Pawar, P.; Vittal, K.P. Design and development of advanced smart energy management system integrated with IoT framework in smart grid environment. *J. Energy Storage* **2019**, *25*, 100846. [CrossRef]
3. Sun, Q.; Li, H.; Ma, Z.; Wang, C.; Campillo, J.; Zhang, Q.; Wallin, F.; Guo, J. A Comprehensive Review of Smart Energy Meters in Intelligent Energy Networks. *IEEE Internet Things J.* **2016**, *3*, 464–479. [CrossRef]
4. Alahakoon, D.; Yu, X. Smart Electricity Meter Data Intelligence for Future Energy Systems: A Survey. *IEEE Trans. Ind. Inform.* **2016**, *12*, 425–436. [CrossRef]
5. Rodrigues Junior, W.L.; Borges, F.A.; da S. Veloso, A.F.; de A. L. Rabêlo, R.; Rodrigues, J.J. Low voltage smart meter for monitoring of power quality disturbances applied in smart grid. *Measurement* **2019**, *147*, 106890. [CrossRef]
6. Avancini, D.B.; Rodrigues, J.J.; Martins, S.G.; Rabêlo, R.A.; Al-Muhtadi, J.; Solic, P. Energy meters evolution in smart grids: A review. *J. Clean. Prod.* **2019**, *217*, 702–715. [CrossRef]
7. Morello, R.; De Capua, C.; Fulco, G.; Mukhopadhyay, S.C. A Smart Power Meter to Monitor Energy Flow in Smart Grids: The Role of Advanced Sensing and IoT in the Electric Grid of the Future. *IEEE Sens. J.* **2017**, *17*, 7828–7837. [CrossRef]
8. Spanò, E.; Niccolini, L.; Pascoli, S.D.; Iannaccone, G. Last-Meter Smart Grid Embedded in an Internet-of-Things Platform. *IEEE Trans. Smart Grid* **2015**, *6*, 468–476. [CrossRef]
9. Kuzlu, M.; Pipattanasomporn, M.; Rahman, S. Review of communication technologies for smart homes/building applications. In Proceedings of the 2015 IEEE Innovative Smart Grid Technologies-Asia, ISGT ASIA, Bangkok, Thailand, 3–6 November 2015; pp. 1–6. [CrossRef]
10. Kong, X.; Zhang, S.; Sun, B.; Yang, Q.; Li, S.; Zhu, S. Research on Home Energy Management Method for Demand Response Based on Chance-Constrained Programming. *Energies* **2020**, *13*, 2790. [CrossRef]
11. Avancini, D.B.; Rodrigues, J.J.; Rabêlo, R.A.; Das, A.K.; Kozlov, S.; Solic, P. A new IoT-based smart energy meter for smart grids. *Int. J. Energy Res.* **2021**, *45*, 189–202. [CrossRef]
12. Javaid, N.; Khan, I.; Ullah, M.; Mahmood, A.; Farooq, M. A Survey of Home Energy Management Systems in Future Smart Grid Communications. In Proceedings of the 2013 Eighth International Conference on Broadband and Wireless Computing, Communication and Applications, Compiegne, France, 28–30 October 2013; pp. 459–464. [CrossRef]
13. Qureshi, K.N.; Alhudhaif, A.; Hussain, A.; Iqbal, S.; Jeon, G. Trust aware energy management system for smart homes appliances. *Comput. Electr. Eng.* **2022**, *97*, 107641. [CrossRef]
14. Metzger, I.; Kandt, A.; VanGeet, O. *Plug Load Behavioral Change Demonstration Project*; Technical Report; National Renewable Energy Lab. (NREL): Golden, CO, USA, 2011.
15. Klein, L.; Kwak, J.-Y.; Kavulya, G.; Jazizadeh, F.; Becerik-Gerber, B.; Varakantham, P.; Tambe, M. Coordinating occupant behavior for building energy and comfort management using multi-agent systems. *Autom. Constr.* **2012**, *22*, 525–536. [CrossRef]
16. Sehar, F.; Pipattanasomporn, M.; Rahman, S. Integrated automation for optimal demand management in commercial buildings considering occupant comfort. *Sustain. Cities Soc.* **2017**, *28*, 16–29. [CrossRef]
17. Ahmed, M.S.; Mohamed, A.; Khatib, T.; Shareef, H.; Homod, R.Z.; Abd, J. Real time optimal schedule controller for home energy management system using new binary backtracking search algorithm. *Energy Build.* **2017**, *138*, 215–227. [CrossRef]
18. Haq, E.U.; Lyu, C.; Xie, P.; Yan, S.; Ahmad, F.; Jia, Y. Implementation of home energy management system based on reinforcement learning. *Energy Rep.* **2022**, *8*, 560–566. [CrossRef]
19. Rocha, H.R.; Honorato, I.H.; Fiorotti, R.; Celeste, W.C.; Silvestre, L.J.; Silva, J.A. An Artificial Intelligence based scheduling algorithm for demand-side energy management in Smart Homes. *Appl. Energy* **2021**, *282*, 116145. [CrossRef]
20. Kang, S.; Park, K.; Shin, S.; Chang, K.; Kim, H. Zero standby power remote control system using light power transmission. *IEEE Trans. Consum. Electron.* **2011**, *57*, 1622–1627. [CrossRef]
21. Metzger, I.; Cutler, D.; Sheppy, M. *Plug-Load Control and Behavioral Change Research in GSA Office Buildings*; Technical Report; National Renewable Energy Lab. (NREL): Golden, CO, USA, 2012.
22. Park, S.; in Choi, M.; Kang, B.; Park, S. Design and Implementation of Smart Energy Management System for Reducing Power Consumption Using ZigBee Wireless Communication Module. *Procedia Comput. Sci.* **2013**, *19*, 662–668. [CrossRef]
23. Tsai, K.L.; Leu, F.Y.; You, I. Residence Energy Control System Based on Wireless Smart Socket and IoT. *IEEE Access* **2016**, *4*, 2885–2894. [CrossRef]
24. Phangbertha, L.N.; Fitri, A.; Purnamasari, I.; Muliono, Y. Smart Socket for Electricity Control in Home Environment. *Procedia Comput. Sci.* **2019**, *157*, 465–472. [CrossRef]
25. Muralidhara, S.; Hegde, N.; PM, R. An internet of things-based smart energy meter for monitoring device-level consumption of energy. *Comput. Electr. Eng.* **2020**, *87*, 106772. [CrossRef]
26. Karthick, T.; Charles Raja, S.; Jeslin Drusila Nesamalar, J.; Chandrasekaran, K. Design of IoT based smart compact energy meter for monitoring and controlling the usage of energy and power quality issues with demand side management for a commercial building. *Sustain. Energy Grids Netw.* **2021**, *26*, 100454. [CrossRef]
27. Tekler, Z.D.; Low, R.; Blessing, L. User perceptions on the adoption of smart energy management systems in the workplace: Design and policy implications. *Energy Res. Soc. Sci.* **2022**, *88*, 102505. [CrossRef]
28. Tekler, Z.D.; Low, R.; Yuen, C.; Blessing, L. Plug-Mate: An IoT-based occupancy-driven plug load management system in smart buildings. *Build. Environ.* **2022**, *223*, 109472. [CrossRef]

29. Ziegler, S.; Woodward, R.C.; Iu, H.H.C.; Borle, L.J. Current sensing techniques: A review. *IEEE Sens. J.* **2009**, *9*, 354–376. [CrossRef]
30. ADE7758 Datasheet. Available online: https://www.analog.com/media/en/technical-documentation/data-sheets/ade7758.pdf (accessed on 27 March 2022).
31. CLAMP ON POWER LOGGER PW3360—Hioki. Available online: https://www.hioki.com/en/products/detail/?product_key=5589 (accessed on 27 March 2022).
32. Bahga, A.; Madisetti, V. *Internet of Things: A Hands-on Approach*, 1st ed.; Vijay Madisetti: Kharagpur, India, 2014; pp. 20–44.
33. Angelis, G.F.; Timplalexis, C.; Krinidis, S.; Ioannidis, D.; Tzovaras, D. NILM applications: Literature review of learning approaches, recent developments and challenges. *Energy Build.* **2022**, *261*, 111951. [CrossRef]

sensors

MDPI

Article

A Low-Cost IoT System for Real-Time Monitoring of Climatic Variables and Photovoltaic Generation for Smart Grid Application

Gustavo Costa Gomes de Melo [1,*], Igor Cavalcante Torres [2], Ícaro Bezzera Queiroz de Araújo [1], Davi Bibiano Brito [1] and Erick de Andrade Barboza [1]

1 Computing Institute, A. C. Simões Campus, Federal University of Alagoas—UFAL, Maceió, AL 57072-970, Brazil; icaro@ic.ufal.br (Í.B.Q.d.A.); davi@ic.ufal.br (D.B.B.); erick@ic.ufal.br (E.d.A.B.)

2 Center of Agrarian Sciences, Engineering and Agricultural Sciences Campus, Federal University of Alagoas—UFAL, Rio Largo, AL 57100-000, Brazil; igor.torres@ceca.ufal.br

* Correspondence: gustavocosta@ic.ufal.br; Tel.: +55-82-999891568

Abstract: Monitoring and data acquisition are essential to recognize the renewable resources available on-site, evaluate electrical conversion efficiency, detect failures, and optimize electrical production. Commercial monitoring systems for the photovoltaic system are generally expensive and closed for modifications. This work proposes a low-cost real-time internet of things system for micro and mini photovoltaic generation systems that can monitor continuous voltage, continuous current, alternating power, and seven meteorological variables. The proposed system measures all relevant meteorological variables and directly acquires photovoltaic generation data from the plant (not from the inverter). The system is implemented using open software, connects to the internet without cables, stores data locally and in the cloud, and uses the network time protocol to synchronize the devices' clocks. To the best of our knowledge, no work reported in the literature presents these features altogether. Furthermore, experiments carried out with the proposed system showed good effectiveness and reliability. This system enables fog and cloud computing in a photovoltaic system, creating a time series measurements data set, enabling the future use of machine learning to create smart photovoltaic systems.

Keywords: monitoring; data acquisition systems; renewable energy

Citation: Melo, G.C.G.d.; Torres, I.C.; Araújo, Í.B.Q.d.; Brito, D.B.; Barboza, E.d.A. A Low-Cost IoT System for Real-Time Monitoring of Climatic Variables and Photovoltaic Generation for Smart Grid Application. *Sensors* **2021**, *21*, 3293. https://doi.org/10.3390/s21093293

Academic Editor: Antonio Cano-Ortega

Received: 31 March 2021
Accepted: 3 May 2021
Published: 10 May 2021

1. Introduction

The share of renewable energies in electricity generation has been growing worldwide. In 2019, there was an increase of 200 gigawatts of renewable energy in the world energy matrix, with photovoltaic energy being responsible for 57.5% of this increase according to [1]. Small and medium-sized distributed photovoltaic generation systems were the ones that grew the most. In Brazil, at the end of 2020, distributed generation represented 59% of installed photovoltaic sources, with a 107% growth compared to 2019, while centralized generation had an increase of only 24% [2].

The expansion of distributed renewable energies presents several benefits, such as less environmental impact, reduced emission of carbon dioxide, and less degradation of fauna and flora. Regarding social impacts, this type of generation system can be employed in remote locations that do not have access to the power grid, enabling and improving access to communication, education, and agricultural production. Renewable energies, especially solar energy, tend to generate more jobs than non-renewable energy generation, and less centralized systems can create more opportunities.

A primary feature of photovoltaic (PV) systems is the correlation between the climatic conditions and the performance of its generation. The availability of sunlight, temperature and various other climatic factors directly affect energy production. In large and medium-sized centralized photovoltaic systems, many of the efforts and resources are used in monitoring and acquiring data, which is essential to recognize the renewable resources available on-site, evaluate the efficiency of electrical conversion, detect failures and optimize electrical production.

On the other hand, in small photovoltaic systems, the high monitoring cost generally makes its implementation inaccessible. The high cost can lead to situations in which the system operators do not detect failures such as loss of efficiency, peaks and falls in voltage, and insertion of harmonics in the power grid [3,4]. These failures can disrupt the operation of the photovoltaic system or even cause damage to the power grid [5].

The introduction of the Internet of Things (IoT) concept to monitoring devices can bring several benefits, such as access to real-time data, remote device management, cost reduction, and system scalability. Moreover, it allows the integration of devices into the smart grid, enabling improvements in the photovoltaic system's processing, fault recovery, and reliability skills [6]. Furthermore, wireless communication decreases the distance range limitation and costs typical of wired communication.

This paper presents the design, development, and validation of a low-cost IoT data acquisition system that focuses on real-time monitoring of photovoltaic energy generation and the main meteorological factors that influence the generation. The principal motivation is to provide an alternative solution for commercial systems that are usually expensive and closed to adjustments and modifications. The proposal consists of three main elements: (1) two data logger devices for data acquisition, one for meteorological data and other for PV generation data; (2) an IoT cloud system that processes and stores the data obtained; (3) and a web application that displays the real-time data and the previous data collected.

The proposal includes improvements in software, hardware, and system architecture (IoT). To the best of our knowledge, considering the similar works found in the literature, this is the first proposed PV monitoring system that aggregates all of the following features:

- Measurements of all the relevant meteorological variables;
- Open software implementation;
- LoRa (Long Range) as the data transmission technology and connection with the internet without cables;
- Data storage locally and in the cloud;
- Network time protocol (NTP) to synchronize the devices' clocks;
- PV generation variables measured directly from the plant, not from the inverter.

The focus of hardware development was flexibility and cost reduction. Furthermore, with the application of the IoT cloud system, the proposed system allows remote control, local and cloud storage of data, real-time access to data, and scalability. Due to these features, the system is more oriented to small/medium operators than distribution system operators (DSOs). Furthermore, it can be of interest to researchers because it provides an enabling technological system at an affordable price. Moreover, it can be of high interest to professionals working in developing countries where the limited diffusion of solar technology can be attributed to lack of funding and research and development activities [7].

It is worth mentioning that this is an enabling system for creating intelligent photovoltaic systems. It provides an IoT architecture that enables machine learning techniques to be executed using cloud or fog computing paradigms. Moreover, the data sets generated by the system can be used to train machine learning algorithms for fault detection and power generation forecasting, for example.

The rest of the paper is structured as follows. Section 2 describes the proposed system with details about the device's hardware, the operation of the devices, the LoRa protocol developed, the IoT architecture used, and the web application developed. Section 3 presents a review of related works in the literature. Section 4 presents the results and discussions with a technical comparison of the proposed system with previous systems, experimental results for the validation of the proposed system, and cost analysis of the proposed system. Finally, Section 5 provides the conclusions.

2. Proposed System Description

2.1. System Overview

Figure 1 shows an overview of the system. The data logger devices are responsible for collecting, conditioning, storing, and transmitting data from all sensors. We developed two different devices. The first one is responsible for collecting the meteorological data from the solarimetric station sensors. The second one is a data logger to monitor the PV generation.

The data logger devices use LoRa wireless communication to send the data obtained to a LoRa gateway, and Wi-Fi (IEEE 802.11) connects the gateway to the internet. The gateway is the intermediary between the devices and the cloud system. It is responsible for redirecting the monitored data to the cloud, storing it, or directing commands and settings from the cloud to the devices.

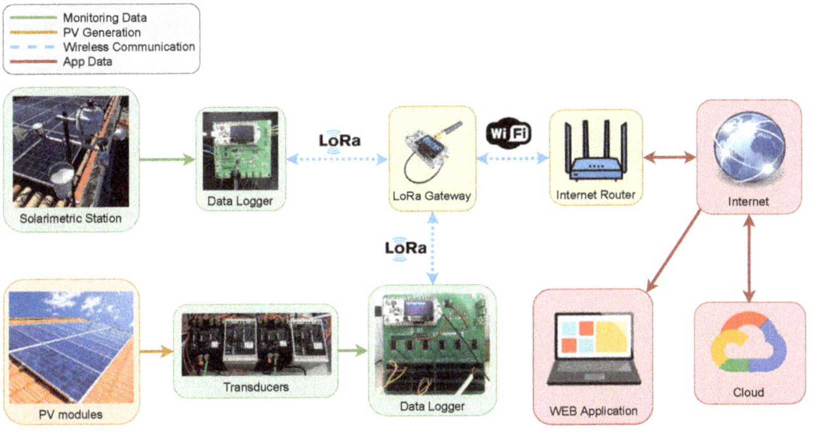

Figure 1. Simplified diagram of the proposed system.

All data acquired by the system can be accessed easily through the web application, allowing real-time viewing of data or querying data stored in the cloud database. A remote server hosts the web application, enabling users to access it from any browser.

2.2. Data Logger Devices

When designing the data logger devices, the goal was to achieve low production costs, provide wireless communication, and be flexible for software and hardware changes. The main component of these devices is the Heltec Wi-Fi LoRa 32 (V2) IoT dev-board [8], which features the ESP32 dual-core microcontroller (MCU) [9] and integrates Wi-Fi, LoRa, Bluetooth (IEEE 802.15.1), onboard OLED display, and micro-USB connector.

The solarimetric station data logger requires a robust MCU to operate in harsh environments since the station will be exposed to different climatic conditions. The ESP32 was built for use in industrial environments. It can work in temperatures between −40 and 125 °C and adapt dynamically to external condition changes.

Each data logger has an SD card (secure digital card), where the obtained data are saved temporarily before sending it to the gateway. This local storage ensures that data are not lost if communication with the gateway or the cloud is not available. The 74HC125D buffer [10] performs communication between the MCU and the SD through a serial peripheral interface (SPI) bus.

Furthermore, a real-time clock (RTC) was used to monitor the date and time of each measurement, providing an accurate time when the MCU will read the sensors' data. The RTC communicates directly with the MCU through an inter-integrated circuit (I2C) bus and has a dedicated battery allowing the time to be tracked continuously even if the data logger is without power.

2.2.1. Solarimetric Station Data Logger and the Meteorological Variables

Figure 2 shows the solarimetric station data logger's main components and the connections between them. This device is responsible for monitoring the following variables: irradiance, PV module temperature, wind speed and direction, ambient temperature, humidity, and rain.

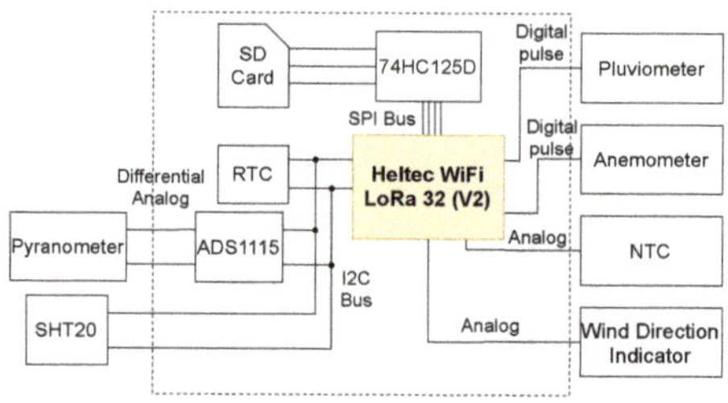

Figure 2. Simplified diagram of the solarimetric station data logger, with emphasis on the components and connections.

Solar irradiance is one of the essential meteorological variables. The energy generated by a photovoltaic system is directly proportional to the irradiance that reaches the photovoltaic modules. A low-cost pyranometer measures solar irradiance. The relationship between pyranometer input and output is given by Equation (1). G is the irradiance incident, K is the calibration constant of the pyranometer, and mV is the output voltage in millivolts. Since the pyranometer generates a very low voltage, the high precision ADS1115 analog-to-digital converter (ADC) [11] was used to read the measurements.

$$G = \frac{mV}{K} \tag{1}$$

The photovoltaic module temperature influences its photovoltaic conversion efficiency. It was estimated at [12] that 0.5% PV module efficiency is reduced with an increase of 1 °C in its temperature. The data logger uses a 10 kΩ negative temperature coefficient (NTC) sensor to measures the PV module temperature. The relationship between the resistance of the NTC (R) and its temperature (T_m) is provided by Equation (2), where R_{25} is the resistance of the NTC at the reference temperature, T_{25} is the reference temperature (25 °C), and β is the NTC constant.

$$T_m(R) = \frac{1}{\frac{ln\left(\frac{R}{R_{25}}\right)}{\beta} + \frac{1}{T_{25}}} \tag{2}$$

The ambient temperature influences the PV module temperature, and air humidity can absorb or reflect solar energy, decreasing the irradiance that reaches the photovoltaic module. The SHT20 sensor [13] measures these two factors. Rain can affect many factors at the same time, such as reducing solar irradiance and panel temperature. The data logger obtains the rainfall index through a rain gauge.

The wind can help reduce the PV module temperature, improving its efficiency. An anemometer measures wind speed, which generates a digital pulse at each turn around itself. The accumulated pulses (CP) during a given period (P) are used to calculate the revolutions per minute (RPM), as shown in Equation (3). The revolutions per minute are then converted to km/h using Equation (4), where r is the radius of the anemometer. The wind direction measurement is performed by a wind direction indicator with an analog output that varies according to the direction the indicator is pointing.

$$RPM = \frac{CP \times 60 \times 1000}{P} \tag{3}$$

$$W_s = \frac{4 \times \pi \times r \times RPM}{60 \times 1000} \times 3.6 \quad (4)$$

2.2.2. PV Generation Data Logger

The PV generation data logger is responsible for monitoring the direct current (DC) voltage and current of multiple PV strings and active alternating current (AC) power at the inverter output. To achieve this, six ADC ADS8668 [14] were used. Each ADS8668 has eight channels of differential analog input and communicates with the MCU through the SPI bus. To establish the communication of the six ADCs with the MCU using only one chip selector pin, a unique topology called daisy-chain was used, as illustrated in Figure 3.

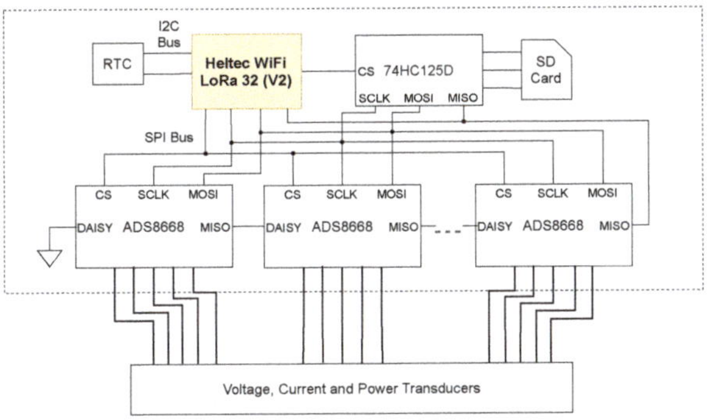

Figure 3. Simplified diagram of the PV generation data logger, with emphasis on the components and connections.

During the experiment and validation of the system, five SECON transducers [15] were used, two for current, two for voltage, and one for power. The current transducer proportionally converts an input current between 0 and 10 A into a voltage between 0 and 5 V. The two voltage transducers have different measurement ranges; the first reads between 0 and 400 V and the second between 0 and 500 V. Both have a proportional output between 0 and 10 V.

$$P = V_{out} \times 3600 - 9000 \quad (5)$$

A ±9 kW bidirectional three-phase transducer measures the AC power. This transducer can receive power up to 9000 W as input and provides an output voltage between 0 and 5 V. Because the transducer is bidirectional, the relationship between its output and measured power is provided by Equation (5), where P is power, and V_{out} is the transducer output voltage.

2.3. Data Loggers Operation

Figure 4 shows a flowchart illustrating the main steps in the operation of the devices; both data loggers operate similarly. When powered on, the data logger first initializes the LoRa radio, RTC, and SD card. Then, the existence of the data and settings files on the SD card is verified, and if these files do not exist, they are created in a comma-separated values (CSV) format. The data file is created with a header that informs the data type in each file column. The configuration file is created with the default sensor settings.

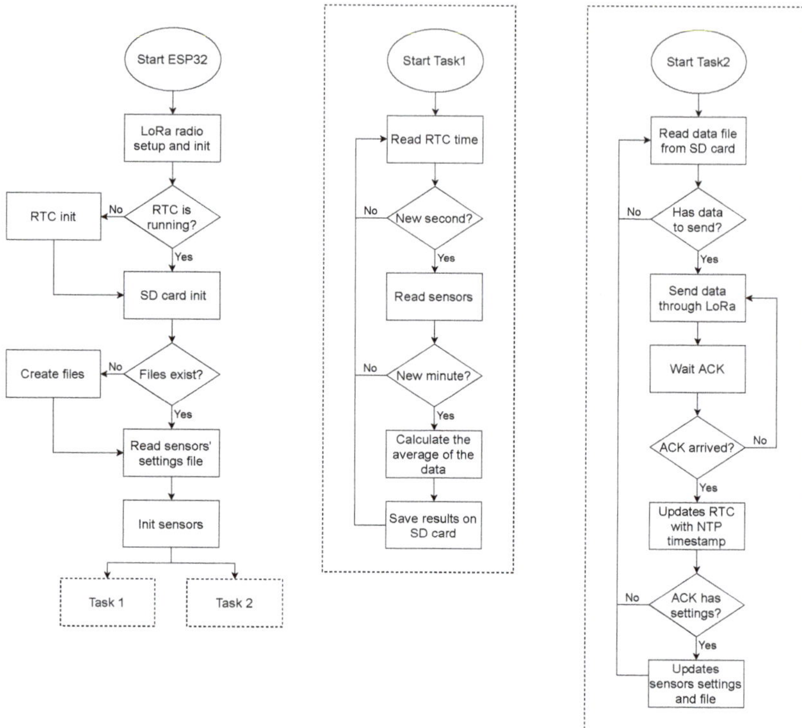

Figure 4. Simplified diagram representing the operation of the data logger devices.

Next, all connected sensors are initialized, and the saved settings are applied to some of those sensors (transducers and the pyranometer). After the initialization of the sensors, two tasks are assigned to each of the ESP32 cores to run in parallel.

The first task is responsible for reading the sensors' data. It contains a loop that performs polling of the time marked by the RTC. If the time obtained second value is different from the last sampling second value, a new sampling is performed, and each of the obtained data is added to the value stored in its respective variable. When the second equals 00, the values accumulated in the variables are divided by the number of samples, averaging the last-minute data (60 samples). This approach is based on [16].

In short, the data are sampled every second for one minute, and then an average is calculated. This method is not applied to the monitoring of rain, which is the daily accumulation, and the direction of the wind, determined by the most frequent direction in that last minute. After calculating the average, a timestamp is obtained from the RTC. Then, the results and timestamps are saved in the SD card data file, and the storage variables are reset to zero.

The second task is performed every five seconds (using the function delay) and manages LoRa communication. This interval was defined based on tests to obtain a high send frequency without interfering with the execution of the first task. Each time the second task is executed, it is checked if there are data on the SD card to be sent. If data are detected on the SD card, the first data set in the file is compressed, added to a LoRa packet, and sent to the LoRa gateway. After, the task expects to receive an acknowledgment (ACK) that the data have reached the gateway. If confirmation does not arrive within a specific time, the data are resent. When the ACK is received, the data set sent is removed from the SD card, and a new set is prepared for sending.

Therefore, in regular operation, data are sent from the data loggers to the cloud every minute. In the event of a communication failure, upon returning, the accumulated data will be sent every five seconds until all data are sent and the operation is normalized.

The LoRa gateway is always synchronized with the NTP.br server [17]. When an ACK is sent to a device, a timestamp obtained through the NTP is sent with it. The NTP timestamp is used by the second task to update the RTC date and time. Thus, the RTC will always maintain the correct time with great precision. This RTC update also allows the devices' time to be synchronized, causing data sampling to occur roughly at the same time.

Sensors' configurations may also be included in the ACK package. These settings can be set in the cloud system and contain constants used in the initialization and the reading of the transducers and the pyranometer, allowing remote adjustments of the measurements. The second task applies the settings and saves them to the SD card file to be maintained even if the data logger is restarted or loses communication with the cloud.

2.4. LoRa Network Protocol

A custom LoRa network protocol was developed specifically for the proposed system to send the data in the most efficient way possible. The Arduino-LoRa library [18] was used for both devices and at the gateway to transmit and receive LoRa packets. This library exposes the LoRa radio directly and allows it to send data to any radios in range with the same radio parameters, without using compression or addressing.

Two functions offered by the library were used: sync word and cyclic redundancy check (CRC). The sync word function limits data transmission only to devices that share the same sync word value, creating an isolated LoRa network. CRC is an error detection method to detect an accidental change in the transmitted data packets.

A payload structure was developed to maximize data transfer using the smallest number of bits. Figure 5 shows the structure for transmitting sensor data and ACK information. The first byte is a header that contains the information of whom the sender and recipient are, performing the addressing, and also contains bits that inform if the payload is an ACK and if it contains sensor settings information.

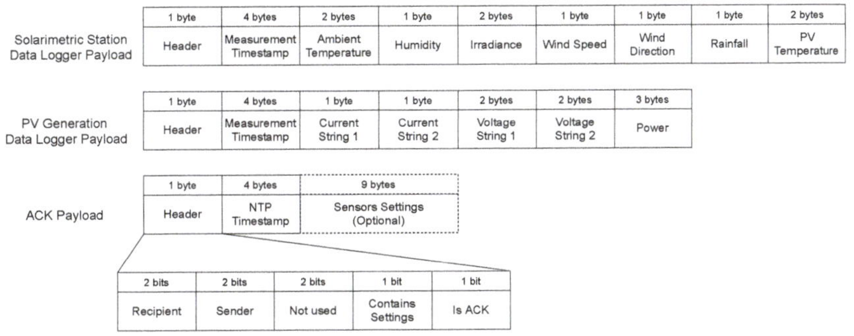

Figure 5. Proposed LoRA payload structure.

The data are compressed using a bit-packing, so its representation uses fewer bits than if they were transmitted as ASCII code. Bit-packing is a simple compression, where the data are first represented in an integer value using the bit packing formula, and then it is represented in binary. Table 1 details this representation for each type of data.

As Arduino-LoRa only supports transmitting char (1 byte in C language), even the pieces of data less than 8 bits were allocated one byte in the structure. This allocation also facilitates the separation of the data in the receiver.

Table 1. Data bit-packing.

Data Type	Bit-Packing Formula	Bits Required	Range	Precision
Timestamp	B = seconds from 1 Jan 1970 00:00:00	32	1 Jan 1970 00:00:00 to Jan 19 2038 03:14:07	1 s
Temperature	B = (T + 40) × 10	16	−40 to 125 °C	0.1 °C
Humidity	B = H	7	0 to 100%	1%
Irradiance	B = I × 10	16	0 to 6553.5 W/m²	0.1 W/m²
Wind Speed	B = S	8	0 to 255 Km/h	1 Km/h
Wind Direction	Maps to a value that represents the direction	3	0, 45, 90, 135, 180, 225, 270, 315°	-
Rainfall	B = R/0.25	8	0 to 63.75 mm	0.25 mm
Voltage	B = V × 10	16	0 to 6553.5 V	0.1 V
Current	B = C × 10	8	0 to 25.5 A	0.1 A
Power	B = (P + 9000) × 10	24	−9000 to 9000 W	0.1 W

2.5. IoT Architecture

The data collected by the proposed system are made available to the user through a cloud system. For this project, the Google Cloud Platform (GCP) [19] was used to implement the IoT architecture. However, Amazon Web Service, Microsoft Azure, or any other cloud system could also be used. Figure 6 illustrates the proposed IoT architecture.

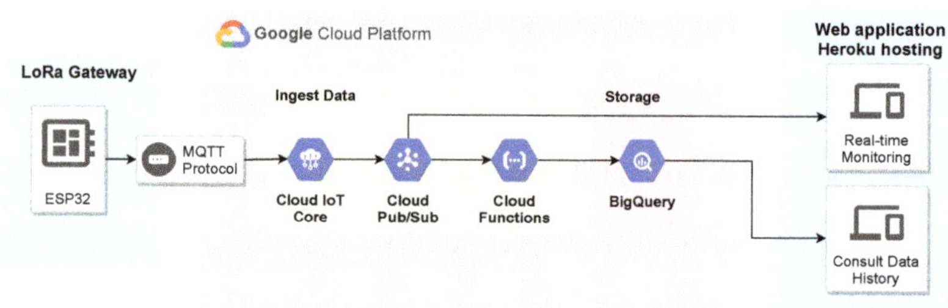

Figure 6. Diagram representing the IoT architecture.

The GCP IoT Core module is responsible for managing the devices and defining which communication protocol they can use. IoT Core provides the options to register, update, and monitor the devices' status as a device manager. In this proposal, the LoRa gateway was registered as an IoT device, and a key pair was generated to perform the device authentication and secure communication. The generated public key was registered in the IoT Core while the private key was implemented in the LoRa gateway.

For communication between the GCP and LoRA gateway, the protocol "message queuing telemetry transport" (MQTT) is used. MQTT is a machine-to-machine communication protocol based on the publish/subscribe pattern to exchange asynchronous messages. The Pub/Sub module is the GCP MQTT broker, responsible for managing topics and subscriptions. Pub/Sub offers temporary message storage and real-time message delivery with high availability and consistent performance on a large scale.

Two MQTT topics were created, one to receive data sent from the station and the other to receive PV generation data. Each of these topics has two subscribers. The first subscriber is the real-time monitoring page of the web application, allowing a real-time data display to the user. The second subscriber is the cloud functions module of GCP, which transfers the data to BigQuery for storage.

The GCP BigQuery module is a standard query language (SQL) database for large data sets. A table was created to store the data for each data logger device. The web application can make queries to BigQuery based on dates and obtain a data history for display.

The IoT Core automatically creates two topics for each registered device, a configuration topic and a command topic. The configuration topic is used to transmit the sensor settings of the data logger devices, and the command topic can be used to reset the LoRa gateway remotely. Both topics are accessible only through the IoT Core web page.

2.6. Web Application

The web application provides a simple and easy way for the user to view and obtain the data acquired by the proposed system. The application was developed using the Django Web framework, which uses Python to manage and render web pages. In Python, four endpoints were created to access the cloud services using the GCP SDK: two endpoints for subscription in the MQTT topics of each data logger and two others to query the BigQuerry tables.

Three web pages were implemented using HTML and JavaScript: home page, real-time monitoring, and consult data history. Figure 7 shows the applications home page that contains some information about the work developed. Each device has a real-time monitoring page and a consult data history page, accessed from the top menu. When the cursor passes through the devices' names in the top menu, a drop-down sub-menu is displayed to select the desired page.

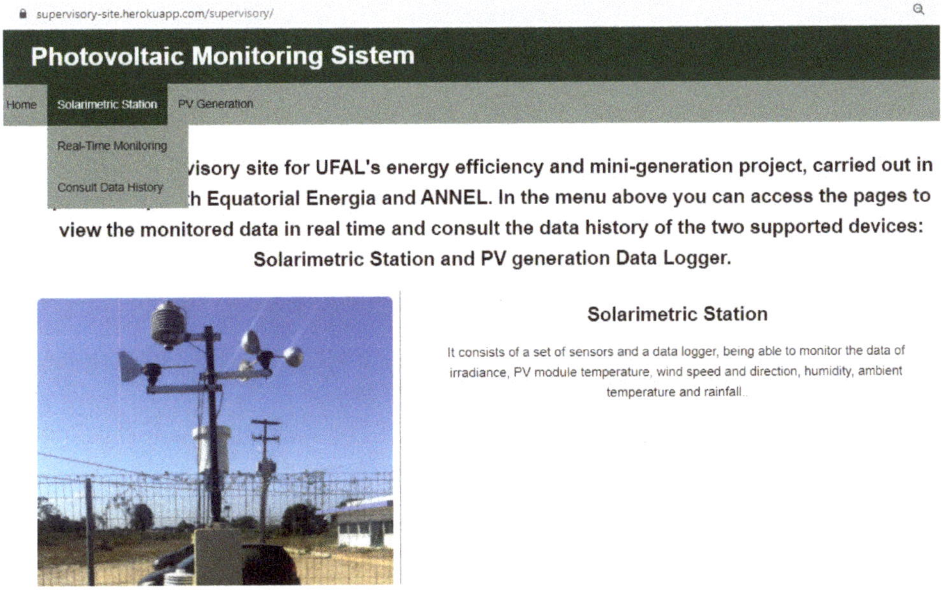

Figure 7. Web application home page, displaying the drop-down sub-menu.

The real-time monitoring page uses one of the Python endpoints to subscribe to the device's data topic. As soon as the data are received through the MQTT, they are displayed in two sections of this page. The first section shows the latest data set that arrived, including the date and time it was obtained. In the second section, each monitored variable is displayed in a different line chart. The charts begin to display the data received from the moment the page is opened and can display up to 1440 points simultaneously (24 h of monitoring). When the maximum point limit is reached, the oldest points are removed as new data are received. Figure 8 shows the real-time monitoring page of the solarimetric station, with the first section and the first chart of the second section.

On the consult data history page (Figure 9), the user can choose a day, manually typing in MM/DD/YYYY format or using an interactive calendar to consult the data saved in BigQuery. When choosing the date and pressing the "Consult" button, the Python endpoint queries the table in BigQuery, and the data obtained, if any, are displayed on the page in table format. The "Download" button converts the data displayed in the table into a CSV file transferred to the user's computer.

The web application is hosted on Heroku [20], a cloud platform as a service with support for different profile languages, including the Django framework.

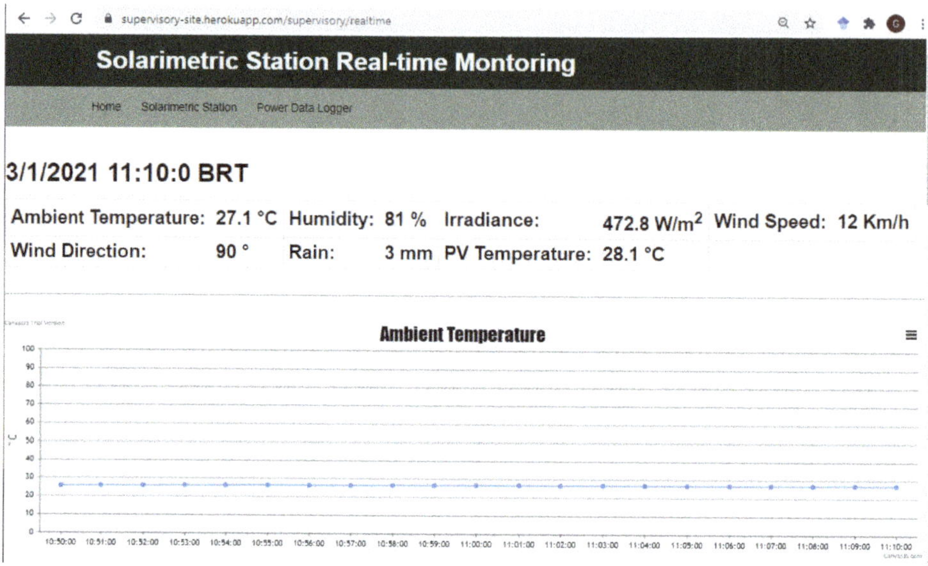

Figure 8. Web application page for real-time monitoring of the solarimetric station.

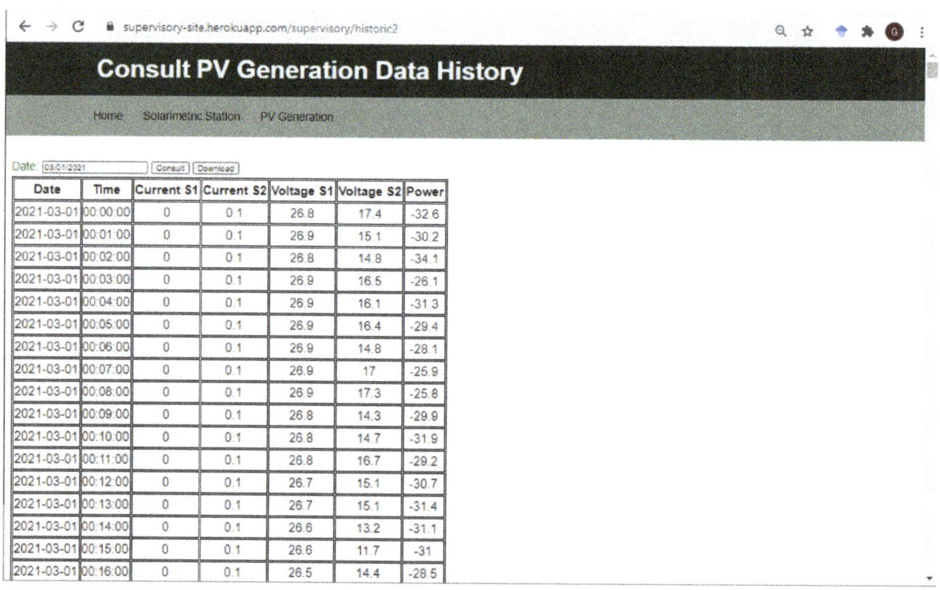

Date	Time	Current S1	Current S2	Voltage S1	Voltage S2	Power
2021-03-01	00:00:00	0	0.1	26.8	17.4	-32.6
2021-03-01	00:01:00	0	0.1	26.9	15.1	-30.2
2021-03-01	00:02:00	0	0.1	26.8	14.8	-34.1
2021-03-01	00:03:00	0	0.1	26.9	16.5	-26.1
2021-03-01	00:04:00	0	0.1	26.9	16.1	-31.3
2021-03-01	00:05:00	0	0.1	26.9	16.4	-29.4
2021-03-01	00:06:00	0	0.1	26.9	14.8	-28.1
2021-03-01	00:07:00	0	0.1	26.9	17	-25.9
2021-03-01	00:08:00	0	0.1	26.9	17.3	-25.8
2021-03-01	00:09:00	0	0.1	26.8	14.3	-29.9
2021-03-01	00:10:00	0	0.1	26.8	14.7	-31.9
2021-03-01	00:11:00	0	0.1	26.8	16.7	-29.2
2021-03-01	00:12:00	0	0.1	26.7	15.1	-30.7
2021-03-01	00:13:00	0	0.1	26.7	15.1	-31.4
2021-03-01	00:14:00	0	0.1	26.6	13.2	-31.1
2021-03-01	00:15:00	0	0.1	26.6	11.7	-31
2021-03-01	00:16:00	0	0.1	26.5	14.4	-28.5

Figure 9. Web application page to consult the history of PV generation data.

3. Literature Review and Comparison of Monitoring Systems Applied to PV Plants

Several related works can be found in the literature. Despite the advantages and advances of the literature systems presented below, they all have at least one limitation. Table 2 presents a comparison of some of the technical characteristics of the systems available in the literature and the system proposed in this work. In the rest of this section, we will discuss the similarities and differences between the proposed system and the related ones.

The monitoring system proposed in [21] consists of two types of devices: smart meter and main brain. Smart meters are the devices responsible for monitoring the voltage and current data of the PV system in real-time, while the main brain is the center where the data collected by the devices smart meter will be stored. Both are based on the ATmega 328P-PU MCU and communicate via a radio frequency (RF) wireless network operating at 315 MHz. The data can be accessed by a mobile application that communicates with the main brain via Bluetooth.

In [22], a two-level sensor network was developed for monitoring PV systems. The first level of the network consists of sensor nodes that monitor the voltage and temperature of each PV module. In contrast, the second level consists of sensor nodes that monitor irradiance, ambient temperature, voltage, and current of each string. In addition, the second-level nodes merge their monitored data with the data obtained by the first level and send it to a data center. The communication between the levels is via a radio frequency wireless network, and the second level uses ZigBee [23] to send all the collected data to the center.

Internet connection is crucial to provide real-time monitoring of data and to allow remote access to the system. In [21,22], the data are only available locally. Local-only availability would also make it difficult for future integration of these systems into a smart grid network. Internet-connected monitoring systems can be configured in two different topologies: the data logger devices connect directly to the internet or intermediate devices between the internet and the data loggers.

The authors of [24] developed a wireless sensor network based on the ESP8266 MCU to monitor the photovoltaic system. Each network node monitors the current and voltage data of a set of photovoltaic modules and connects to the internet via Wi-Fi to send the collected data to an IoT cloud platform. The system is also capable of monitoring the humidity and temperature of the solar plant.

In [25], the proposed system is based on ESP32 and ESP8266, which communicate with an unspecified cloud system via Wi-Fi. Data for temperature, irradiance, humidity, wind speed, and DC generation are collected every 47 s and are made available through a web application.

Initially, the proposal's data loggers were configured to communicate with the internet via Wi-Fi, similarly to [24,25]. However, this approach presented a limitation in the positioning of data logger devices due to the Wi-Fi range. Thus, the second topology using LoRa and Wi-Fi was adopted in the current version.

Aghenta and Iqbal [26] present an IoT approach that focuses on monitoring PV generation without performing meteorological data acquisition. An Arduino performs sensor data acquisition and communicates with a Raspberry Py via a serial bus. The Raspberry Pi is connected to the internet via an Ethernet cable through which the collected data are sent to an IoT platform based on a local server, where it is stored and can be accessed.

Table 2. Technical characteristics of cited data acquisition systems. UN means unspecified and refers to features that are present in the system, but the technology used has not been specified. NM means not mentioned and indicates features that were not mentioned in the articles. The abbreviations of the measured parameters are DC current (Idc), DC voltage (Vdc), DC power (Pdc), AC current (Iac), AC voltage (Vac), AC power (Pac), ambient temperature (Ta), PV module temperature (Tm), irradiance (G), humidity (h), pressure (p), rainfall (rf), wind speed (Ws) and wind direction (Wd).

System	Measured Parameters	Open SW	Data Transmission	Internet Connection	Data Storage	Devices Sync	Dedicated Sensors
Caruso et al. [21]	Idc,Vdc	Yes	315 MHz RF	No Connection	Local SD card	NM	Yes
Su et al. [22]	Ta,Tm,G,h,Idc,Vdc	No	RF and ZigBee	No Connection	Local computer	NM	Yes
Al-Naima and Hamad [24]	Ta,h,Idc,Vdc	Yes	-	Wi-Fi	ThingSpeak cloud database	NM	Yes
Pereira et al. [25]	Ta,Tm,G,h,Ws, Idc,Vdc,Pdc	Yes	-	Wi-Fi	Local flash memory and cloud database	NTP	No
Aghenta and Iqbal [26]	Idc,Vdc,Vb	Yes	Serial	Ethernet	EmonCMS local server	NM	Yes
Zedak et al. [27]	Ta,G,Idc,Vdc	UN	I2C	UN	Local Raspberry Pi and cloud database	NM	Yes
Zago and Fruett [28]	Idc,Vdc	Yes	ZigBee	Wi-Fi	Local Raspberry Pi	NM	Yes
Xia et al. [29]	Idc,Vdc	UN	ZigBee	4G	Cloud server database	NM	No
Paredes-Parra at al. [30]	Ta,Tm,G,Idc,Vdc,Iac,Vac	Yes	LoRa	Ethernet	The Things Network cloud server	NM	Yes
Lazzaretti et al. [31]	Ta,Tm,G,h,Ws,Wd,Idc,Vdc,Iac,Vac	No	-	Ethernet	Local database	LabVIEW	Yes
Moreno-Garcia et al. [32]	Ta,Tm,G,p,rf,Ws,Wd,Idc,Vdc,Iac,Vac	No	UN	Ethernet	Local database	PTP	Yes
Erraissi et al. [33]	Ta,Tm,G,Ws,Wd,Idc,Vdc,Pdc,Iac,Vac,Pac	Yes	Bluetooth	Ethernet	Local SD card	NM	No
Proposed system	Ta,Tm,G,h,rf,Ws,Wd,Idc,Vdc,Pac	Yes	LoRa	Wi-Fi	Local SD card and cloud database	NTP	Yes

The system proposed in [27] monitors voltage, current, temperature, and irradiance. This system is based on an Arduino and uses a Raspberry Py as a gateway. The two devices communicate using the I2C protocol, while the Raspberry Py communicates with a cloud service using the MQTT protocol. In addition to storing and making data available, the cloud service can also send configuration commands to devices.

Regarding data transmission between devices, in [26,27], short-distance wired communication was used, limiting the disposition of devices and making installation more complex.

The system presented in [28] has a structure based on wireless sensor networks, in which each sensor node monitors the current and voltage generated by an individual photovoltaic module. The nodes send data via ZigBee to a Raspberry Py that hosts a web page, allowing access to data locally and over the internet via Wi-Fi.

In [29], ZigBee modules are used to collect and transmit data obtained from the PV plant inverters, building a local sensor network. A 4G gateway is used to connect the local network to the internet, enabling remote data access. Checksum verification is used to ensure the stability of the data transmission and to verify its integrity.

ZigBee technology generally has a range of 10 to 100 m and low energy consumption. The LoRa typically has a range of 2–5 km in urban areas or 15 km in suburban areas and has an even lower energy consumption than ZigBee. These were the main reasons for the adoption of LoRa in the proposal. However, LoRa has a lower data transfer rate than ZigBee.

LoRa is also used in the system implemented in [30] for data transmission. The system can monitor DC and AC electrical data, the temperature of the PV modules, irradiance, ambient temperature, and humidity. A LoRa gateway is responsible for making the data available on a local network to be accessed from a computer.

A few studies have reported a method for synchronizing the clocks of the devices that make up the system. The clock synchronization is essential for systems that use a single device and systems composed of multiple devices, allowing the measurements to have a correct timestamp and accurately represent the events in the PV plant.

A system for fault detection in PV systems is presented in [31]. The National Instruments CompactRIO (cRIO) controller is used to obtain the solar irradiance and ambient temperature data from a weather station and the DC and AC voltage and current data from the PV system. The collected data are then used in techniques for detecting and classifying faults in the PV system. The cRIO clock is updated through the LabVIEW software.

In [32], precision time protocol (PTP) (IEEE 1588) was used to synchronize the timestamps of the slides that make up a wireless sensor network. The network comprises wireless sensors that monitor irradiance, ambient temperature, the temperature of the PV modules, rainfall index, wind speed and direction, atmospheric pressure, DC and AC electrical data.

Network time protocol was used in the proposed system due to its easy access to information and because it is widely used in applications that require a precise timestamp. In [25], the NTP was also used, as the devices connect directly to the internet, it is only necessary to access the date and time information using the IP address of the NTP server. In the proposal, for the NTP data to be transmitted to the data logger devices, it was necessary to integrate it into the LoRa payload.

The system proposed in [33] is based on PcDuino (discontinued), which combines Arduino with Raspberry Py operating on Linux, being able to monitor temperature, irradiance, wind speed and direction, and AC and DC electrical data. The data are stored locally on an SD card and accessed over the internet. Storing data only locally on the system can create difficulties and a greater complexity when providing remote access. This form of storage is performed in [21,22,26,28,31,32].

Storing data only on remote servers can cause data loss if there is a communication failure. This is done in [24,29,30]. Performing both types of storage can prevent these problems and make the system more reliable, as was done in [25,27] and in our proposal.

Commercial software that requires a license is used in [22,31,32]; in addition to making changes to the system challenging, it also makes it more expensive. In the proposed system, the device software is developed in C++ using the Arduino framework, which is open-source and widely used and supported by the community.

Finally, for complete monitoring of a PV system, it is necessary to monitor: (1) the meteorological factors to which the system is subjected; (2) the DC electrical generation of the PV modules; (3) and the AC output of the inverter. This is accomplished in [30–33]. Our proposal also involves acquiring these three types of data. Furthermore, dedicated sensors are used without relying on data provided by the inverter. Dedicated sensors allow the proposed system to be applied to any PV system and independent of the sampling of data provided by the inverters. The systems in [25,29,33] are dependent on the inverter.

4. Results and Discussions

4.1. Experiment

The proposed system was applied to monitor a PV microsystem consisting of 19 polycrystalline silicon (Si-p) modules. Each module has a nominal power of 270 Wp (watt-peak) and can be associated in series with a nominal 5130 Wp. The PV system was configured in two strings due to the maximum voltage limitations of the inverter used. String 1 has ten panels associated in series, and String 2 has nine panels also associated in series. The connection of the strings is made through a DC/AC inverter with a nominal power of 5000 W, reaching a maximum peak of 6500 W.

Figure 10 shows the system installed on-site. Figure 10a shows the solarimetric station discussed in Section 2.2.1, with highlight 1 showing the pyranometer, anemometer, rain gauge, and wind direction indicator. The SHT20 is positioned inside a weather shelter in the middle of the station's structure. The NTC is fixed to the back of one of the PV modules. The solarimetric station data logger is contained in an airtight box for protection and is shown in highlight 2.

The cabinet shown in Figure 10b is located next to the inverter and has devices for protection and sectioning of the PV plant. Highlight 3 shows the DC voltage transducers (white) and DC current (black). Next to them is the AC power transducer. As detailed in Section 2.2.2, the data from the transducers are acquired by the PV generation data logger (highlight 4). The following results refer to the data monitored from 27 February 2021, to 14 March 2021.

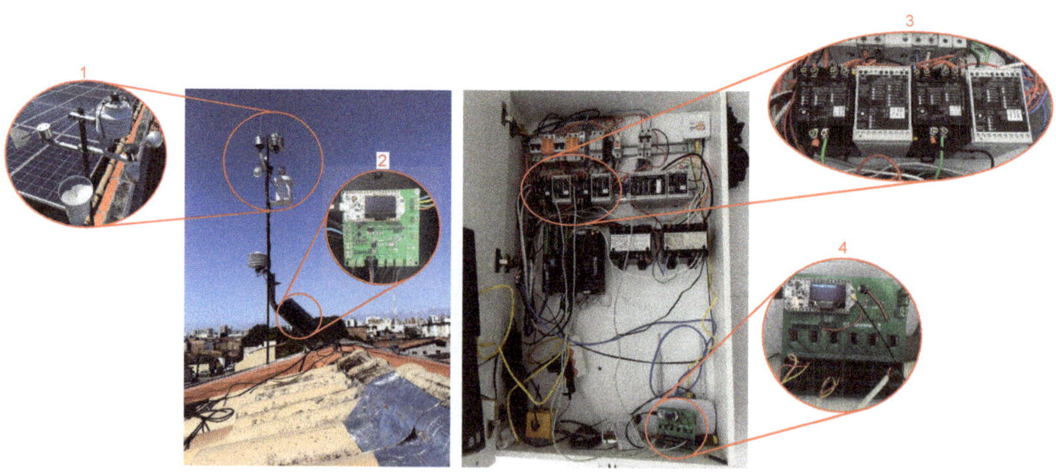

(a) (b)

Figure 10. Proposed system installed in a PV plant. (**a**) Solarimetric station, with emphasis on its data logger and sensors. (**b**) Cabinet with the transducers and the PV generation data logger.

4.2. Proposed System Operation

The architecture of the proposed system has good reliability and was effective in displaying collected data in real-time. During the 16-day experimental period, 23,040 data sets from each data logger were expected to be collected and sent. A total of 99.13% of the data sets from the solarimetric station, and 99.40% of the PV generation data sets, reached the cloud, demonstrating the system's reliability.

All data that arrived at GCP were successfully saved in BigQuery and sent to the web application via MQTT. Regarding the effectiveness of being in real-time, calculating the average of the data obtained by task 1 until these data are displayed in the web application is fast, with a delay of at most five seconds. Most of this delay is introduced by executing task 2 of the data loggers every five seconds, which is necessary to maintain the consistency of the device's operation.

The use of dedicated sensors allows the proposed system to be applied to other photovoltaic systems, regardless of the inverter used. Furthermore, the sampling of the proposal does not depend on the sampling of PV generation data provided by the inverter. Sampling every second for one minute, followed by the average of the data obtained in that interval, provides accurate measurements, keeping the transmission, storage, and computation of data in low complexity [16].

The proposed LoRa protocol reduces the size of the payload allowing a more efficient transmission, which reduces the transmission time and energy consumption. Table 3 shows a comparison with the load sizes transmitted by the proposed structure, by LoRaWAN using Cayenne low power payload (LPP) [34] and as a text string.

Table 3. Payload size in bytes considering different protocols.

Protocol	Station Data	PV Generation Data	ACK (min)
Proposed	15 B	14 B	5 B
Cayenne LPP [34]	33 B	28 B	9 B
Text String	61 B	56 B	27 B

The use of NTP to synchronize RTCs brought significant advantages. First, it allowed RTCs to have the correct date and time. Before using NTP, the RTC had a small precision error that could accumulate, generating an error of several minutes. Second, it allowed a simple way to synchronize the measurements between the two data loggers.

4.3. Measured Data Validation

A comparison with the data collected by a second monitoring system was performed to validate the data collected by the proposed system. This second monitoring system is based on the CR1000 data logger from Campbell Scientific [35] and has external sensors for measuring the ambient temperature, solar irradiance, and temperature of a PV module. The PV generation data are obtained by the same transducers used by the proposed system.

Before performing the comparisons and statistical calculations, a preprocessing of the data was performed. Regarding the data obtained by the proposed system, some of the PV generation values had errors in their measurements (e.g., being outside the expected range). The last valid reading replaced these values. Some current data obtained by CR1000 showed the not-a-number (NAN) value. These values were also removed and replaced by the last valid read value.

After preprocessing, the following statistical metrics were calculated between the data obtained from the two systems: mean absolute error (MAE) (Equation (6)), root mean square deviation (RMSD) (Equation (7)) and weighted absolute percent error (WAPE) (Equation (8)).

$$MAE = \frac{1}{n}\sum_{n}^{i=1}|y_i - x_i| \tag{6}$$

$$RMSD = \left[\frac{1}{n}\sum_{n}^{i=1}(y_i - x_i)^2\right]^{0.5} \tag{7}$$

$$WAPE = \frac{\sum_{n}^{i=1}|y_i - x_i|}{\sum_{n}^{i=1}|y_i|} \tag{8}$$

where n is the number of data samples, y_i is the i-th sample of data collected by CR1000, and x_i is the i-th sample of data collected by the proposed system. Table 4 shows the results.

MAE measures the average magnitude of the errors between the two data sets without considering their direction. Similarly, RMSD expresses average error, but as the errors are squared before they are averaged, the RMSD gives a relatively high weight to large errors. MAE and RMSD provide the error in units of the variable of interest, which can generate a misleading comparison between the errors of the different measurements. For this reason, WAPE was also calculated, showing the errors as a percentage.

Table 4. Statistical comparison between the measures of the proposed system and the CR1000 data logger considering the 16 days of the experiment and three types of metrics: MAE, RMSD and WAPE.

Data type	MAE	RMSD	WAPE
Ambient Temp.	1.21 °C	1.46 °C	4.13%
Irradiance	37.05 W/m²	68.12 W/m²	13.54%
PV Module Temp.	5.67 °C	6.73 °C	17.56%
DC Current String 1	0.12 A	0.41 A	5.37%
DC Current String 2	0.17 A	0.41 A	7.29%
DC Voltage String 1	21.46 V	48.02 V	15.15%
DC Voltage String 2	14.51 V	38.23 V	11.99%
AC Power	74.01 W	210.90 W	6.60%

The factors that presented WAPE below 10% were considered acceptable, including ambient temperature (4.13%), currents of the two strings (5.37% and 7.29%), and the AC power (6.6%). Considering the units and magnitudes of these factors, they also presented a low MAE that should not impact the measurement quality. The RMSD of the power (210.9 W) and currents were between 2.4 and 3.2 times higher than the MAE. This relationship between errors may indicate the presence of outliers in these measurements. The irradiance (13.54%), the temperature of the PV module (17.56%), and the voltage measurements (15.15% and 11.99%) showed WAPE greater than the acceptable value. Although, these values can be improved with some adjustments discussed below.

Figures 11 and 12 shows graphical comparisons of measurements over seven days (1 March to 7 March 2021). The blue line represents the measurements returned by our proposed system, and the red line represents the measurements returned by the CR1000. The CR1000-based system does not monitor data on humidity, wind speed, wind direction, and rain, so they have not been compared.

One can see that the error in the measurement of irradiance is mainly present when this factor reaches its maximum value (Figure 11c). This error can be caused due to the difference in the pyranometers' installation location. The pyranometer of the proposed system is installed on top of the structure of the solarimetric station (Figure 10a). In contrast, the pyranometer of the CR1000 system is installed at a lower level, next to the PV modules. Furthermore, the irradiance signal of the proposed system also presents noise. The application of a filter can reduce this noise and the error.

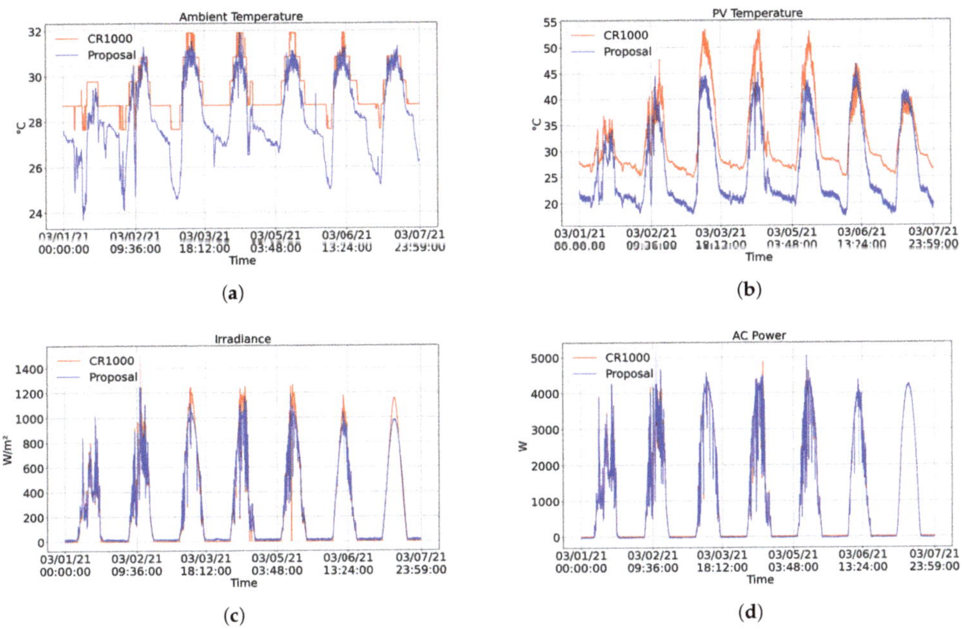

Figure 11. Graphical comparison of the data obtained during one week (1 March to 7 March 2021) by our proposed system (blue) and the CR1000 (red). The graphs show the following measurements: (**a**) ambient temperature, (**b**) PV module temperature, (**c**) irradiance, (**d**) AC power.

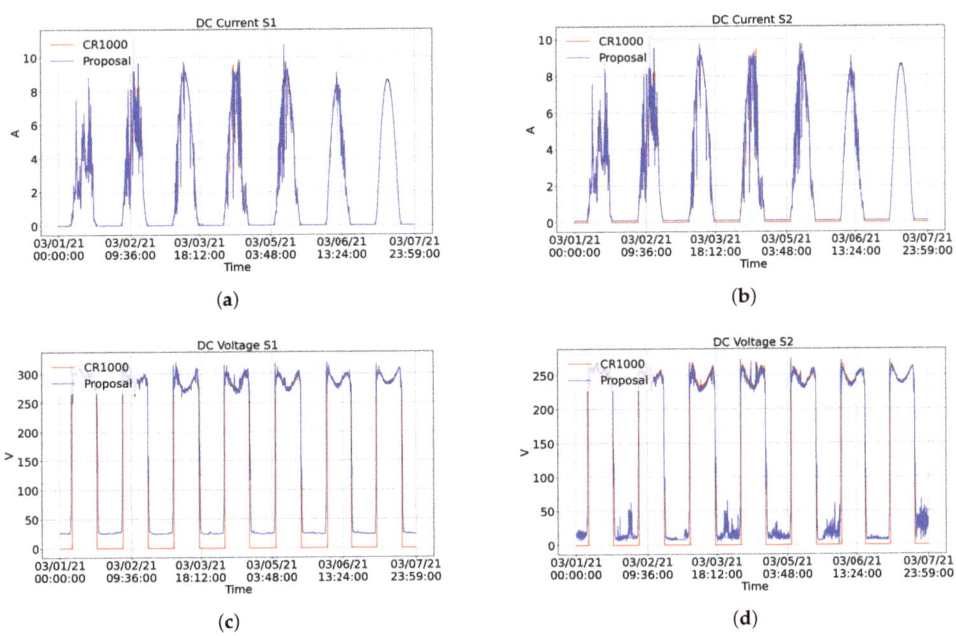

Figure 12. Graphical comparison of the data obtained during one week (1 March to 7 March 2021) by our proposed system (blue) and the CR1000 (red). The graphs show the following measurements: (**a**) string 1 current, (**b**) string 2 current, (**c**) string 1 voltage, (**d**) string 2 voltage.

Most of the error in voltage measurements is generated during the night, where the proposed station measured values above 0 V when it should be zero (Figure 12c,d). Zeroing the reading below a threshold value would reduce these errors in the voltage readings. The temperature of the PV module showed the most significant error among all factors when using an NTC to monitor these data. In the equation used to convert the resistance presented by the NTC to temperature, expected constants were applied for an ideal 10 kΩ NTC. Performing a calibration to find the specific constants of the NTC used would reduce the error presented.

4.4. Cost Description and Comparison

The cost of producing the solarimetric station data logger was USD 65.42, and this value includes the printed circuit board and electronic components, such as resistors, capacitors, voltage regulator, ADS1115, RTC, SD card, and Heltec Wi-Fi LoRa 32 (V2). The production of the PV generation data logger was USD 109.11. The higher price is due to the ADCs that the device contains. The LoRa gateway is a Heltec Wi-Fi LoRa 32 (V2), costing USD 20.80.

The cost of the sensors used are: pyranometer—USD 279.45, NTC 10K—USD 3.14, SHT20 (with waterproof protection)—USD 27.97, anemometer—USD 37.10, wind direction indicator—USD 37.10, pluviometer—USD 48.44, voltage transducer—USD 55.93, current transducer—USD 82.96 and power transducer—USD 236.24. Adding the costs of the data loggers and their sensors, we have a total cost of USD 498.62 for the solarimetric station and USD 623.13 for the PV generation monitors, so the total hardware cost of the proposed system is USD 1142.55, including the gateway.

The production cost can be lower when considering only the components used in the experiment and if the purchase of the components is optimized. For example, only one of the ADCs of the PV generation data logger was used during the experiments. Thus, the remaining ADCs can be removed, reducing USD 44.4, for a total cost of USD 1098.15. Furthermore, as the system is flexible, any sensors or transducers can be easily replaced with cheaper alternatives.

A comparison can be made with the work developed in [33] since it is one of the most complete of the literature and provides the cost of its development. The authors reported a cost of USD 25,000.00 to develop 20 units of the system. Therefore, each unit has a value of around USD 1250.00. To monitor meteorological and PV generation factors, two units of this system are required. Thus, our system is two times cheaper than this one. Another comparison can be made with the CR1000 data logger, the cost of which is about USD 1354.32 (average of eBays offers). The CR1000 does not include any sensors or transducers. Adding these devices to the CR1000, forming a PV monitoring system that monitors the proposed system's same factors, would be more expensive. Hypothetically, applying the sensors and the transducers used in the proposed system, which cost USD 926.42, to the CR1000 would result in a system with a total cost of approximately USD 2280.74 (CR1000 + sensors).

Regarding the proposed system software, all the code used in the developed devices was open source, adding no extra cost. In relation to GCP, the monthly cost is USD 0.20, based on the amount of data obtained during the month of March 2021 (31 days) and without considering the free monthly use of some of the services. We intend to implement in the future the same IoT infrastructure based on the open-source messaging agent Mosquitto [36], offering a free alternative to GCP. Heroku offers 1000 h per month to run free applications at no cost, so the web application does not add costs to the system.

5. Conclusions

In this work, an IoT system was developed for the real-time monitoring of photovoltaic systems. The IoT system comprises two data logger devices, a cloud system, and a web application. It can monitor weather and PV generation data.

The proposed system differential is that it measures all the relevant meteorological variables, is implemented using open software, uses LoRa as the data transmission technology, connects with the internet without cables, storages data locally and in the cloud, uses network time protocol to synchronize the devices' clocks, and measures PV generation variables directly from the plant (not from the inverter). To the best of our knowledge, no work reported in the literature presents these features altogether.

Moreover, experimental results showed the correct effectiveness of real-time data display and good reliability of the proposed system. The cost of production proved to be low, being almost twice as cheap as a system based on a commercial data logger and one of the complete systems found in the literature. Therefore, the proposed system can be an excellent alternative to micro and mini PV systems. Nevertheless, since it is an open system, it is scalable and easily modified, enabling it to be used in PV systems of different topologies and sizes.

Some of the future works are:

- Implement filters and perform sensor calibration on the proposed system to improve the accuracy of PV module temperature, DC voltages, and irradiance measurements;
- Implement a Mosquitto message broker in a dedicated server to avoid the need for a paid cloud service, which will decrease the data cost;
- We intend to integrate the proposed system with machine learning techniques to forecast photovoltaic generation based on meteorological data and automatically detect failures, allowing optimization of the electrical production process and increase the reliability of the PV plant.

Author Contributions: Conceptualization, G.C.G.d.M. and D.B.B.; Data curation, G.C.G.d.M. and I.C.T.; formal analysis, G.C.G.d.M.; funding acquisition, D.B.B.; investigation, G.C.G.d.M., D.B.B., E.d.A.B. and I.C.T.; methodology, G.C.G.d.M.; project administration, D.B.B.; resources, D.B.B.; software, G.C.G.d.M.; supervision, E.d.A.B.; validation, G.C.G.d.M. and I.C.T.; visualization, E.d.A.B.; writing—original draft, G.C.G.d.M.; writing—review and editing, D.B.B., E.d.A.B., I.C.T. and Í.B.Q.d.A. All authors have read and agreed to the published version of the manuscript.

Funding: This work is part of the project "Projeto de eficiência enegética e minigeração na UFAL" financed by Agência Nacional de Energia Elétrica (ANEEL) and the electric company Equatorial Energia.

Institutional Review Board Statement: Not applicable.

Informed Consent Statement: Not applicable.

Data Availability Statement: The data collected by the proposed system are openly available. The data for the solarimetric station can be found in Meteorological Data Proposed System at 10.6084/m9.figshare.14216969. Photovoltaic generation data are available in PV Generation Data Proposed System at 10.6084/m9.figshare.13953845. The data from the CR1000-based system were provided by Igor Cavalcante Torres and can be found in CR1000 Data at 10.6084/m9.figshare.14225111. The software codes of the two data loggers and the of LoRa gateway, as well as the complete schematics of the data logger devices, can be found in the Photovoltaic Monitoring System repository at https://github.com/gustavo95/Photovoltaic-Monitoring-System. All the developed code is authorial, except for the code used to access the Google cloud. Some modifications made to the Google code are documented at the beginning of each modified file, as requested by the Apache 2.0 license. Moreover, the hardware project, design, and assembly were done by the authors.

Acknowledgments: The authors would like to thank Equatorial Energia, ANEEL and Edge Innovation Center for the support.

Conflicts of Interest: The authors declare no conflict of interest.

Abbreviations

The following abbreviations are used in this manuscript:

PV	Photovoltaic
IoT	Internet of Things
LoRa	Long Range
NTP	Network Time Protocol
DSO	Distribution System Operator
MCU	Microcontroller Unit
RF	Radio Frequency
cRIO	Compact RIO
DC	Direct Current
AC	Alternating Current
PTP	Precision Time Protocol
I2C	Inter-Integrated Circuit
MQTT	Message Queuing Telemetry Transport
SD	Secure Digital
SPI	Serial Peripheral Interface
RTC	Real-Time Clock
ADC	Analog-to-Digital Converter
ACK	Acknowledgment
CRC	Cyclic Redundancy Check
GCP	Google Cloud Platform
SQL	Standard Query Language
CSV	Coma-Separated Values
UN	Unspecified
NM	Not Mentioned
Idc	DC Current
Vdc	DC Voltage
Pdc	DC Power
Iac	AC Current
Vac	AC Voltage
Pac	AC Power
Ta	Ambient Temperature
Tm	PV Module Temperature
G	Irradiance
h	Humidity
rf	Rainfall
p	Pressure
Ws	Wind Speed
Wd	Wind Direction
USD	United States Dollar
LPP	Low Power Payload
NAN	Not A Number
MAE	Mean Absolute Error
RMSD	Root Mean Square Deviation
WAPE	Weighted Absolute Percent Error

References

1. Renewables 2020 Global Status Report. Technical Report, REN21 Secretariat: Paris, France, 2020. Available online: https://www.ren21.net/reports/global-status-report/ (accessed on 5 May 2021).
2. Infográfico ABSOLAR. Available online: http://www.absolar.org.br/infografico-absolar.html (accessed on 14 January 2021).
3. Begovic, M.M.; Kim, I.; Novosel, D.; Aguero, J.R.; Rohatgi, A. Integration of photovoltaic distributed generation in the power distribution grid. In Proceedings of the 45th Hawaii International Conference on System Sciences, Maui, HI, USA, 4–7 January 2012; pp. 1977–1986.
4. Agüero, J.R.; Steffel, S.J. Integration challenges of photovoltaic distributed generation on power distribution systems. In Proceedings of the IEEE Power and Energy Society General Meeting, Detroit, MI, USA, 24–28 July 2011; pp. 1–6.
5. López-Vargas, A.; Fuentes, M.; Vivar, M. IoT application for real-time monitoring of solar home systems based on Arduino with 3G connectivity. *IEEE Sens. J.* **2018**, *19*, 679–691. [CrossRef]
6. Ghasempour, A. Internet of things in smart grid: Architecture, applications, services, key technologies, and challenges. *Inventions* **2019**, *4*, 22. [CrossRef]
7. Shahsavari, A.; Akbari, M. Potential of solar energy in developing countries for reducing energy-related emissions. *Renew. Sustain. Energy Rev.* **2018**, *90*, 275–291. [CrossRef]
8. WiFi LoRa 32 (V2). Available online: https://heltec.org/project/wifi-lora-32/ (accessed on 16 March 2021).
9. ESP32. Available online: https://www.espressif.com/en/products/socs/esp32 (accessed on 16 March 2021).

10. 74HC125 Datasheet. Available online: https://assets.nexperia.com/documents/data-sheet/74HC_HCT125.pdf (accessed on 16 March 2021).
11. ADS111x Datasheet. Available online: https://www.ti.com/lit/ds/symlink/ads1115.pdf (accessed on 16 March 2021).
12. Luque, A.; Hegedus, S. *Photovoltaic Science and Engineering*, 2nd ed.; John Wiley & Sons Ltd.: Hoboken, NJ, USA, 2010.
13. SHT20 Datasheet. Available online: https://www.mouser.com/datasheet/2/682/Sensirion_Humidity_Sensors_SHT20_Datasheet-1274196.pdf (accessed on 16 March 2021).
14. ADS866x Datasheet. Available online: https://www.ti.com/lit/ds/symlink/ads8668.pdf (accessed on 16 March 2021).
15. Current, Tension, and Frequency Transducers/Transmitters. Available online: http://www.secon.com.br/english/grupo.php?id=16 (accessed on 16 March 2021).
16. Riley, D.M.; Cameron, C.P.; Jacob, J.A.; Granata, J.E.; Galbraith, G.M. Quantifying the effects of averaging and sampling rates on PV system and weather data. In Proceedings of the 2009 34th IEEE Photovoltaic Specialists Conference, Philadelphia, PA, USA, 7–12 June 2009; pp. 456–461.
17. NTP.br. Available online: https://ntp.br/ (accessed on 16 March 2021).
18. arduino-LoRa. Available online: https://github.com/sandeepmistry/arduino-LoRa (accessed on 16 March 2021).
19. Google Cloud Platform. Available online: https://cloud.google.com/ (accessed on 16 March 2021).
20. Heroku: Cloud Application Platform. Available online: https://www.heroku.com/ (accessed on 16 March 2021).
21. Caruso, M.; Miceli, R.; Romano, P.; Schettino, G.; Spataro, C.; Viola, F. A low-cost, real-time monitoring system for PV plants based on ATmega 328P-PU microcontroller. In Proceedings of the IEEE International Telecommunications Energy Conference, Osaka, Japan, 18–22 October 2015; pp. 1–5.
22. Su, F.P.; Chen, Z.C.; Zhou, H.F.; Wu, L.J.; Lin, P.J.; Cheng, S.Y.; Li, Y.F. A distributed monitoring system for photovoltaic arrays based on a two-level wireless sensor network. In Proceedings of the IOP Conference Series: Earth and Environmental Science, Kunming, China, 22–25 September 2017; Volume 93, p. 012077.
23. Zigbee. Available online: https://zigbeealliance.org/solution/zigbee/ (accessed on 18 February 2021).
24. Al-Naima, F.; Hamad, A. A Low-cost Solar Farm Monitoring System Based on Cloud Database. In Proceedings of the International Conference on Innovation and Intelligence for Informatics, Computing, and Technologies, University of Bahrain, Zallaq, Bahrain, 18–20 November 2018; pp. 1–5.
25. Pereira, R.I.; Jucá, S.C.; Carvalho, P.C. IoT embedded systems network and sensors signal conditioning applied to decentralized photovoltaic plants. *Measurement* **2019**, *142*, 195–212. [CrossRef]
26. Aghenta, L.O.; Iqbal, M.T. Development of an IoT Based Open Source SCADA System for PV System Monitoring. In Proceedings of the IEEE Canadian Conference of Electrical and Computer Engineering, Edmonton, Alberta, AB, Canada, 5–8 May 2019; pp. 1–4.
27. Zedak, C.; Lekbich, A.; Belfqih, A.; Boukherouaa, J.; Haidi, T.; El Mariami, F. A proposed secure remote data acquisition architecture of photovoltaic systems based on the Internet of Things. In Proceedings of the 6th International Conference on Multimedia Computing and Systems, Rabat, Morocco, 10–12 May 2018; pp. 1–5.
28. Zago, R.M.; Fruett, F. A low-cost solar generation monitoring system suitable for internet of things. In Proceedings of the 2nd International Symposium on Instrumentation Systems, Circuits and Transducers, Fortaleza, Ceará, Brazil, 28 August–1 September 2017; pp. 1–6.
29. Xia, K.; Ni, J.; Ye, Y.; Xu, P.; Wang, Y. A real-time monitoring system based on ZigBee and 4G communications for photovoltaic generation. *CSEE J. Power Energy Syst.* **2020**, *6*, 52–63.
30. Paredes-Parra, J.M.; García-Sánchez, A.J.; Mateo-Aroca, A.; Molina-García, Á. An alternative Internet-of-Things solution based on LoRa for PV power plants: Data monitoring and management. *Energies* **2019**, *12*, 881. [CrossRef]
31. Lazzaretti, A.E.; Costa, C.H.D.; Rodrigues, M.P.; Yamada, G.D.; Lexinoski, G.; Moritz, G.L.; Oroski, E.; Goes, R.E.; Linhares R.R.; Stadzisz, P.C.; et al. A monitoring system for online fault detection and classification in photovoltaic plants. *Sensors* **2020**, *20*, 4688. [CrossRef] [PubMed]
32. Moreno-Garcia, I.M.; Palacios-Garcia, E.J.; Pallares-Lopez, V.; Santiago, I.; Gonzalez-Redondo, M.J.; Varo-Martinez, M.; Real-Calvo, R.J. Real-time monitoring system for a utility-scale photovoltaic power plant. *Sensors* **2016**, *16*, 770. [CrossRef] [PubMed]
33. Erraissi, N.; Raoufi, M.; Aarich, N.; Akhsassi, M.; Bennouna, A. Implementation of a low-cost data acquisition system for "PROPRE. MA" project. *Measurement* **2018**, *117*, 21–40. [CrossRef]
34. Cayenne Low Power Payload. Available online: https://developers.mydevices.com/cayenne/docs/lora/#lora-cayenne-low-power-payload (accessed on 16 March 2021).
35. CR1000X. Available online: https://www.campbellsci.com/cr1000x (accessed on 16 March 2021).
36. Eclipse Mosquitto. Available online: https://mosquitto.org/ (accessed on 16 March 2021).

sensors

MDPI

Article

Influence of Data Sampling Frequency on Household Consumption Load Profile Features: A Case Study in Spain

J. C. Hernandez [1], F. Sanchez-Sutil [2], A. Cano-Ortega [2,*] and C. R. Baier [3]

1 Center for Advanced Studies in Earth Sciences, Energy and Environment, University of Jaén, 23071 Jaén, Spain; jcasa@ujaen.es
2 Department of Electrical Engineering, University of Jaen, 23071 Jaen, Spain; fssutil@ujaen.es
3 Department of Electrical Engineering, University of Talca, Camino los Niches Km. 1, Curicó 3344158, Chile; cbaier@ieee.org
* Correspondence: acano@ujaen.es; Tel.: +34-953-212343

Received: 13 September 2020; Accepted: 20 October 2020; Published: 23 October 2020

Abstract: Smart meter (SM) deployment in the residential context provides a vast amount of data of high granularity at the individual household level. In this context, the choice of temporal resolution for describing household load profile features has a crucial impact on the results of any action or assessment. This study presents a methodology that makes two new contributions. Firstly, it proposes periodograms along with autocorrelation and partial autocorrelation analyses and an empirical distribution-based statistical analysis, which are able to describe household consumption profile features with greater accuracy. Secondly, it proposes a framework for data collection in households at a high sampling frequency. This methodology is able to analyze the influence of data granularity on the description of household consumption profile features. Its effectiveness was confirmed in a case study of four households in Spain. The results indicate that high-resolution data should be used to consider the full range of consumption load fluctuations. Nonetheless, the accuracy of these features was found to largely depend on the load profile analyzed. Indeed, in some households, accurate descriptions were obtained with coarse-grained data. In any case, an intermediate data-resolution of 5 s showed feature characterization closer to those of 0.5 s.

Keywords: smart meter; temporal data granularity; electric load profile; time slices; time series; advanced metering infrastructure

1. Introduction

Large temporal datasets for household electricity consumption, provided by smart meters (SMs), offer significant potential for energy time-series scientists. These datasets permit increased resolution and analysis at the level of individual households. Recent studies by Zhou and co-workers [1,2] focus on the challenges and opportunities that SMs provide for smarter energy management where SM data are an essential component.

Traditionally, consumption load metering in the residential context has been conducted at a low time resolution. Thus, consumption profiles are generally gathered for different dwelling types, based on a sampling frequency that provides a data granularity from 1 to 30 min [3]. However, current SCADA systems can sample consumption data at a higher frequency (typically 1 Hz) though standard practice is to store averaged values of 1 min or higher [4].

Reference [5] highlights the importance of taking into account both a wide time slice and frequency spectrum for an accurate description of the load profile features of household consumption.

More specifically, the choice of a temporal data granularity (data sampling frequency) for specifying consumption load profile features has a crucial impact on the results of any action or assessment, as discussed in the literature [6–48], see Table 1. This table summarizes for each potential action or assessment the time resolution (data granularity) and time horizon (time slice) envisaged for the works related to load profiles in households. The impact of temporal data granularity is important because the consumption profile is known to fluctuate at a high temporal resolution (i.e., interval of 0.01–5 Hz [16,49,50]). Therefore, when a longer time resolution is envisaged, the profile dynamics become increasingly biased. This means that the profile should be sampled at a more fine-grained level, which will more accurately describe its behavior. Nonetheless, there is a trade-off between the computational burden and accuracy of any action or assessment, which is determined by this discrete time resolution [51]. Moreover, the effective measurement of the electrical variables –root-mean-square (RMS) value– requires at most a 5-Hz sampling frequency [52]; a higher frequency would provide just punctual values.

The dynamic nature and frequently high variation typical of household consumption load profiles in the residential context has been analyzed for different temporal data granularities, as shown in Table 1. As disclosed, most of the time resolutions were longer than 1 min. Few references focused on granularities lower than 15 s. High data granularity naturally implies larger amounts of data to be locally stored (either hard disk or memory card) or uploaded to the cloud. This has led many researchers and industry practitioners to develop and survey a vast number of analytical tools that could help to segment and cluster SM big data so that they can be analyzed in real time [11,32]. On the other hand, uploading these data to the cloud (data traffic with the cloud) is another important limitation [11,44,53]. High granularity requires a large bandwidth, which is not always available in individual households. Since data cannot be transmitted at such high resolutions, data compression algorithms are required. The compression ratio can be 10:1 or even 1000:1 can be achieved. For example, Kelly [8] performed measurements every 6.25×10^{-5} s(16 kHz) and used the free lossless audio codec (FLAC) compression algorithm that reduced the daily data of 28.8 GB to 4.8 GB with a 6:1 compression ratio. Also, reference [54] developed a new method of data compression via stacked convolutional sparse auto-encoder. These algorithms are usually time consuming, and thus, the gathered data are not available in real time.

The scientific community currently has very limited access to consumption load data in the residential context. The information available for private purposes is either free or must be purchased. Free options are available at different web sites that provide records for homes. Pecan Street [55] is a web site that provides data from 1115 houses with PV and/or EV. Mack [56] developed a web site on SMs for homes, with the aim of saving electricity. Wilcox [53] built a hadoop-scaled SM analytics platform that allows the use of large datasets at a 20 TB scale. Furthermore, there are other web sites that provide household consumption load data with a granularity from 1 s to 1 min [57–64].

Some studies evaluated the feature bias due to the use of coarse-grained data when assessing the consumption load profiles in households. Murray [38] compared time resolution data of 1 min and 15 min and demonstrated the damping effect when working with data with high temporal resolutions in 21 houses in the UK. Naspolini [39] showed that the use of a 15-min data granularity was not well adjusted because of load fluctuations with a period lower than 15 min (5 min) in the operation of electric water heaters. Bucher [40] studied a 1 s and 15-min data granularity in domestic PV-household load profiles. Shi [41] analyzed the accuracy of predictions based on different data granularity. Hoevenaars [64] showed that using a 1-h time step hid the load variability within the hour for models of renewable power systems. Regarding optimization purposes, Van der Meer [42] concluded that a 5-min time resolution provided a good balance between accuracy and the burden of data size, whereas [45] showed that using hourly data led to large biases compared to 1-min data. However, coarser data could be sufficient for household aggregation [48]. A shorter fluctuation at a granularity of only 4 s was investigated in [65].

Table 1. Literature published related to load profiles in households.

Time Resolutions (Data Granularity -Sampling Frequency)	Actions or Assessments							Time Horizon (Time Slice)		
	Adjust Tariff Electrics	Energy Efficiency & Comfort	Recognize Activities	Optimization Size Renewable Generation	Electrical Load Forecasting	Power Consumption Characterization	Applying Probabilistic Techniques	A Few Minutes	A Few Hours	A Few Days
<0.5 s	–	–	–	–	–	[8]	–	–	–	–
1 s	–	[8,9,12]	[23]	[13,42]	[12]	[7–9,11]	–	[7,13]	[9,23]	[13]
>1 s and <10 s	–	[8,38]	–	[13]	–	[8,38,41]	–	[13]	–	[13]
15 s	–	–	–	[13,14]	[14]	–	–	[13]	–	[13]
1 min	–	[9,12]	[6,12,26]	[13,45]	[12,24–27,29]	[6,9,27]	–	[19,26,27]	[6,9]	–
5 min	[15]	[15,17]	[26]	[7,16]	[17,18,26,27]	[7,27]	[18]	[15–18,26,27]	[17]	[17]
15 min	[16]	–	[6,22,26]	[16]	[22,25,26,30,34,37,47]	[6,11,32]	–	[16,26]	[6,33,34,47]	[22]
30 min	–	–	[26]	[42,45]	[22,27,30,33–35,42]	[35]	[42]	[26,27,35,42]	[33,34]	[22,35,42]
1 h	–	–	[26,28,36]	[45,48]	[25,26,30,34,36]	[28]	–	[26]	[34,36]	[22]
1 day	–	–	[36]	[43]	[36,37,43]	[37]	–	[37]	[36]	[43]
No available information	[14]		–	[4,40,49]	[19–21,29]	[7]	[19,20]		[20]	[19,21]

In this context, it was found that the availability of household consumption load data at a fine granularity (high sampling frequency) obtained from SM measurements allows users to do the following actions or assessments (see Table 1):

- To dynamically adjust electricity tariffs as fine temporal resolutions make it possible to quickly adapt to changes in consumption and thus reduce the electricity bill.
- To improve energy efficiency, comfort, and safety in households with an intelligent automation system.
- To recognize activities (i.e., analysis of energy consumption through observation) that are potentially more meaningful to users.
- To optimally size renewable generation system and storage systems.
- To forecast household load profiles with different time horizons, from a short-term load forecast (hourly and daily) to long-term forecast-based planning studies.
- To perform studies of angle stability, transient analysis, and frequency control in electrical energy systems.
- To characterize household consumption load profile features.
- To perform load forecasting by applying probabilistic techniques.

To date, the major obstacle to accurately describing household consumption load profile features is the fact that this type of profile has not been sampled at temporal low-resolution (high frequency) that gathers its intrinsic dynamics. Thus, most of the examples mentioned in Table 1 used an hourly or even a 30-min or 15-min time resolution, but they do not include enough information to accurately provide actions or assessments. Furthermore, even though the current trend is to use more temporally granular data sets in household applications, the influence of temporal granularity has not as yet been analyzed using a comprehensive and high-resolution data set. Quite a few papers [38–42,45,48,65] evaluated the bias due to the use of coarse-grained data, but never compared resolutions lower than 1 to 5 min. Still another shortcoming is the fact that most studies cover short time slices. These involve a reduced timespan chosen to characterize key aspects of temporal variability, for example, covering weekdays and weekends, different times of day, and different seasons. The time horizon (time slice) envisaged in Table 1 was typically restricted to minutes, a few hours or a few days. However, the resulting description is not accurate since it does not take seasonality into account. Lastly, SM data only reflect a few electrical variables, which means that very little information regarding electrical behavior can be derived (usually energy or power).

To fill this gap, the new methodology presented in this paper makes two major contributions. It first proposes periodograms along with autocorrelation and partial autocorrelation analyses and an empirical distribution-based statistical analysis, which are able to describe household consumption profile features with greater accuracy. This type of analysis reveals key issues about the granularity impact on the load fluctuation, such as the accurate description of its constituent signals. In contrast, the temporal analysis usually found in literature only offers information regarding the granularity impact on the change in the magnitude of the peak and trough load. Secondly, it proposes a framework for data collection in households at a high sampling frequency (>4 Hz) that provides data to be used in the proposed methodology.

A case study of four households in Spain, using thirteen data granularities, from a half-second to 30 min (0.5, 1, 2, 5, 10, 15, 30 s, and 1, 2, 5, 10, 15, and 30 min), provided valuable insights into the influence of data granularity on the description of consumption load profile features. The data set selected, during almost two years, had different consumption features, namely with varying characteristics in terms of the relation between the peak and base load and load fluctuations, which made it possible to take the heterogeneity of real-world load profiles into account. We acknowledge that conducting our analysis with a data sample from four households is a limitation of this study. However, this data sample was adequate to achieve our primary objective of demonstrating the usefulness of the methodology proposed, which was to highlight the information loss regarding the profile features when using coarse-grained data.

The remainder of the paper is organized as follows: Section 2 describes the methodology implemented in this study. Section 3 discusses the results that reflect the influence of data granularity (sampling frequency) on consumption load profile features. Finally, Section 4 presents the conclusions that can be derived from this research.

2. Methodology

This section first presents a set of variables for a full description of features for the stochastic dataset derived from household consumption load profiles. The variables are defined from estimated time series models. This is followed by an explanation of the concepts of granularity and time slices. An outline is then provided of the framework for consumption data collection in households at high sampling frequency and its post-processing.

2.1. Time-Series Theory

For stationary stochastic data, the theory of time series models provides estimated models, which include the description of the probability mass function (PMF), power spectral density function, and autocorrelation function [66,67].

The features of a stationary stochastic dataset are fully described by the joint probability density function of the observations [66,67]. The joint probability density function of the observations are fully depicted by a stationary stochastic dataset. If the density could be calculated on the basis of observations, then this density would provide all of the information pertaining to the signal. Nevertheless, this is usually not feasible without a great deal of additional knowledge about how such observations were obtained. Features that can always be estimated include the power spectral density and the autocorrelation function. In addition, knowledge of the spectrum (power spectral density function) or autocorrelation (autocorrelation function) along with the first two statistical moments makes it possible to accurately describe the joint probability density function of normally distributed observations [66,67]. Even when assumptions regarding the normal distribution and strict stationarity are not confirmed, previous estimators still provide a sound basis for further research [67]. Nonetheless, in the case of other distributions, higher-order moments provide more information.

2.1.1. Stationary

A time series x (and thus the underlying stochastic process) is considered stationary if the process is in a certain state of statistical equilibrium. Accordingly, the properties of a stochastic processes are assumed to be invariant during the translation trough time. This signifies that the joint probability distribution associated with m observations $(x_1, x_2, \ldots, x_m)$, for any set of time measurements $(t_1, t_2, \ldots, t_m)$, is the same as that for m observations $(x_{1+k}, x_{2+k}, \ldots, x_{m+k})$, at times $(t_{1+k}, t_{2+k}, \ldots, t_{m+k})$. Therefore, the joint distribution must not change when all of the observation times are shifted backward or forward by any integer amount k.

Household consumption load profiles are usually not stationary; there is usually daily, weekly, and monthly seasonality and an upward trend as the number of appliances in the household rises. However, as stationary datasets are easier to analyze, there are numerous techniques that can be applied on time series to make it stationary, i.e., transformations, deseasonalisation, and differencing [68].

There are many methods to check whether a time series is stationary or non-stationary: (i) look at plots; (ii) summary statistics; and (iii) statistical tests. The most rigorous approach to detecting stationarity in time series data is using statistical tests developed to detect specific types of stationarity, such as simple parametric models that generate the stochastic process. Among them, it is important to mention the following: (i) Augmented Dickey-Fuller (ADF) test [69]; (ii) Kwiatkowski, Phillips, Schmidt, and Shin (KPSS) test [70,71]; (iii) Variance ratio test [72]; (iv) Leybourne-McCabe (LMC) test [73]; and (v) Phillips-Perron (PP) test [74].

The Dickey-Fuller test was the first statistical test developed to check the null hypothesis that a unit root is present in an autoregressive model of a given time series, and that the process is thus not stationary. An extension of this test (the ADF test) was developed to accommodate more complex models and data. The KPSS test assumes the null hypothesis of stationarity in relation to the linear or average trend, as opposed to the ADF test. While the alternative is the presence of unitary root. The unitary hypothesis represents the stationary nature of the process. The variance ratio test is included in the semi-parametric tests, unlike the two previous ones, which are parametric. The null hypothesis implies non-stationary, and the alternative hypothesis indicates that the process is stationary. El LCM test allows for additional autoregressive lags similar to the ADF test. Although both tests have the same asymptotic distribution, the statistics from the LMC test converge at a higher rate. The PP test states the null hypothesis as the non-stationarity in time series data and the rejection of the unit-root null in favor of the alternative model.

2.1.2. Metrics for Statistics and Probability Analysis

In probability theory, the moments of a stochastic dataset (random variable) $\boldsymbol{x}$ consist of the expected values of certain functions $\boldsymbol{x}$. Such variables are a set of descriptive measurements that correspond to the probability distribution of $\boldsymbol{x}$ and determine whether all the moments of $\boldsymbol{x}$ are known.

Let $\boldsymbol{x}$ be a discrete univariate dataset with a finite number of outcomes $(x_1, x_2, \ldots, x_m)$ occurring with probabilities $p_x(x_i)$, i.e., with PMF $p_x(x_i)(= P(\boldsymbol{x} = x_i)])$, its moment of order one, two, and r can be specified as follows [75,76]:

$$\begin{gathered} m_x^1 = E[\boldsymbol{x}];\ m_x^2 = E\left[\boldsymbol{x}^2\right];\ m_x^r = E[\boldsymbol{x}^r] \\ m_x^r = \sum_{i=1}^{\infty} x_i^r \cdot p_x(x_i) \end{gathered} \tag{1}$$

The cumulants of a stochastic dataset are variables that constitute an alternative to the moments described in the previous paragraph [76]. Unlike moments, these cumulants cannot be directly obtained by summatory or integrative processes, such as (1). Cumulants can only be found by identifying the moments and applying relationship formulas [75,76]. Accordingly, the first cumulant is the expected value; the second cumulant is the variance; the third cumulants measures asymmetry; and the fourth cumulant measures the tailedness of the probability distribution.

Equation (1) describes how to find the expected value, variance, skewness and kurtosis for discrete random variables according to probability theory. However, some of these variables such as the expected value and variance might strike as very similar to the sample mean and sample variance, respectively, in descriptive statistics. The sample mean and sample variance are random variables because their values depend on what the particular random sample happens to be. In other words, if we know the frequency distribution, or how many times a data value is repeated in the dataset, the following formula can be used to determine the statistics sample mean, $\overline{x}$, and sample variance, s^2:

$$\begin{gathered} \overline{x} = \frac{\sum f \cdot x_i}{\sum f} \\ s^2 = \frac{\sum f \cdot (x_i - \overline{x})^2}{\sum f} \end{gathered} \tag{2}$$

Notice that $m = \sum f$ is the dataset sample size.

The distinction between variables in statistics and probability analysis is that the statistics vary with each sample dataset, whereas probability variables are fixed when you know the dataset's probability distribution. The law of large numbers states that as the sample size grows to infinity, statistics provide a more accurate picture of moments of distribution.

2.1.3. Autocorrelation and Spectrum

The covariance between two observations x_n and x_{n+k} of a stationary stochastic dataset is formulated as follows:

$$r(k) = \operatorname{cov}(x_n, x_{n+k}) = E[(x_n - \mu_x)(x_{n+k} - \mu_x)] \tag{3}$$

where μ_x is the mean of the dataset.

The quantity $r(k)$ is specified for each integral value of k, and the combination of all of these quantities is known as the autocovariance function of x_n. It quantifies the covariance between pairs at a distance or lag k, for all values of k. This signifies that it is a function of lag k.

The autocovariance function expresses all knowledge pertaining to Gaussian stochastic data. In combination with the first two statistical moments, it fully characterizes the joint probability distribution function of the data. Only when the distribution significantly departs from normal, it is interesting to study higher-order moments or other features.

In the same way as the covariance between two variables, it is also possible to normalize the autocovariance function $r(k)$ and thus obtain the autocorrelation function $\rho(k)$:

$$\rho(k) = \frac{r(k)}{r(0)} = \frac{r(k)}{\sigma_x^2} \tag{4}$$

where σ_x^2 is the variance of the dataset.

The autocorrelation function reveals how rapidly a signal can change over a period of time. At lag 0, the autocorrelation value is 1. In the case of most physical processes, there is an autocorrelation function that progressively diminishes for greater lags. Accordingly, the relation at a short temporal distance is greater than the relation for longer distances. A long lag value in the autocovariance function is indicative of the slow variation of the data. In contrast, a short lag value signifies that the data at short distances are correlated. Nonetheless, a high value in the autocorrelation function signifies a repetition pattern, and thus reveals a constituent signal in the analyzed dataset. Therefore, the resulting variable that includes the set of high values of the autocorrelation function can provide a means for describing the features of a stationary stochastic dataset.

The partial autocorrelation function, $\alpha(k)$ represents the autocorrelation between x_n and x_{n+k} is an indication of x_n on x_{n+k} through x_{n+k-1} removed. In the same way, it denotes the autocorrelation between x_n and x_{n+k}, which is not explained by lags 1 through *k-1*, inclusively [77]:

$$\begin{gathered} \alpha(1) = corr[x_{n+1}, x_n] \text{ for } k = 1 \\ \alpha(k) = corr\Big[x_{n+k} - P_{n,k}(z_{n+k}), x_n - P_{n,k}(x_n)\Big] \text{ for } k \geq 2 \end{gathered} \tag{5}$$

where $P_{n,k}(x)$ is the surjective operator of orthogonal projection of x onto the linear subspace of the Hilbert space spanned by $x_{n+1}, \ldots, x_{n+k-1}$.

There are algorithms for estimating the partial autocorrelation based on the sample autocorrelations [78].

The Discrete Fourier Transform (DFT) of the autocovariance function constitutes the spectrum or power spectral density function $h(\omega)$. The Wiener-Khintchine theorem [79,80] defines conditions in which valid autocovariances have a transform that is always non-negative in all contexts; see [81].

$$\begin{aligned} h(\omega) &= \tfrac{1}{2\pi} \sum_{k=-\infty}^{\infty} r(k) e^{-j\omega k}, \qquad -\pi \leq \omega \leq \pi \\ r(k) &= \textstyle\int_{-\pi}^{\pi} h(\omega) e^{j\omega k} d\omega, \qquad k = 0, \pm 1, \pm 2, \ldots \end{aligned} \tag{6}$$

The reason why this is known as the 'power spectral density function' is evident in the integral for the value $k = 0$:

$$r(0) = \int_{-\pi}^{\pi} h(\omega) d\omega = r(0) = \sigma_x^2 \tag{7}$$

The variance represents the total power in the signal. It is the power spectral density function that provides the distribution of the total power over the frequency range. When the data are characterized by a strong quasi-periodicity with a specific period, they show a narrow peak in the power spectral density instead of one exact frequency, thus revealing a constituent signal in the analyzed dataset.

Therefore, the resulting variable that includes the set of narrow peaks of the power spectral density function can provide a means for describing the features of a stationary stochastic dataset.

The fast Fourier transform (FFT) is an algorithm used for computing the DFT more efficiently and faster, described in Equation (6) as an infinite sum. The FFT algorithm systematizes the redundant calculations in a very efficient way, taking advantage of the algebraic properties of the Fourier matrix. The FFT applied to a signal allows thus obtaining its power spectrum, namely, the periodogram of the signal. When a high number of observations (N) is involved, the FFT results assure a high accuracy.

When (4) is divided by the variance of the signal, this results in the normalized autocorrelation function $\rho(k)$ and the normalized power spectral density $\varphi(\omega)$:

$$\begin{aligned} \varphi(\omega) &= \frac{1}{2\pi}\sum_{k=-\infty}^{\infty} \rho(k)e^{-j\omega k}, \qquad -\pi \leq \omega \leq \pi \\ \rho(k) &= \int_{-\pi}^{\pi} \varphi(\omega)e^{j\omega k}d\omega, \qquad k = 0, \pm 1, \pm 2, \ldots \end{aligned} \tag{8}$$

2.2. Temporal Granularity and Time Slices

Granularity is the temporal resolution of a recorded measurement set, which is used to obtain the variability of the data set, pertaining to a given measurement interval. The time slice is the temporal framework for a given study. The quality of the results depends on the choice of the appropriate granularity and time slice for establishing the household consumption load profile features.

The temporal framework for analyzing the electrical energy system differs, depending on the action or assessment performed [19], as shown in Figure 1. Very short-term analysis involves a temporal framework from 0.5 s to 1 min, and mainly includes transient analysis, demand response in real time, angle stability, and frequency control. Short-term analysis is associated with the system operation from various seconds to thirty minutes. It includes day-to-day system operation, hour-ahead scheduling, studies of probabilistic load forecasting, and seasonal prediction of demand. A mid-term analysis involves a temporal framework from various days to one year. This includes maintenance of system assets, unit commitment, energy trading, and energy sales. Finally, long-term analyses, which range from 3 years to more than a decade [19], cover new capacity addition [82], system planning, and energy policies covering future demand growth.

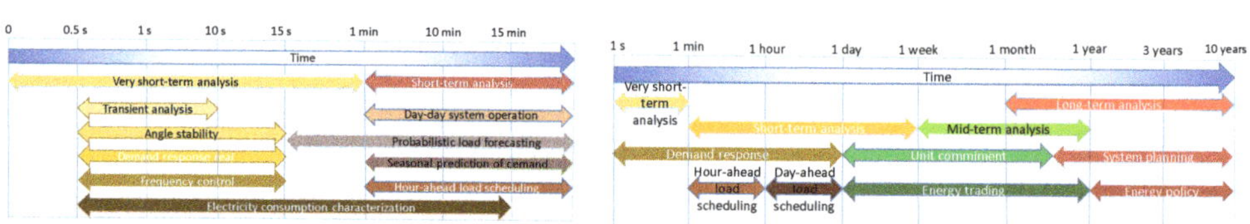

Figure 1. Temporal framework for analyzing the electrical energy system.

2.3. Planned Framework for Consumption Data Collection in Households

Figure 2 shows the planned framework for the remote real-time collection of consumption data at high sampling frequency (>4 Hz) in households with SMs and the uploading of this information to the cloud. This development is one of the most important issues targeted by the SEREDIS project ('Nuevos servicios de red para microredes renovables inteligentes. Contribución a la generación distribuida residencial': Grant No. ENE 2017-83860-R [44,49,83,84]). In this framework, the SM designed in [44] is installed in the general protection box, and the data gathered are dumped directly into two data storage solutions: (i) the cloud and (ii) a local storage in SD card.

2.3.1. SM

As shown in Figure 3, the SM is composed of a data acquisition block and a data-to-cloud upload block (see [44] for a more in-depth explanation). This SM was calibrated and tested to ensure its reliability and accuracy [44]. This SM has two Arduino boards: (i) the Arduino Uno Rev3 (AUR3 [85]); (ii) the Wemos Arduino D1R1 (AD1R1 [86]). The AUR3 board was used for the measurement process whereas the AD1R1 board uploaded data to the cloud. This reduced the time needed to process data and upload them to the cloud.

The SM simultaneously performed two processes. In the first process, the AUR3 microcontroller software determined the fundamental and derived electrical variables, sent data via serial port to AD1R1, and then stored information in the local data logger. The AD1R1 software read from the serial port and uploaded the data to the cloud via Wi-Fi in a parallel process.

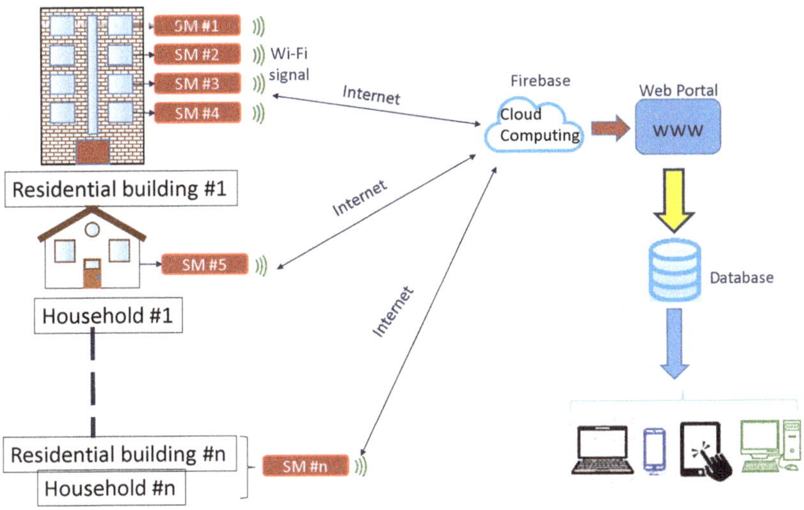

Figure 2. Remote real-time data collection schematic.

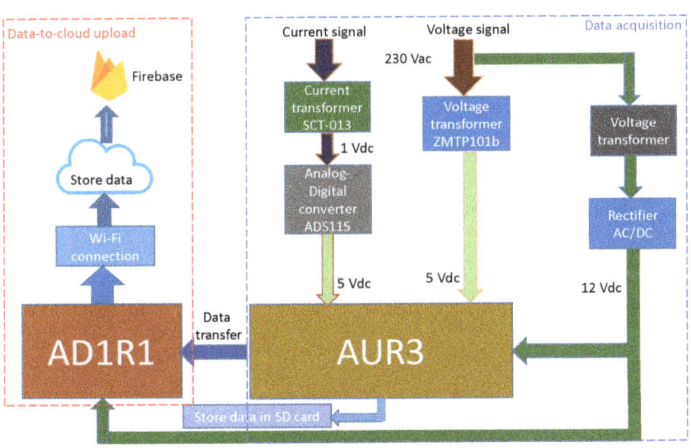

Figure 3. Block diagram for the data acquisition and uploading.

The timeline of the processes is shown in Figure 4. In the first process, the required time for measuring electrical variables is a 10-cycle time interval for a 50-Hz power system (Class-A performance [52]). In order to be accurately measured, the frequency of sampling must be at least twice the frequency of the signal. Therefore, 200 samples were used for the 10-cycle interval with a sampling frequency of 1 kHz. Derived variables are then calculated, which takes about 30 ms. The transmission of the information to the serial part only takes 1 ms, and data storage in SD memory, 9 ms. This leaves

10 ms for the waiting time In the second process, the SM reads the data received in 1 ms and uploads the data in 150 ms. About 50 ms are required to confirm the data upload, which leaves 49 ms for the waiting time.

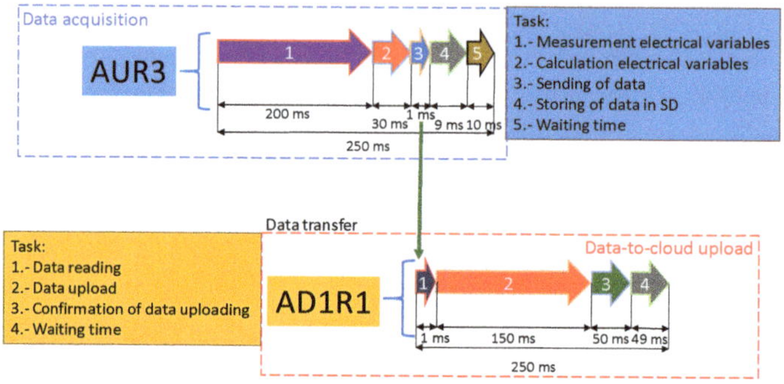

Figure 4. Process timeline for the SM.

Data Acquisition

Figure 3 shows the hardware for the data acquisition process. Analogical voltage and current sensors measured the electrical variables, which were then processed in the Arduino AUR3 [85] microprocessor. Once the fundamental variables were obtained, the derived variables were computed. More specifically, the current sensor STC-013 [87] (of the non-invasive type) together with the voltage sensor ZMPT101b were used for this purpose [88]. A digital/analogue converter ADS1115 was planned to increase the 1V DC output of the current sensor to the 5 V analogue input of AUR3 [89].

Data-To-Cloud Upload

The cloud provides a cost-effective method of supporting big data analytics. Therefore, the cloud data storage solution is suitable in scenarios where a real-time response from a given stream of SM data is required. This real-time data availability aids in personalizing applications that benefit both household owners and the scientific community when analyzing consumer profiles.

When data with finer granularities are gathered, the amount of information involved is high and requires data compression algorithms. These algorithms are usually time consuming; thus, a delay between the measurement and the data availability in the cloud appears. To enable a real-time response, this research does not apply compression algorithms, and the time of data-to-cloud upload is set to 0.25 s (see Figure 4), the same as the data acquisition time.

According to Figure 3, the base of the wireless communication module of the SM was the AD1R1 board [88] that acted as the interface between the microcontroller and cloud data storage (i.e., Firebase). The board used the ESP8266 platform as the operation core, which permitted Wired Equivalent Privacy (WEP) or Wi-Fi Protected Access (WPA)/WPA2 () authentication for secure Wi-Fi communication. In addition, it operated with 802.11 b/g/n wireless systems, which were compatible with the majority of the routers and modems on the market. This framework used the platform Firebase [90] to store huge amounts of data from households monitored with IoT technology and cloud computing. Alternatively, wireless communication systems such as 4G and 5G networks can be used. This implies a more expensive data service contract for data-to-cloud uploading.

Local Data Storage

A SM is equipped with a SD card mounted on a data logger shield, which is used as a backup to avoid data lost because of data-to-cloud upload problems. The memory size required per household in a year is about 2.2 GB.

2.4. Data Post-Processing

In an asynchronous way, data collected can be used for different actions or assessments. In our study, the assessment aims to show the influence of data granularity on the description of consumption load profile features. This required post-processing the data that involved the extraction of data from the planned storage solution and its adaptation for different granularities.

Cloud data were stored in *json* format and could be downloaded at any time. Also, SD card data were stored in *CSV* format and could be downloaded anytime. Therefore, firstly the data of each house in Figure 2 could be downloaded and converted to *CSV* format, if required, resulting in a daily *CSV* file of all electrical measurements. This format could be recognized by applications used for the processing and analysis of data, such as MS Excel and MatLab.

Secondly, the adaptation for different granularities from raw data on a 0.25 s-basis was carried out by the up-sampling method of the RMS value. The data size was reduced in the up-sampling operation. This allowed obtaining data at thirteen resolutions of data granularity, from a half-second to 30 min (0.5, 1, 2, 5, 10, 15, 30 s, and 1, 2, 5, 10, 15, and 30 min).

3. Results and Discussion

This section shows the results of the influence of data granularity on the description of household consumption load profile features based on the methodology presented in Section 2. This framework highlighted the information loss regarding the profile features when coarse-grained data were used. We first focus on the temporal results for different time slices from a sub-hourly to a monthly analysis, including daily and weekly analyses. However, the global influence assessment was based on a yearly analysis. This first involved a statistical analysis. After this, periodograms were compared to autocorrelation and partial autocorrelation analyses to highlight significant outcomes regarding profile features.

3.1. Case Study

The case study focused on four real-world households in the city of Jaen in southern Spain. As explained, this research is part of the SEREDIS project, which characterized load profiles for household consumption, electric vehicles (EVs), and PV systems in the residential context. Nonetheless, this study is limited to the consumption load in households. The consumption load data came from SM readings as described in Section 2.3, being post-processed as stated in Section 2.4.

The household set selected had different consumption features, namely with varying characteristics in terms of the relation between the peak and base load and load fluctuations, which made it possible to take the heterogeneity of real-world load profiles into account (see Table 2). In addition, important issues determined the household selection, such as the power contracted from the electric mains and the type of supply at home. Single-phase systems were designed at households #1, #2, and #4, whereas household #3 had a three-phase system. The contracted power in Spanish legislation reflects how the household is equipped with different electrical appliances (see Table 3). In this study, the limitation in the number of households was due to a combination of limited financial budget, limited number of households that included EVs and PV systems, and a low number of families who voluntarily cooperated on the research project.

In particular, household #1 was a family flat with three children. Household #2 was a semi-detached house with only two inhabitants. Household #3 was a detached house with two children and their parents. Household #4 was a terraced house of four components, inhabited by two adults and two teenagers.

The city of Jaén has a Mediterranean climate with hot summers but cool winters. Therefore, the houses in our study were all equipped with climate control systems. More specifically, household #1 had a heating system for the whole building with a gas boiler. The flat also had an air-conditioning system for summer. Household #2 had an individual gas boiler for heating, and a two-split air conditioner system (living room and master bedroom). Household #3 was equipped with a central electrical air-thermal system for heating and cooling. Household #4 had an air-conditioning/heat pump system. Each household had different electrical appliances, depending on the age and behavior of the family members, see Table 3.

Table 2. Key features of four households in Jaén (southern Spain). Case study.

	Household #1	Household #2	Household #3	Household #4
Total annual consumption (kWh/year)	3033	2626	22,058	4139
Total surface (m^2)	100	125	210	140
Number of family members	5	2	4	4
Is there at least an adult during the morning at home?	No	No	Yes	Yes
Electric heating	No	No	Yes	Yes
Electric air conditioned	Yes	Yes	Yes	Yes
Building type	Flat	Semi-Detached house	Detached house	Terraced house
Contracted power from the electric mains (kW)	3.45	2.3	5.75	4.6
Number of phases	1-phase	1-phase	3-phase	1-phase

Table 3. Appliances and lighting installed in every household case study.

Item	Power (W)	Household #1	Household #2	Household #3	Household #4
Oven	1200–2200	√	√	√	√
Electric cooker	900–2000	√	√	√	√
Extractor hood	70–200	√	√	√	√
Microwave oven	900–2500	√	√	√	√
Dishwasher	1500–2200	√	—	√	√
Refrigerator	250–350	√	√	√	√
Washing machine	1500–2200	√	√	√	√
Electric water heater	1500–5500	—	—	√	—
Vacuum cleaner	1100–2000	√	—	√	—
Dryer	1000–2500	√	√	√	√
Clothes dryer	1500–3000	√	—	√	—
Desktop computer	150–300	√	—	√	—
Laptop	100–250	√	√	√	√
Smart phone	15–25	√	√	√	√
Tablet	20–30	√	—	√	—
LED TV	150–550	√	√	√	√
BlueRay-DVD player	50–75	—	√	—	—
Stereo system	100–150	√	√	—	√
Video games console	25–150	√	—	√	√
Low energy bulbs	5–20	√	√	√	—
Florescent lamps	18–58	√	√	—	√
LED lamps	4–12	√	—	√	√
Halogen lamps	25–60	—	—	—	√

All of the households had the usual appliances installed in the kitchen. Household #2 did not have a dishwasher. The computer equipment in each household was quite heterogeneous (desktop computer, laptop, smart phone, tablet), and depended on the profession of the occupants, and whether certain members of the household were still students. Moreover, the ages of the occupants also determined the entertainment equipment (TV, stereo system, video games console, etc.). Thus, households #1, #3 and #4 had video game consoles for the children and adolescents that lived there. In contrast, the occupants of household #2 preferred listening to music on a CD/disc or playing movies on DVD/BlueRay. The lighting system in all households was low consumption and high efficiency. Low consumption lamps were planned in households #1, #2 and #3. Households #1, #2 and #4 also had fluorescent lamps. LED lighting was present in households #1, #3 and #4. Only household #4 used halogen lamps.

3.2. Reliability of the Planned Framework to Provide Data

Figure 5 shows the data availability for the data-to-cloud upload in each household within the SEREDIS project up to July 2020. The gaps in the figure represent the days when some information was lost because of data-to-cloud upload problems, particularly when the data loss exceeded 25% in one day. Data collection started in July 2018 (household #3) and is still going on for all of the households. On September 2018, SMs were added to households #1 and #2. Finally, household #4 was equipped with an SM in October 2018.

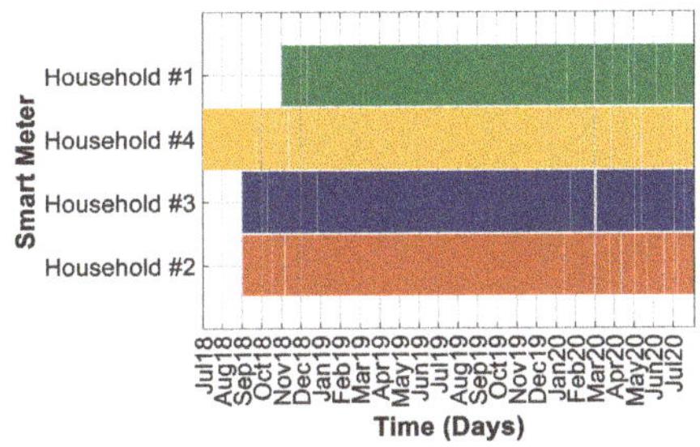

Figure 5. Data availability for every household.

As data were collected on a 0.25 s basis, 345,600 measurements for six electrical variables (voltage, current, active, reactive and apparent power, power factor) were stored each day. This provided a total data set of 2,073,600 per day, 62,208,000 per month, and 746,496,000 per year.

The reliability of the data-to-cloud upload (Firebase [90]) during 2019 was greater than 99% because of the high quality of the fiber optic Internet connection in all of the households. Figure 6 depicts the percentages of successful data-to-cloud upload.

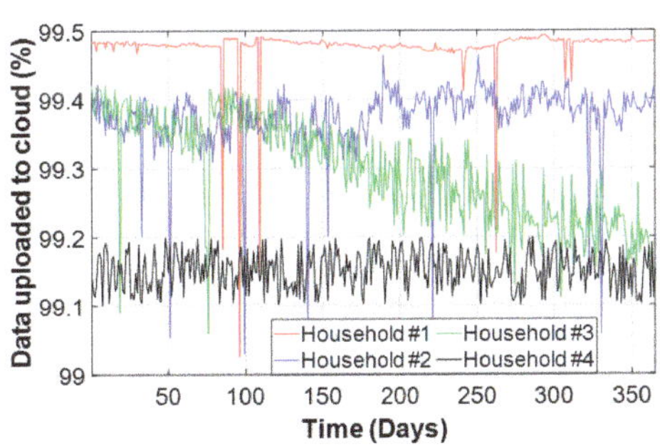

Figure 6. Percentage of successful data-to-cloud upload in 2019.

As explained, the SM also included a local 8-GB memory card that served as a data backup. This permitted an operational autonomy of 1.88 years. This local storage guaranteed 100% data availability; thus, the assessment shown hereafter is based on data of this storage solution.

3.3. Sub-Hourly Time Slice Analysis

The influence of the data granularity on the temporal change features of the household consumption load profile at sub-hourly level is analyzed in this section. Figure 7 shows the measurements for all of the households and thirteen resolutions of data granularity from 0.5 s to 30 min on a working day in September. Data with finer granularities, from 0.5 s to 30 s, are shown on a 5-min timespan whereas coarser resolutions, from 0.5 min to 30 min, cover a 2-h timespan.

For household #1, Figure 7a1 highlights the smoothing of peaks and troughs in the consumption load because of the use of coarse-grained data. For example, the 1.77 kW load peak at 22:26:26 h was only achieved at a 0.5-s granularity. The granularity of 1 and 2 s gave a load peak of 1.67 kW and 1.61 kW, which meant a decrease of 5.64% and 9.04%, respectively. However, the flattening of this peak was very pronounced for the 30-min granularity where a decrease of 27.12% was observed.

Regarding household #2, the increase in the maximum load peak for the finest granularity in regard to the coarse resolution occurred at 23:42:52 h and was 56.44%. The comparison of the consumption load in household #3 at 12:07:00 h, as an example, shows a reading of 1.83 kW for a 0.5-s granularity, 0.53 kW for a 15 min granularity and 0.81 kW for a 30 min granularity, which meant a decrease of 71.03% and 55.73%, respectively. For household #4, the reading of 1.81 kW for a 0.5-s granularity at 14:30:00 h was reduced to 0.88 kW for the 30-min granularity, which meant a decrease of 51.38%.

Coarse data granularities tend to flatten the peaks and troughs in the consumption load. Since there is a considerable loss of information, the consumption thus evaluated does not conform to reality. The smaller the granularity used, the smoother the load peaks and troughs will be. As a result, the reduction in the measured power is greater, and is thus a less accurate reflection of reality. One problem with reducing the temporal resolution of consumption data sampling is the loss of variability in regard to the intra-temporal steps, which has a particularly high effect on the actions or assessments. The loss of detail observed in Figure 7, does not justify the reduction of the sampling resolution for the consumption load data.

3.4. Daily Time Slice Analysis

This section extends the timeframe of Section 3.3 to one day when analyzing the influence of the data granularity. This provides information pertaining to the daily consumption in the households. Figure 8 shows the six samples of data granularity for all of the households on the previously mentioned day of September. This figure highlights the dual nature of the consumption load profile, namely, the continuity of the rough base-load and the intermittent spikes of the peak-load.

For household #1, the maximum daily load peak occurred at 22:00:00 h with a 0.5-s reading of 4.13 kW. However, the readings for the 10 and 30-min granularity dropped to 2.65 and 1.69 kW, respectively, which signified a decrease of 36.07% and 59.09%. As can be observed, the consumption load for household #2 had a sequence of peaks throughout the day for the finest data granularity. These peaks were strongly attenuated for a 5-s data granularity, and much more reduced for the coarse data resolutions. These repeated peaks were caused by the operation of the refrigerator. As an example, the 0.5-s reading at 23:49:53 h was 2.27 kW, after which it dropped to 0.95 kW (a 58.14% decrease) for the 30-min granularity. The central electrical air-thermal system in household #3 originated peaks at a stable timespan. The maximum daily load peak occurred at 19:23:51, with a reading of 7.47 kW for the 0.5-s granularity. In contrast, the 30-min granularity reading dropped to 5.95 kW, a value that was 20.34% lower. In household #4 the maximum daily load peak occurred at 9:17:23 h with a value of 3.14 kW for a 0.5-s granularity, whereas for a 30-min granularity, it decreased by 57.64% (1.33 kW). Repeated peaks were only identified at 0.5 and 5-s granularities during hours when the consumption was lower. These were caused by equipment disconnection. In summary, the greatest loss of information occurred in household #1 (59.07%), followed by #2 (58.14%), #4 (57.64%) and #3 (20.34%).

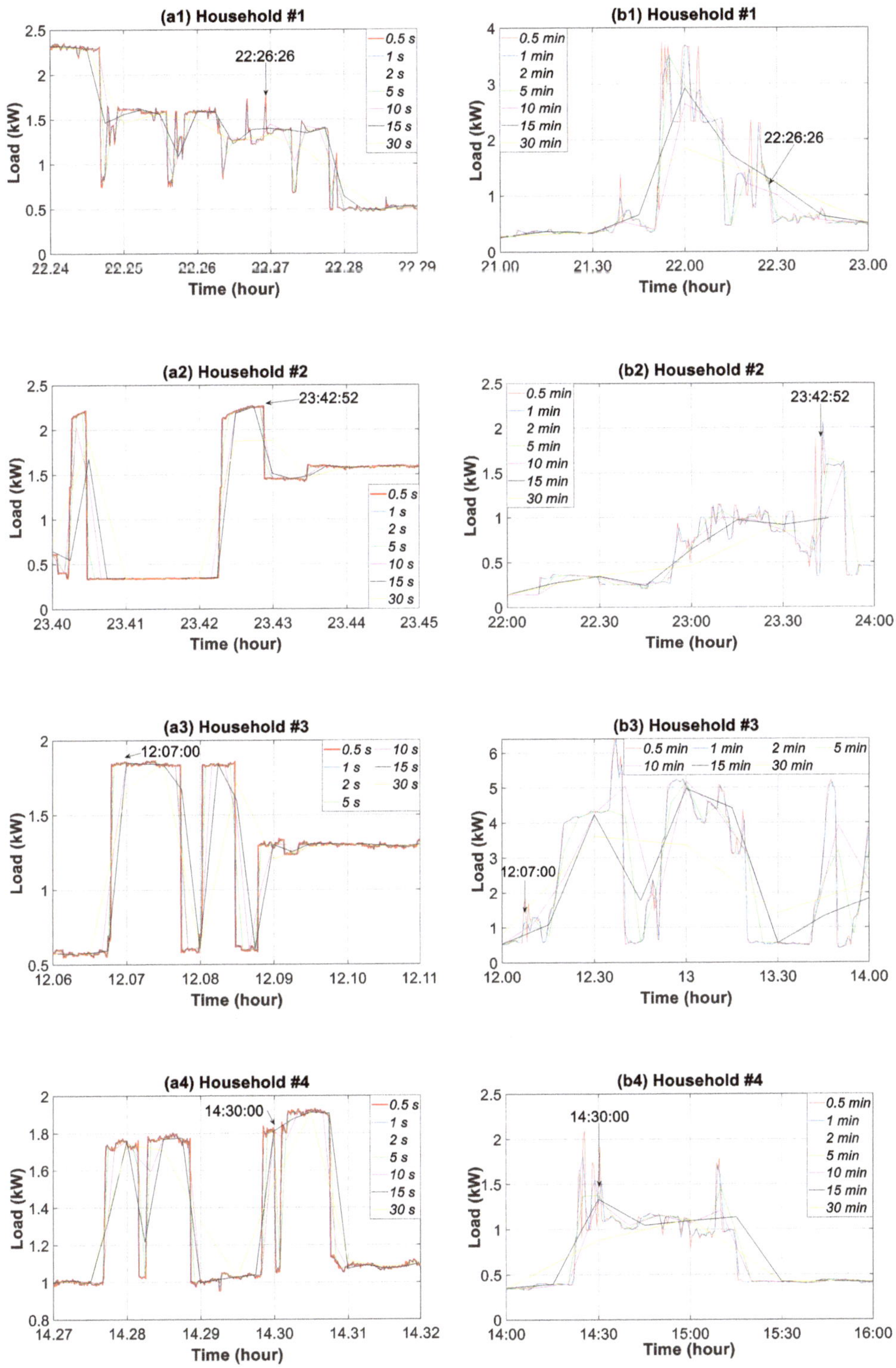

Figure 7. Consumption load profile for household #1 to #4: (**a1–a4**) 0.5 to 30 s granularities (5-min timespan); (**b1–b4**) 0.5 to 30-min granularities (2-h timespan).

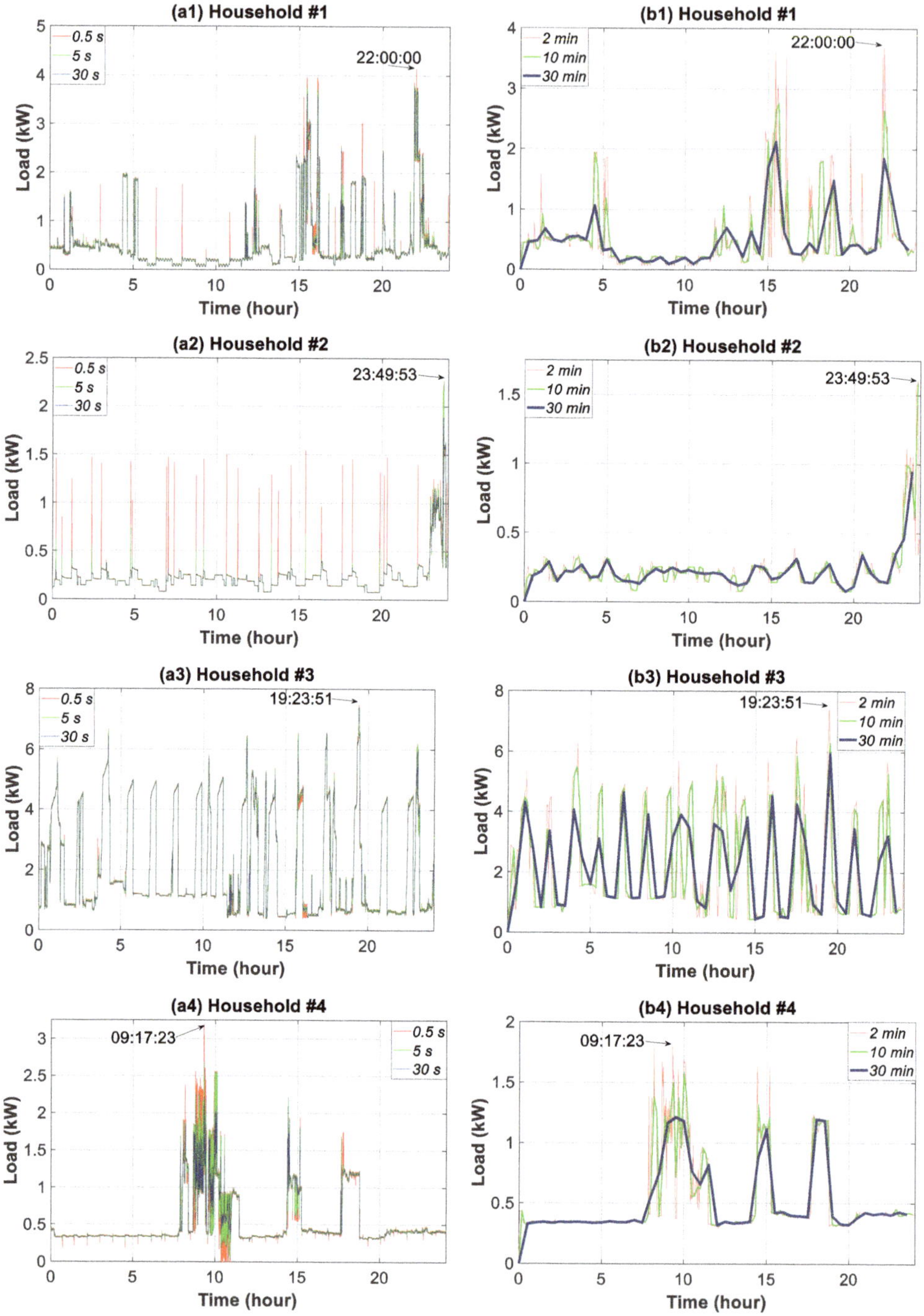

Figure 8. Daily consumption load profile for household #1 to #4: (**a1–a4**) 0.5 to 30 s granularities; (**b1–b4**) 2 to 30-min granularities.

3.5. Weekly Time Slice Analysis

This section discusses the influence of the three samples of data granularity on weekly consumption features, Figure 9. For household #1, the maximum load peaks occurred on Thursday and Friday, with 5.48 kW for the 0.5-s data granularity. These peaks were reduced by 12.77% and 34.67% at the granularities of 30 s and 30 min, respectively. Repeated peaks throughout the day in household #2 were also observed during the weekly analysis. The reduction of peaks by using coarse data granularity, (up to 75%), was the most pronounced in the households analyzed. Intermediate data granularity decreased the weekly load peaks by 37%. In household #3, the highest reduction occurred at the end of Wednesday, which signified a reduction of 30.62% and 72.76% in the temporal resolution of 30 s and 30 min, respectively. The reduction in household #4 was lower, and came to a 62.07% droop for the 30-min data granularity.

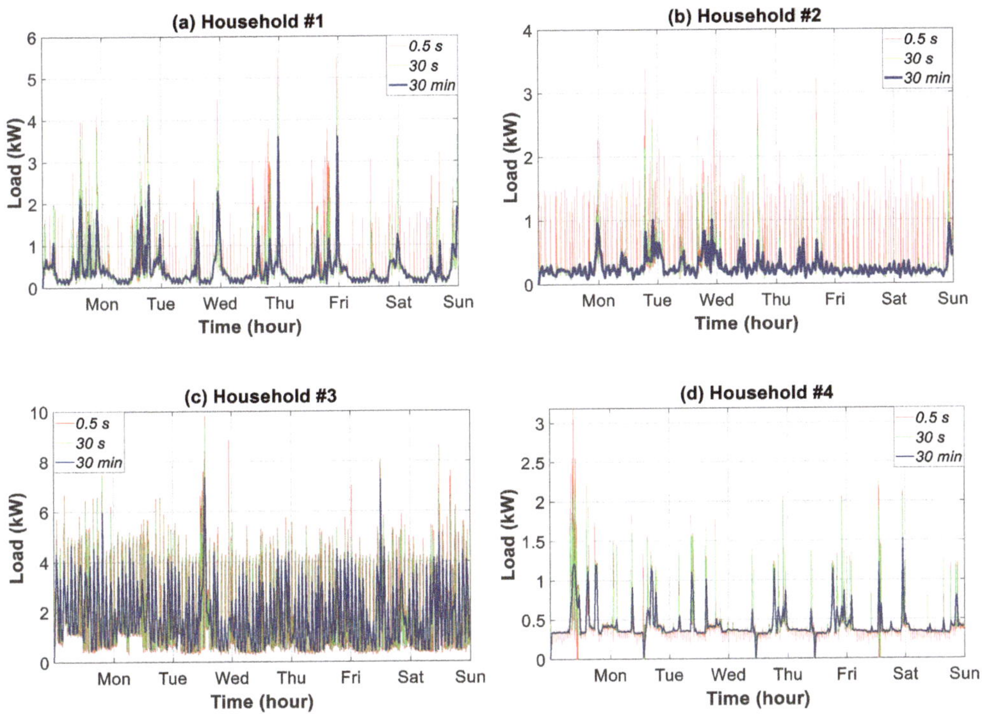

Figure 9. Weekly consumption load profile in households #1 to #4 for three data granularities.

3.6. Monthly Time Slice Analysis

This section examines the daily smoothing of the highest peak and deepest trough when using coarse-grained data to underline their accuracy. The analysis focused on data from January.

Unlike the remaining sections where the temporal framework and the analysis were very short-term, the results of this section are applicable to medium-term analysis where knowledge of day-to-day operation is required. For this purpose, the ratio between the daily peak or trough load and the daily mean load is used as a metric.

Figure 10 shows the ratio for the peak load and all of the households for a 0.5–30 min data granularity. The load profiles show a widening spread in the daily mean load for households #3, #1, #2, and #4. This reveals an increase in daily variability as shown in Figure 9.

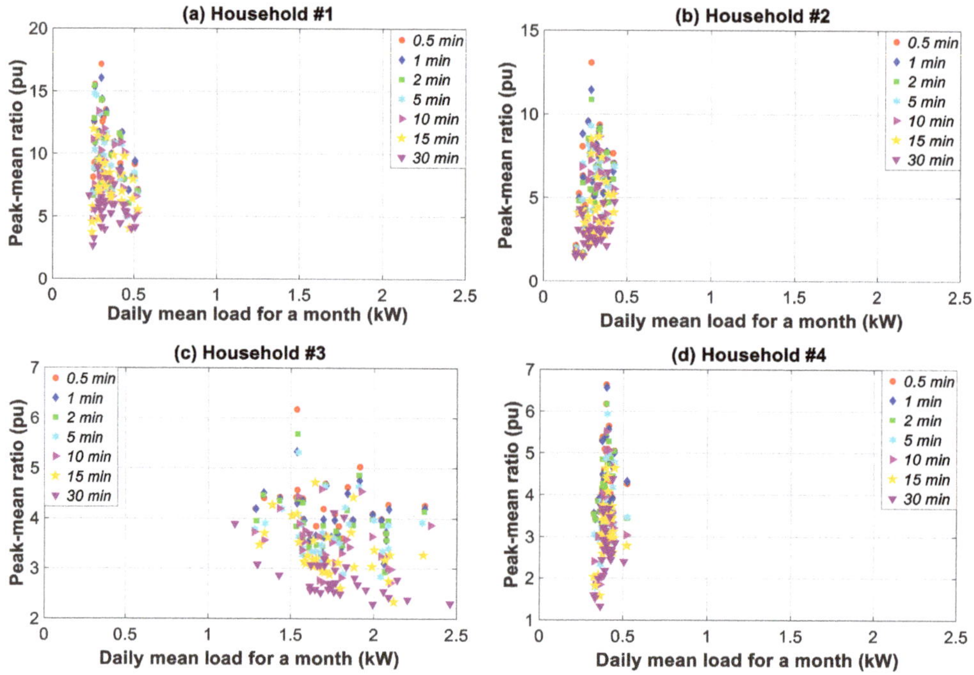

Figure 10. Ratio of daily peak load vs. daily average load in households #1 to #4 for 1.66–30 min data granularity in January.

Table 4 summarizes the highest peak-mean and deepest trough-mean ratios for the thirteen granularities and four households. As can be observed, the monthly maximum peak-mean ratios achieved the highest values for a data granularity of 0.5 s. The ratio decrease for the 30-min data granularity was 50.02%, 39.66%, 35.07% and 36.89% with regard to the 0.5-s granularity, respectively for households #1, #2, #3, and #4. The monthly minimum trough-mean ratio also achieved the highest value for a 0.5-s data granularity. The decrease in percentages with coarser granularity, which was greater than the one for maximum peak-mean ratios, was the following: 64.07%, 75.22%, 27.67% and 59.64%. This lack of accuracy indicates the need to adjust the data granularity at 0.5 s for the medium-term analysis.

Table 4. Monthly maximum/minimum ratio of daily peak-mean load and of daily trough-mean load in January.

Data Granularity	Household #1		Household #2		Household #3		Household #4	
	Maximum Peak-Mean	Minimum Trough-Mean	Maximum Peak-Mean	Minimum Trough-Mean	Maximum Peak-Mean	Minimum Trough-Mean	Maximum Peak-Mean	Minimum Trough-Mean
0.5 s	17.7714	7.4346	13.2673	6.1029	6.3413	3.1745	6.9320	3.2970
1 s	17.4875	7.4233	13.2562	5.4252	6.3114	3.1726	6.9318	3.2951
2 s	17.5749	7.2256	13.2559	4.7885	6.2524	3.1693	6.8127	3.2876
5 s	17.4989	7.0880	13.1816	3.6465	6.2348	3.1668	6.7546	3.2712
10 s	17.3746	6.8418	13.1445	2.3266	6.2191	3.1642	6.7268	3.2671
15 s	17.1912	6.5573	13.0965	2.0043	6.2258	3.1601	6.6947	3.2524
30 s	17.1558	6.5692	13.0793	1.6689	6.1782	3.1554	6.6404	3.2491
1 min	16.0662	6.5542	11.4579	1.6763	5.3288	3.0967	6.5740	3.2495
2 min	15.4650	6.0329	10.8880	1.7029	5.6834	2.9352	6.1811	3.0951
5 min	14.8479	5.0117	9.3116	1.6465	5.3142	2.8401	5.9447	1.7821
10 min	13.4116	4.7214	8.6840	1.6023	4.5818	2.7117	5.5095	1.8488
15 min	12.0069	3.7204	8.6675	1.5882	4.7191	2.3409	4.6373	1.5840
30 min	8.8820	2.6709	8.0044	1.5119	4.1170	2.2958	4.3747	1.3305

3.7. Yearly Time Slice Analysis

This section highlights the influence of data granularity on the description of household consumption load profile features by means of different complementary analyses. Firstly, the consumption pattern of each household is justified. Then, based on the use of coarse-grained data, a statistical analysis underlines the change in the annual empirical distribution shape. Finally, periodograms and autocorrelation analyses are used to focus on the loss of information pertaining to profile features, caused by the use of coarse-grained data. This was based on the knowledge of the main constituent signals of the load fluctuations.

3.7.1. Temporal Consumption Pattern

The yearly consumption pattern in a household reflects the energy behavior of the occupants during all seasons, and is strongly influenced by temperature, wind speed, relative humidity, etc. [16,34]. It also takes vacation and holiday periods into account. Accordingly, Figure 11 shows the daily average consumption load during the year for five samples of data granularity in all of the households.

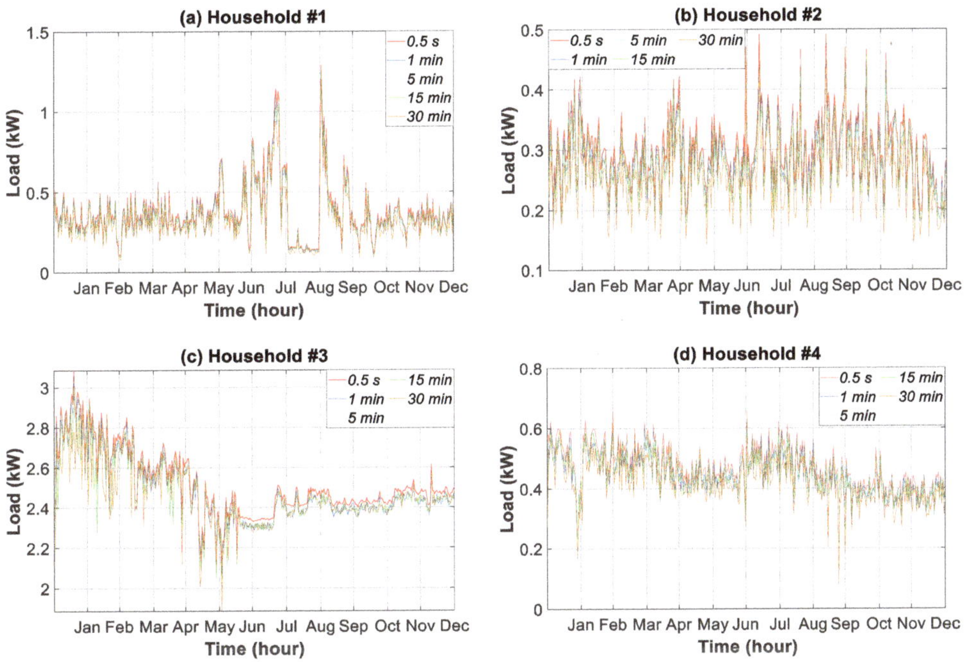

Figure 11. Daily average consumption load profile in households #1 to #4 for five data granularities.

Consumption pattern #1 was stable during spring, autumn, and winter. The heating system in the household was for the whole building, and thus did not influence electricity consumption. However, the air-conditioning system from May to September strongly increased consumption, except for the month of August when the occupants were away on holiday. During the summer (June to September), the children spent more time at home, which increased consumption.

Consumption pattern #2 was stable throughout the year because the occupants were at work all day, and were only at home at night. Nonetheless, there were days between June and October when consumption peaked because of the use of the split air conditioner system.

The central electrical air-thermal system for heating and cooling in household #3 operated the whole year. When temperatures were lower (i.e., January to April), consumption was greater.

Regarding household #4, electricity consumption decreased in January and August because the family moved to their second residence.

3.7.2. Statistical Analysis

This section presents a statistical analysis of the datasets monitored in the course of a year (see Section 3.2 for a detailed explanation of the timespan) for the four households in the case study (see Section 3.1), once these datasets were post-processed for different granularities according to Section 2.4. This analysis shows the impact of data granularity on the description of household consumption load profile features. Accordingly, Figure 12 represents the annual empirical distributions (PMFs of the discrete variables) of the consumption load for all of the households and for four data granularities. Specific zooms were included for a better understanding of results.

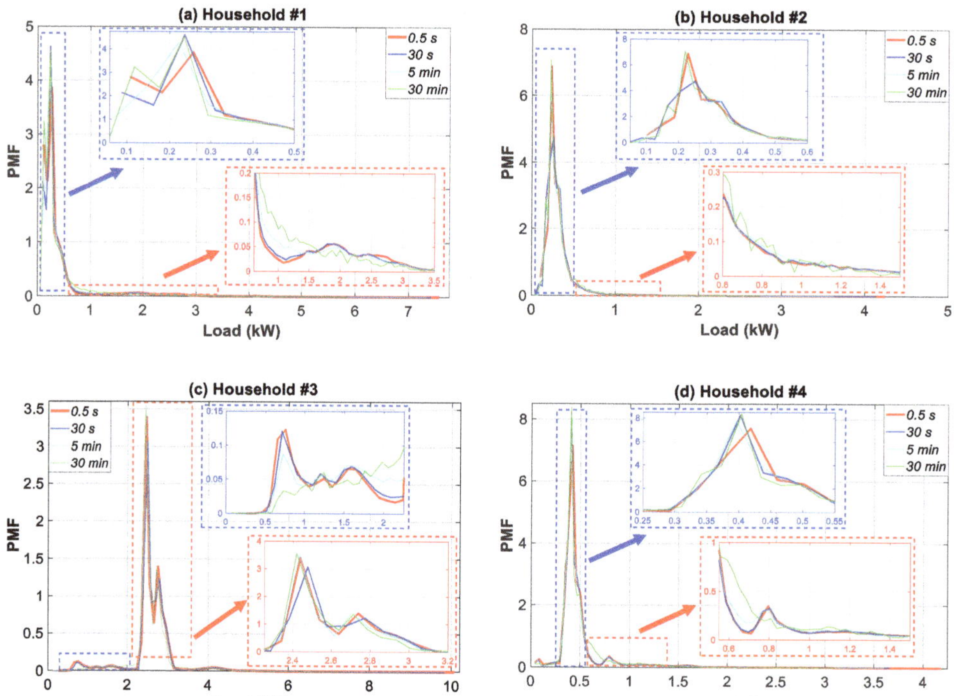

Figure 12. Annual empirical distribution of the consumption load profile in households #1 to #4 for four data granularities.

As can be observed, the PMFs of the household consumption load data are clearly not normal or Gaussian. The most non-Gaussian behavior is evident for household #1, followed by household #3. This result is in consonance with the dual nature of the consumption load profile in Figure 8. In addition, the use of coarser temporal granularity, ranging from 0.5 s to 30 min, substantially affected the PMF shapes. Thus, opposite behaviors were observed. The PMF shape either moved further away from a Gaussian distribution (households #1 and #3) or began to show a more Gaussian behavior (households #2 and #4). In general, the shape was more frequently skewed near those hours with a lower load, which removed many of the extremes. The extreme ends of the PMF were of potential interest as they represented periods of very low or very high consumption.

For profile #1, the PMF moved largely within the intervals of 0.0–0.4 kW and 0.6–3.0 kW. A reduction of occurrences in the 0.0–0.4 kW interval was observed, which moved the higher occurrences towards the 0.6–1.5 kW load interval. In household #3, the lower occurrences in the 0.0–2.5 kW interval were compensated by higher ones in the 2.25–3.20 kW interval. For household #4, a displacement of occurrences from the 0.30–0.55 kW interval to the 0.55–1.50 kW interval was observed.

Table 5 summarizes the statistical results of the annual empirical distributions for all of the data granularities. For purposes of comparison, the values at the 0.5-s data granularity were used as a reference. Using coarser temporal granularity, ranging from 0.5-s to 30-min, led the sample mean of the consumption load to decrease by 17.61%, 17.68%, 17.65%, and 17,74%, respectively for households #1, #2, #3, and #4. In general, the level of variability for all households was significantly reduced as the maximum load values decreased by 15.55%, 18.28%, 13.74%, and 18.49%, respectively. Nonetheless, larger droops were observed for the relevant minimum values, namely, 20.98%, 19.93%, 17.59%, and 16.02%. The reduction in percentage for the variance of all households was in the 27.54–31.29% interval. This again confirmed the drop in the variability of the consumption load for the households.

Table 5. Descriptive statistics of the consumption load profile throughout the year.

Household	Data Granularity	Sample Mean (kW)	Maximum Value (W)	Minimum Value (kW)	Sample Variance (kW^2)	Sample Skewness (kW^3)	Sample Kurtosis (kW^4)
#1	0.5 s	0.3771	1.2884	0.1021	0.0326	1.7935	7.8044
	1 s	0.3752	1.2821	0.1016	0.0323	1.7932	7.8035
	2 s	0.3733	1.2756	0.1011	0.0320	1.7938	7.8060
	5 s	0.3658	1.2497	0.0991	0.0307	1.7930	7.8024
	10 s	0.3628	1.2397	0.0982	0.0302	1.7931	7.8048
	15 s	0.3619	1.2369	0.0980	0.0301	1.7959	7.8179
	30 s	0.3582	1.2247	0.0971	0.0295	1.7940	7.8073
	1 min	0.3545	1.2112	0.0960	0.0288	1.7950	7.8130
	2 min	0.3506	1.1994	0.0950	0.0283	1.7988	7.8334
	5 min	0.3430	1.1722	0.0918	0.0270	1.7963	7.8290
	10 min	0.3388	1.1616	0.0926	0.0264	1.8027	7.8813
	15 min	0.3285	1.1191	0.0884	0.0248	1.7985	7.8953
	30 min	0.3106	1.0882	0.0801	0.0224	1.8782	8.5657
#2	0.5 s	0.2998	0.4912	0.1922	0.0030	0.4923	3.6786
	1 s	0.2983	0.4888	0.1912	0.0028	0.4920	3.6774
	2 s	0.2968	0.4863	0.1903	0.0029	0.4934	3.6812
	5 s	0.2908	0.4764	0.1864	0.0028	0.4927	3.6803
	10 s	0.2884	0.4725	0.1849	0.0028	0.4930	3.6810
	15 s	0.2878	0.4715	0.1845	0.0028	0.4906	3.6731
	30 s	0.2848	0.4668	0.1827	0.0027	0.4913	3.6795
	1 min	0.2818	0.4622	0.1809	0.0027	0.4947	3.6930
	2 min	0.2788	0.4573	0.1790	0.0026	0.4876	3.6675
	5 min	0.2728	0.4488	0.1748	0.0025	0.4886	3.6615
	10 min	0.2701	0.4417	0.1732	0.0025	0.4827	3.6079
	15 min	0.2606	0.4229	0.1683	0.0023	0.4702	3.5331
	30 min	0.2469	0.4014	0.1539	0.0021	0.3548	3.3360
#3	0.5 s	2.5180	3.0848	2.0955	0.0276	0.8327	3.4692
	1 s	2.5054	3.0690	2.0844	0.0273	0.8323	3.4684
	2 s	2.4928	3.0545	2.0745	0.0270	0.8332	3.4696
	5 s	2.4425	2.9928	2.0321	0.0260	0.8329	3.4723
	10 s	2.4224	2.9683	2.0174	0.0255	0.8328	3.4685
	15 s	2.4173	2.9608	2.0108	0.0254	0.8317	3.4663
	30 s	2.3922	2.9311	1.9906	0.0249	0.8312	3.4660
	1 min	2.3670	2.9016	1.9720	0.0244	0.8309	3.4660
	2 min	2.3420	2.8627	1.9485	0.0239	0.8324	3.4470
	5 min	2.2913	2.7917	1.9029	0.0229	0.8126	3.4436
	10 min	2.2669	2.8123	1.8726	0.0225	0.8498	3.5525
	15 min	2.1899	2.6639	1.8024	0.0211	0.7854	3.5506
	30 min	2.0736	2.6612	1.7269	0.0200	1.0132	4.1224

Table 5. *Cont.*

Household	Data Granularity	Sample Mean (kW)	Maximum Value (W)	Minimum Value (kW)	Sample Variance (kW^2)	Sample Skewness (kW^3)	Sample Kurtosis (kW^4)
#4	0.5 s	0.4726	0.6576	0.0880	0.0059	−0.5149	4.8532
	1 s	0.4702	0.6544	0.0876	0.0058	−0.5147	4.8532
	2 s	0.4678	0.6510	0.0871	0.0057	−0.5154	4.8545
	5 s	0.4584	0.6379	0.0854	0.0055	−0.5152	4.8533
	10 s	0.4546	0.6323	0.0847	0.0054	−0.5175	4.8579
	15 s	0.4537	0.6314	0.0845	0.0054	−0.5147	4.8602
	30 s	0.4489	0.6251	0.0836	0.0053	−0.5129	4.8537
	1 min	0.4442	0.6179	0.0828	0.0052	−0.5167	4.8618
	2 min	0.4396	0.6133	0.0819	0.0051	−0.5198	4.8590
	5 min	0.4302	0.5983	0.0803	0.0049	−0.5133	4.8487
	10 min	0.4248	0.5972	0.0797	0.0048	−0.5042	4.8242
	15 min	0.4111	0.5617	0.0773	0.0044	−0.5558	4.9013
	30 min	0.3888	0.5360	0.0739	0.0041	−0.4903	4.7124

The sample skewness [91] was positive in households #1, #2, and #3, which underlines that the right tail of the consumption load distribution was longer than the left. On the contrary, household #4 had a negative sample skewness. For coarser data granularities, households #1 and #3 increased their sample skewness value whereas the behavior in households #2 and #4 was exactly the opposite. This was confirmed by the displacement of PMFs in Figure 12.

The sample kurtosis values indicate that all households were leptokurtic [91], which means that the consumption loads were concentrated around the sample mean as household profiles had values greater than 3. Households #1 and #3 had higher sample kurtosis values for coarser granularities, which revealed that the consumption load tended to be closer to the sample mean. This outcome was more pronounced in household #3. However, the behavior in households #2 and #4 was the opposite. Once again, the change in the sample kurtosis values was confirmed by the displacement of PMFs in Figure 12.

3.7.3. Periodogram, Autocorrelation, and Partial Autocorrelation Analyses

This section presents a set of complementary analyses that were performed to explain the influence of data granularity on the description of load profiles. A periodogram analysis, along with an autocorrelation analysis and a partial autocorrelation analysis made it possible to obtain the main periods (or frequencies) of the constituent signals of consumption load fluctuations. The results highlighted the loss of information when using coarse-grained data to describe the load profile features. The analyses were split into two time slices, namely, the 1–100 s interval and the 100 s (1.66 min)–30 min interval. This showed the influence of the aggregation data on these intervals of data granularity. This section concludes by highlighting the daily, weekly, and monthly seasonality of the household consumption load profiles.

Since household consumption load profiles contain more than one source of seasonality, as previously described, our approach at the very beginning removed the trend and seasonal components altogether through differencing and seasonal differencing [68].

To confirm the stationary of the datasets used in periodograms and autocorrelation and partial autocorrelation analyses, Table 6 shows the results the statistical tests described in Section 2.1.1. These tests were applied on the datasets of the annual consumption load profiles for different data granularities. We interpret results using the p-value from the test. A p-value below a threshold (such as 5%) suggests that we reject the null hypothesis, whereas a p-value above the threshold suggests that we fail to reject the null hypothesis.

Table 6. Statistical tests to check stationary of the consumption load profiles.

Household	Data Granularity	ADF Test		KPSS Test		Variance Ratio Test		LMC Test		PP Test	
		Null Hypothesis (*H*)	*p*-Value	Null Hypothesis (*H*)	*p*-Value	Null Hypothesis (*H*)	*p*-Value	Null Hypothesis (*H*)	*p*-Value	Null Hypothesis (*H*)	*p*-Value
#1	0.5 s	0	0.0374	1	0.0100	0	1.5166×10^{-9}	0	0.0100	0	0.0374
	1 s	0	0.0363	1	0.0100	0	6.8206×10^{-8}	0	0.0100	0	0.0363
	2 s	0	0.0472	1	0.0100	0	5.6420×10^{-7}	0	0.0100	0	0.0472
	5 s	0	0.0455	1	0.0100	0	6.2874×10^{-11}	0	0.0100	0	0.0455
	10 s	0	0.0436	1	0.0100	0	1.4630×10^{-7}	0	0.0100	0	0.0436
	15 s	0	0.0433	1	0.0174	0	7.1719×10^{-6}	0	0.0100	0	0.0433
	30 s	0	0.0499	1	0.0100	0	2.6792×10^{-40}	0	0.0100	0	0.0399
	1 min	0	0.0399	1	0.0100	0	4.9340×10^{-10}	0	0.0100	0	0.0399
	2 min	0	0.0379	1	0.0100	0	0.0001	0	0.0100	0	0.0379
	5 min	0	0.0279	1	0.0100	0	0.0458	0	0.0100	0	0.0279
	10 min	0	0.0456	1	0.0478	0	0.0215	0	0.0181	0	0.0456
	15 min	0	0.0158	1	0.0100	0	0.0036	0	0.0100	0	0.0158
	30 min	0	0.0183	1	0.0213	0	0.0184	0	0.0100	0	0.0183
#2	0.5 s	0	0.0010	1	0.0100	0	0.0410	0	0.0100	0	0.0010
	1 s	0	0.0224	1	0.0100	0	0.0328	0	0.0100	0	0.0224
	2 s	0	0.0010	1	0.0100	0	0.0321	0	0.0100	0	0.0010
	5 s	0	0.0152	1	0.0100	0	0.0320	0	0.0100	0	0.0152
	10 s	0	0.0446	1	0.0100	0	0.0306	0	0.0100	0	0.0446
	15 s	0	0.0470	1	0.0100	0	0.0341	0	0.0100	0	0.0470
	30 s	0	0.0387	1	0.0100	0	0.0412	0	0.0100	0	0.0387
	1 min	0	0.0401	1	0.0100	0	0.0152	0	0.0100	0	0.0401
	2 min	0	0.0408	1	0.0100	0	0.0117	0	0.0100	0	0.0408
	5 min	0	0.0434	1	0.0100	0	0.0399	0	0.0100	0	0.0434
	10 min	0	0.0446	1	0.0310	0	0.0324	0	0.0100	0	0.0446
	15 min	0	0.0475	1	0.0386	0	0.0320	0	0.0100	0	0.0475
	30 min	0	0.0470	1	0.0100	0	0.0483	0	0.0100	0	0.0470

Table 6. *Cont.*

Household	Data Granularity	ADF Test		KPSS Test		Variance Ratio Test		LMC Test		PP Test	
		Null Hypothesis (*H*)	*p*-Value	Null Hypothesis (*H*)	*p*-Value	Null Hypothesis (*H*)	*p*-Value	Null Hypothesis (*H*)	*p*-Value	Null Hypothesis (*H*)	*p*-Value
#3	0.5 s	0	0.0302	1	0.0100	0	0.0223	0	0.0100	0	0.0302
	1 s	0	0.0299	1	0.0100	0	0.0332	0	0.0100	0	0.0299
	2 s	0	0.0361	1	0.0100	0	0.0135	0	0.0100	0	0.0361
	5 s	0	0.0357	1	0.0100	0	0.0442	0	0.0100	0	0.0357
	10 s	0	0.0331	1	0.0100	0	0.0443	0	0.0100	0	0.0331
	15 s	0	0.0330	1	0.0100	0	0.0447	0	0.0100	0	0.0330
	30 s	0	0.0329	1	0.0100	0	0.0477	0	0.0100	0	0.0329
	1 min	0	0.0314	1	0.0100	0	0.0206	0	0.0100	0	0.0314
	2 min	0	0.0338	1	0.0100	0	0.0379	0	0.0100	0	0.0338
	5 min	0	0.0417	1	0.0100	0	0.0464	0	0.0100	0	0.0417
	10 min	0	0.0265	1	0.0333	0	0.0426	0	0.0100	0	0.0265
	15 min	0	0.0151	1	0.0100	0	0.0150	0	0.0100	0	0.0151
	30 min	0	0.0415	1	0.0100	0	0.0408	0	0.0100	0	0.0415
#4	0.5 s	0	0.0298	1	0.0100	0	1.6278×10^{-30}	0	0.0100	0	0.0298
	1 s	0	0.0263	1	0.0100	0	7.7177×10^{-20}	0	0.0100	0	0.0263
	2 s	0	0.0429	1	0.0100	0	5.5778×10^{-219}	0	0.0100	0	0.0429
	5 s	0	0.0491	1	0.0100	0	1.9919×10^{-52}	0	0.0100	0	0.0391
	10 s	0	0.0459	1	0.0100	0	0	0	0.0100	0	0.0459
	15 s	0	0.0444	1	0.0100	0	4.1505×10^{-150}	0	0.0100	0	0.0447
	30 s	0	0.0436	1	0.0229	0	4.8221×10^{-9}	0	0.0100	0	0.0364
	1 min	0	0.0451	1	0.0719	0	4.3992×10^{-22}	0	0.0100	0	0.0418
	2 min	0	0.0449	1	0.0450	0	2.2699×10^{-7}	0	0.0100	0	0.0498
	5 min	0	0.0458	1	0.0100	0	6.6692×10^{-18}	0	0.0100	0	0.0482
	10 min	0	0.0451	1	0.0100	0	0.0155	0	0.0238	0	0.0413
	15 min	0	0.0482	1	0.0100	0	0.0051	0	0.0100	0	0.0421
	30 min	0	0.0389	1	0.0100	0	0.0384	0	0.1000	0	0.0399

As can be seen in Table 6, the null hypothesis was rejected in all the tests ($h = 0$ for the tests ADF, variance ratio, LMC, and PP, whereas $h = 1$ for the KPSS test). In addition, the p-value analysis for the KPSS and LMC tests yielded reliability levels higher than 99% for most granularities and households. Nonetheless, the reliability levels for the remaining cases were higher than 95%. Therefore, results in Table 6 assured that the differenced time series data were stationary.

Figure 13 depicts the periodogram of the four consumption load profiles for the 0.5-s data granularity and a period of load fluctuations from 1 to 100 s. The different curves in the graphs were generated with 4000 observations (N) drawn on logarithmic scales for better visualization [67]. The property of the autocovariance function assured that its accuracy was proportionally improved with $1/N$. The accuracy of the periodogram was thus 2.5×10^{-4}. The finest data granularity of 0.5 s, as described in Section 2, limited the frequency analysis to 1 Hz (10^0). The power spectral density for load fluctuations showed striking behavior differences for each household. The most stable power spectrum was that of household #1, followed by households #2, #4, and #3. In general, the power spectrum for the four load profiles showed two different patterns. For households #1 and #2, the power level for the 15–80 s interval remained stable. This meant that load fluctuations included several constituent signals of equivalent significance. Thus, main constituent signals at periods of 14, 27, 40, 55, and 68 s can be clearly observed. In contrast, although households #3 and #4 showed various peaks in the periodogram, which corresponded to different constituent signals, their relevance strongly increased with the period rise. This was due to the greater relevance of the induced cycling by the climate control system that masked other minor fluctuating cycles. Nonetheless, some main constituent signals can be observed, such as those at periods of 80 s for household #3, and others at periods of 30, 60, and 90 s for household #4.

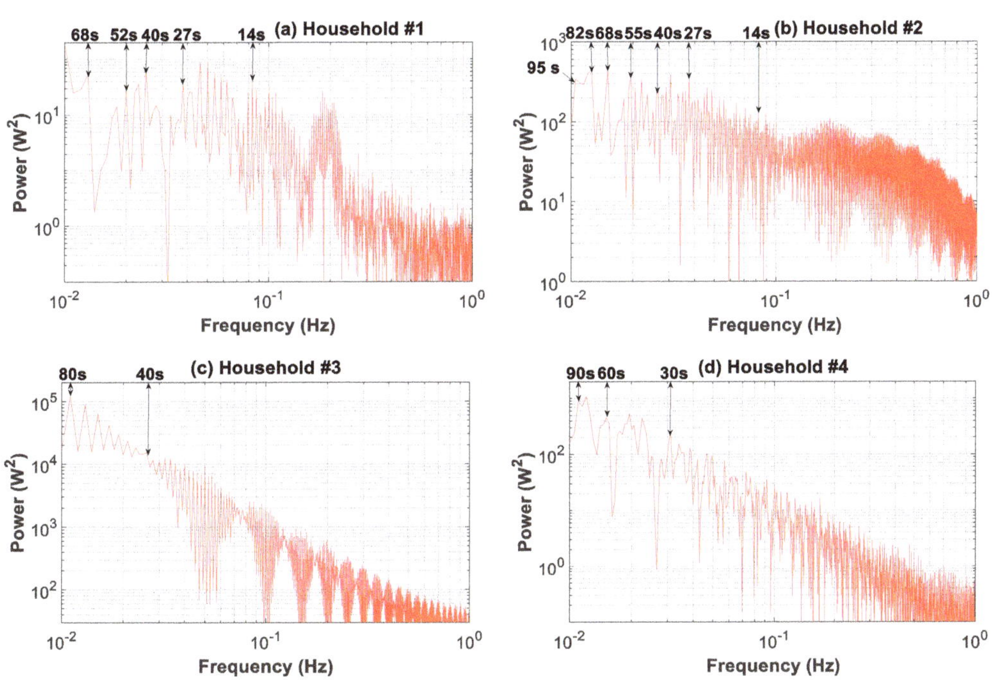

Figure 13. Periodogram of the consumption load profile for households #1 to #4: 1–100 s time slice.

These graphs show that when coarse-grained data were used, from 1 to 100 s, there was a loss of information regarding load profile features. This loss was the greatest for household #1, and in descending order of relevance for households #2, #4, and #3. Furthermore, the main constituent signals that reflect this loss of information for coarse data are evident in these graphs. This analysis confirms the results of the information loss in the daily time-slice analysis (Figure 8), where household #1 also showed the worst behavior, followed by households #2, #4, and #3.

The autocorrelation function, which studies the cross-correlation of a signal with itself, also underlined the constituent signals of load fluctuations. Accordingly, Figure 14 shows the autocorrelation function of the load profiles for four finer levels of data granularity. The curves in the graphs for granularities of 0.5, 2, 5, and 10 s were generated with 200, 50, 20, and 10 observations, respectively. Therefore, the related accuracies were 0.005, 0.02, 0.05, and 0.1, respectively.

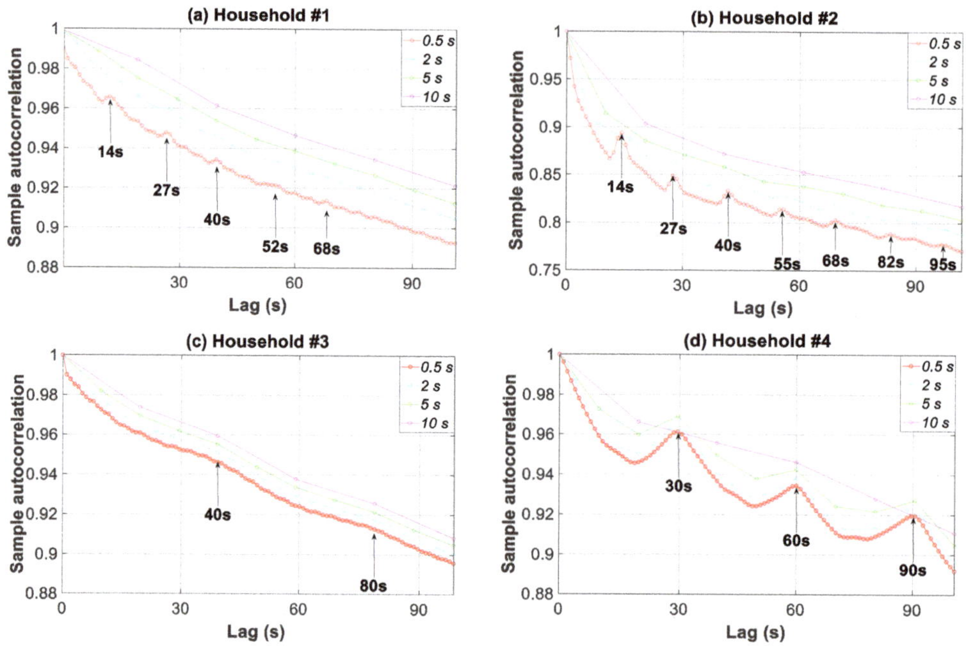

Figure 14. Autocorrelation function of the consumption load profile in households #1 to #4 for four data granularities.

A high value in the autocorrelation function signifies a repetition pattern, and thus reveals a constituent signal in the load fluctuation. The first two households had an autocorrelation function that was much lower than 30. For example, in household #1, both the peak autocorrelation values at lags of 14, 27, 40, 52, and 68 s and very strong dips at 11, 25, 38, 50, and 66 s were quite remarkable. The former lags were in consonance with the periods found for the constituent signals in Figure 13a (the periodogram). This outcome is also striking for the other households in Figure 14 when compared with the results in Figure 13.

Figure 14 also underlines the loss of information for the load profile features when using coarser temporal granularity, ranging from 0.5 to 10 s. As the granularity increased, the autocorrelation value at the specific lags moved closer to unity. This unit value meant that data were fully autocorrelated and no different information (different constituent signals) was involved. The higher the granularity was, the greater the loss of information. As an example for household #1 and a lag of 14 s, the autocorrelation value increased by 0.9%, 1,72%, and 2.21% for a data granularity of 2, 5, and 10 s, respectively. As can be observed, the shifting of the curves for each granularity was the highest in household #1, followed by households #2, #4, and #3. This was a consequence of a much ampler and more stable power spectral density level for the 2–10 s period in the constituent signals for household #1 (Figure 13).

The partial autocorrelation function can also be used to underline the loss of information when coarse-grained data were used. Thus, Figure 15 shows the partial autocorrelation function of the load profiles for four data granularities. As an example, the constituent signal of a 14 s period in household #2 was analyzed. As the granularity increased, the partial autocorrelation value at this specific lag moved closer to unity. This unit value meant that data were fully autocorrelated and no different information (different constituent signals) was involved. More specifically, the value of 0.5 s moved from 0.0166 to 0.1031, 0.2794, and 0.5042 for 2, 5, and 10 s, respectively.

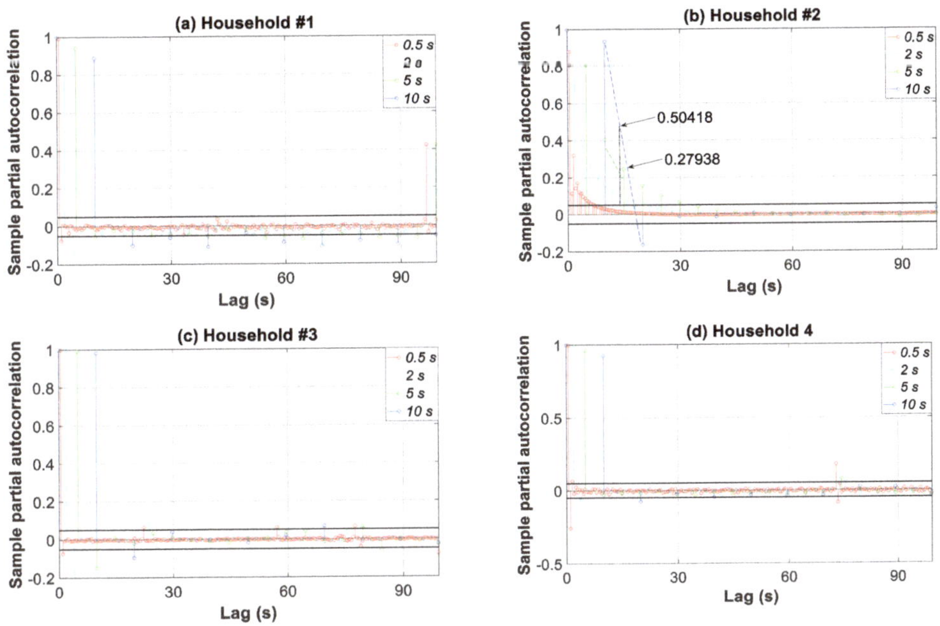

Figure 15. Partial autocorrelation function of the consumption load profile in households #1 to #4 for four data granularities.

Figure 16 broadens the periodogram in Figure 13 covering the period of load fluctuations from 1.66 min to 30 min. The comparison of the power spectral density for the load profiles shows an increasing spread between the base and peak load for profiles #1, #2, #4, and #3. This outcome is evident in Figures 8 and 9, where a drastic reduction of the peaks with higher granularities can be observed. Furthermore, in this interval, the power spectrum level is two orders of magnitude higher compared to that of the 1–100 s interval (Figure 13). Consequently, the contribution to the overall power of the 1–100 s interval was less important and was at least 1% of that of the 1.66–30 min interval.

The power spectrum clearly shows a single constituent signal for household #3, at a period of 25 min, whereas for the remaining households, several main constituent signals are evident. Therefore, in households #1 and #2, there were signals at the 27.45 and 12 min periods, and in household #4, at the 24 and 12.20 min periods. Figure 16 shows that when coarse-grained data from 1.66 to 30 min were used, load profile features were increasingly inaccurate. This lack of accuracy was the greatest for household #1, and in descending order of relevance, for households #2, #4, and #3. Furthermore, the main constituent signals that represent the loss of information for coarse data are shown in Figure 16.

The autocorrelation analysis in Figure 17 pertaining to the 1.66–30 min time slice confirms the constituent signals found in Figure 16 for the different households. In addition, the shifting of the curves (inaccurate profile characterization) for each granularity was in consonance with the power spectral density levels for this 1.66–30 min interval (Figure 16). Thus, the greatest shifting and thus the most inaccurate profile characterization was found in household #1, followed by households #2, #4, and #3.

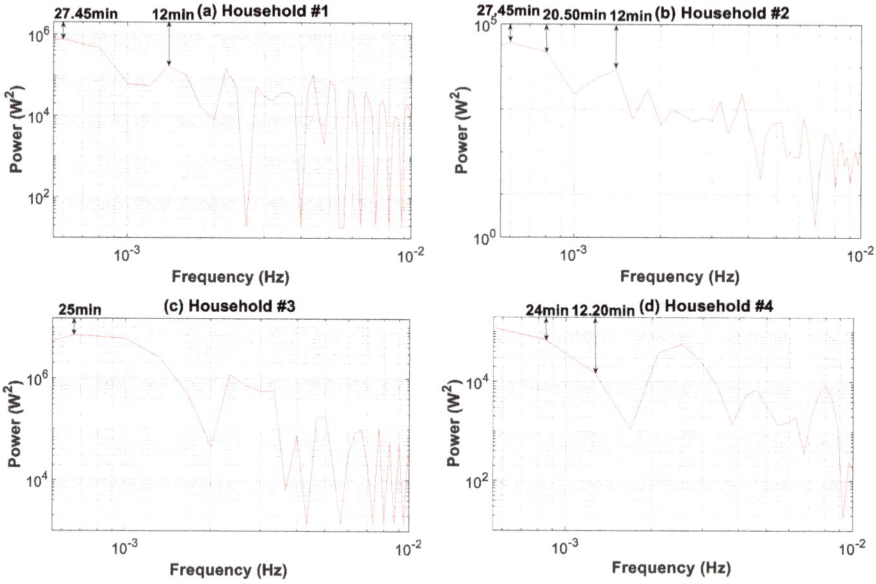

Figure 16. Periodogram of the consumption load profile for households #1 to #4: 1.66–30 min time slice.

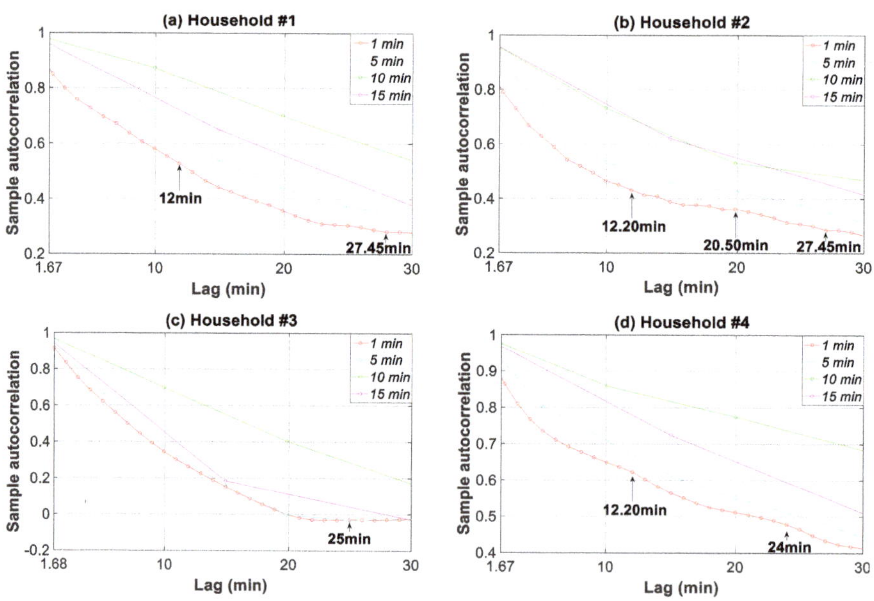

Figure 17. Autocorrelation function of the consumption load profile in households #1 to #4 for four data granularities.

This section concludes by highlighting the seasonality of household consumption load profiles. However, it is important to note that this seasonality, namely daily, weekly, and monthly, is out of the temporal resolution scope of data granularity analyzed in this study (up to 30 min). Accordingly, raw datasets without applying differencing processes were used.

Figure 18 shows the periodograms of the four consumption load profiles for the time horizon daily (a) and weekly/monthly (b). Regarding the daily seasonality in Figure 18a, except for household #2, the power spectrum from the interval of 0.65 h to 24 h had almost the same order of magnitude. For the different households, their main constituent signals can be clearly identified as follows: (i) #1 (0.65, 3.49 and 6 h); (ii) #2 (0.65, 3 and 8 h); (iii) #3 (0.71, 1.50, 1.71, 3, and 12 h); and (iv) #4 (0.92, 1.41, 2.67, and 4.9 h).

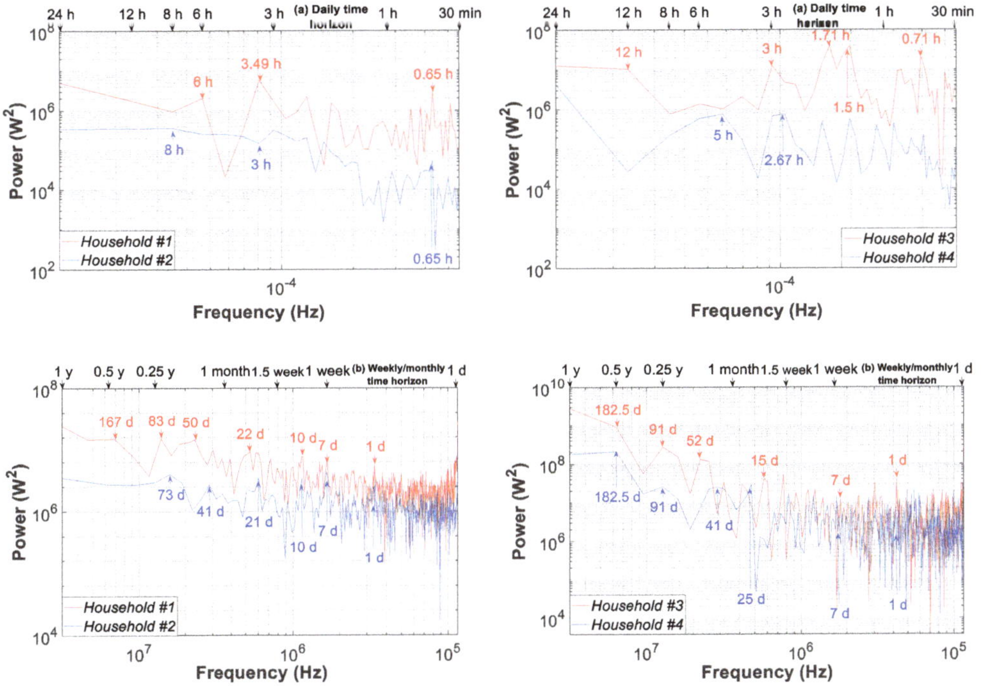

Figure 18. Periodogram of the consumption load profile for households #1 to #4: (**a**) daily time horizon (30 min–1 day time slice); (**b**) weekly/monthly time horizon (1–365 day time slice).

Within the time horizon of one year, Figure 18b, the power spectrum shows that cyclic household power levels differ substantially from one month to the next. In addition, the cyclic power in the interval lower than seven days shows striking behavior differences for each household, decreasing in households #1 and #2 and keeping a more stable value in households #3 and #4. For the different households, their main constituent signals during the weekly and monthly time horizons are as follows: (i) #1 (1, 7, 10, 22, 50, 83, and 167 days); (ii) #2 (1, 7, 10, 21, 41, and 73 days); (iii) #3 (1, 7, 15, 52, 91, and 182.5 days); and (iv) #4 (1, 7, 25, 41, 91, 182.5 days).

3.8. Comparative Study of Granularity Impact in the Literature

The results obtained in this study were compared with other studies in the literature that address granularity impact. It is important to note that this research not only provides temporal results for different time slices from a sub-hourly to a monthly analysis, but it also offers periodograms along with autocorrelation and partial autocorrelation analyses and empirical distribution-based statistical analysis. The temporal analysis offers information regarding the granularity impact on the change in the magnitude of the peak and trough load. In contrast, the second type of analysis reveals additional information such as the constituent signals in the load fluctuation. Since studies in the literature have focused only on temporal analysis, this comparison was limited to the impact on the magnitude change of the peak load because of the lack of data for autocorrelation analyses and empirical distribution-based statistical analysis.

In this study, the values obtained in the temporal analysis (Figures 7–11) and the empirical distribution-based statistical analysis (Figure 12) illustrated that the load peak decreased when the coarse-grained data was compared with the finest data granularity as follows: (i) sub-hourly analysis (between 27.12–56.44%); (ii) daily analysis (between 20.34–59.09%); and (iii) monthly analysis (between 35.07–50.02%).

Wright [27] compared granularities of 30 and 1 min and revealed a peak load reduction between 16–47% for different households. Murray [38] showed peak load changes in the range of 8.9–16% for granularities of 8 s using a meter. Naspolini [39] registered a drop between 18.56–28.36% when 15 min granularity was compared to 5 min granularity. Bucher [40] studied granularities between 5 s and 1 h, which resulted in a reduction between 2–38% as compared to a 1-s based granularity. Shi [41] carried out analysis with granularities of 1, 5, 10, 15, 30, and 60 min and obtained reductions in the peak load of up to 20% (5 min granularity) and up to 80% (60 min granularity) as compared with the 1 min granularity. Widen [48] found drops between 19.19–26.29% for the 60 min granularity with respect to the 10 min. Hoevenaars [64] disclosed a reduction in the peak load when compared to the 1-s based granularity for different granularities as follows: (i) 10 s (between 1.28–7.45% for different sources); (ii) 1 m (between 1.78–15.04%); (iii) 10 min (between 2.46–22.62%); (iv) 60 min (between 8.01–20.65%).

4. Conclusions

Increasing interest in the analysis of household electricity consumption profiles, thanks to the rapid deployment of SMs in the residential context, may significantly change the relevance of such profiles in the near future. To understand profile features and their applicability to any action or assessment, it is necessary to appreciate the full range of consumption load fluctuations.

For this purpose, this paper has presented and discussed a methodology that makes two contributions to the state of the art. Firstly, this research proposed periodograms along with autocorrelation and partial autocorrelation analyses and empirical distribution-based statistical analysis, which were used to describe household consumption load profile features. This type of analysis reveals key issues of the granularity impact on the load fluctuation, such as the accurate description of its constituent signals. Secondly, a framework was developed to collect household consumption data at high sampling frequency (>4 Hz). This methodology allowed us to analyze the influence of data granularity on the description of household consumption load profile features. The effectiveness of this methodology was illustrated in a case study of four households in Spain, using thirteen resolutions of data granularity (0.5, 1, 2, 5, 10, 15, 30 s, and 1, 2, 5, 10, 15, and 30 min). We acknowledge that conducting our analysis with this reduced data sample is a limitation of this study. However, it was adequate for achieving our primary objective of demonstrating the usefulness of the proposed methodology in which the ultimate goal was to highlight the information loss regarding the profile features when using coarse-grained data.

Bearing in mind the limits of applicability of our findings, the main outcomes of the study are detailed below. The influence of data granularity on the results for different time slices from sub-hourly to monthly analysis, including daily and weekly analyses, was discussed. Results from sub-hourly analyses highlight the smoothing of peaks and troughs in the consumption load, based on coarse-grained data from 0.5 s to 30 min. More specifically, peaks decreased by 27.12%, 56.44%, 55.73%, and 51.38%, respectively for households #1, #2, #3, and #4. The daily analysis showed higher peak reductions such as 59.09%, 58.14%, 20.34%, and 57.64%, respectively for the previously mentioned households. The repeated peaks were only identified in the daily and weekly analysis at granularities from 0.5 s to 5 s. The monthly analysis provided data pertaining to the day-to-day load behavior by using the ratio between the daily peak or trough load and the daily mean load. This ratio decreased for coarse data granularity by 50.02%, 39.66%, 35.07%, and 36.89%, respectively for households #1, #2, #3, and #4.

However, the overall influence that data granularity had on the description of household consumption load profile features was performed on an annual basis by using a set of complementary analyses. A statistical analysis based on coarse-grained data underlined the significant change in the empirical distribution shape. The analysis of statistical moments up to the fourth-order reflected the reduction of the level of variability of the consumption load for households when coarse-grained data were used. Periodograms and autocorrelation analyses also indicated the loss of information regarding the profile features caused by the use of coarse-grained data. These analyses were based on the main constituent signals of the load fluctuations. In conclusion, the analyses for different granularities showed that some important loads (e.g., cooling or heating devices, electric water heaters, etc.) produced fluctuations that became increasingly ill-suited for resolutions of 5 s or higher. This confirms that coarse granularities should not be used to collect consumption data because they do not reflect the reality.

The results of our study indicate that it is not necessary to use the finest data granularity, i.e., the 0.5-s resolution. In fact, even for profiles #1 and #2, which showed the greatest fluctuation, a data-resolution of 5 s produced a sufficiently accurate characterization of profile features since the results generated were very close to those of a data-resolution of 0.5 s. Therefore, the use of the 5-s granularity achieves a balance between the computational burden associated with storage data in the cloud and their post-processing, and the loss of information for the consumption profile features.

The results in this research were in line with other studies in the literature that address granularity impact. Since studies in the literature focused only on temporal analysis, this comparison was limited to the impact on the magnitude change of the peak load because of the lack of data for autocorrelation analyses and empirical distribution-based statistical analysis. This review offered a peak load reduction between 1.28–80% with granularities in the interval of 1 s to 1 h.

Future work in the field should take the current limitation of this study into consideration. Further analysis could have been conducted with other households of different characteristics, or the methodology could have been applied to a large set of buildings. It is our hope that this study will spur future work and discussion in the research community regarding the accurate description of household load profile features based on an appropriate data granularity and will ultimately lead to similar work on datasets from other multi-family residential buildings.

Author Contributions: All the authors contributed substantially to this paper. F.S.-S., A.C.-O. and C.R.B. performed the simulations and experimental work, and also wrote the paper. J.C.H. provided the conceptual approach, commented on all the stages of the simulation and experimental work, and revised the manuscript. All authors have read and agreed to the published version of the manuscript.

Funding: This research was funded by the Agencia Estatal de Investigación (AEI) and the Fondo Europeo de Desarrollo Regional (FEDER) aimed at the Challenges of Society (Grant No. ENE 2017-83860-R "Nuevos servicios de red para microredes renovables inteligentes. Contribución a la generación distribuida residencial").

Conflicts of Interest: The authors declare no conflict of interest.

References

1. Zhou, K.; Fu, C.; Yang, S. Big data driven smart energy management: From big data to big insights. *Renew. Sustain. Energy Rev.* **2016**, *56*, 215–225. [CrossRef]
2. Zhou, K.; Yang, S. Understanding household energy consumption behavior: The contribution of energy big data analytics. *Renew. Sustain. Energy Rev.* **2016**, *56*, 810–819. [CrossRef]
3. Torriti, J. A review of time use models of residential electricity demand. *Renew. Sustain. Energy Rev.* **2014**, *37*, 265–272. [CrossRef]
4. Gonzalez, E.; Stephen, B.; Infield, D.; Melero, J.J. Using high-frequency SCADA data for wind turbine performance monitoring: A sensitivity study. *Renew. Energy* **2019**, *131*, 841–853. [CrossRef]

5. Ushakova, S.; Jankin Mikhaylov, S. Big data to the rescue? Challenges in analysing granular household electricity consumption in the United Kingdom. *Energy Res. Social Sci.* **2020**, *64*, 101428. [CrossRef]
6. George, D.; Swan, L.G. A method for distinguishing appliance, lighting and plug load profiles from electricity 'SM' datasets. *Energy Build.* **2017**, *134*, 212–222. [CrossRef]
7. Bianchi, M.; Branchini, L.; Ferrari, C.; Melino, F. Optimal sizing of grid-independent hybrid photovoltaic–battery power systems for household sector. *Appl. Energy* **2014**, *136*, 805–816. [CrossRef]
8. Kelly, J.; Knottenbelt, W. The UK-DALE dataset, domestic appliance-level electricity demand and whole-house demand from five UK homes. *Sci. Data* **2015**, *2*, 150007. [CrossRef] [PubMed]
9. Pipattanasomporn, M.; Kuzlu, M.; Rahman, S.; Teklu, Y. Load profiles of selected major household appliances and their demand response opportunities. *IEEE Trans. Smart Grid* **2014**, *5*, 742–750. [CrossRef]
10. Barker, S.; Mishra, A.; Irwin, D.; Cecchet, E.; Shenoy, P. Smart: An open data set and tools for enabling research in sustainable homes. In Proceedings of the 18th ACM SIGKDD International Conference on Knowledge Discovery and Data Mining, Beijing, China, 12–16 August 2012; pp. 1–6.
11. Cano-Ortega, A.; Sánchez-Sutil, F. Performance optimization LoRa network by artificial bee colony algorithm to determination of the load profiles in dwellings. *Energies* **2020**, *13*, 517. [CrossRef]
12. Tascikaraoglu, A.; Sanandaji, B.M. Short-term residential electric load forecasting: A compressive spati-temporal approach. *Energy Build* **2016**, *111*, 380–392. [CrossRef]
13. Ferdowsi, F.; Mehraeen, S.; Upton, G.B. Integration of behind-the-meter solar into distribution feeders: The importance of time resolution on model results. In Proceedings of the IEEE Green Technologies Conference (GreenTech), Lafayette, LA, USA, 3–6 April 2019; pp. 1–5. [CrossRef]
14. Fridgen, G.; Kahlen, M.; Ketter, W.; Rieger, A.; Thimmel, M. One rate does not fit all: An empirical analysis of electricity tariffs for residential microgrids. *Appl. Energy* **2018**, *210*, 800–814. [CrossRef]
15. Elma, O.; Tascıkaraoglu, A.; Tahir Ince, A.; Selamogullar, U.S. Implementation of a dynamic energy management system using real, time pricing and local renewable energy generation forecasts. *Energy* **2017**, *134*, 206–220. [CrossRef]
16. Hu, M.; Xiao, F.; Jorgensen Bagterp, J.; Wang, S. Frequency control of air conditioners in response to real-time dynamic electricity prices in smart grids. *Appl. Energy* **2019**, *242*, 92–106. [CrossRef]
17. Ahmad, T.; Huanxin, C.; Zhang, D.; Zhang, H. Smart energy forecasting strategy with four machine learning models for climate-sensitive and non-climate sensitive conditions. *Energy* **2020**, *198*, 117283. [CrossRef]
18. Bassamzadeh, N.; Ghanem, R. Multiscale stochastic prediction of electricity demand in smart grids using Bayesian networks. *Appl. Energy* **2017**, *193*, 369–380. [CrossRef]
19. Hong, T.; Fan, S. A university of probabilistic electric load forecasting: A tutorial review. *Int. J. Forecast.* **2016**, *32*, 914–938. [CrossRef]
20. De Felice, M.; Alessandri, A.; Catalano, F. Seasonal climate forecasts for medium-term electricity demand forecasting. *Appl. Energy* **2015**, *137*, 435–444. [CrossRef]
21. Andersen, F.M.; Larsen, H.V.; Gaardestrup, R.B. Long term forecasting of hourly electricity consumption in local areas in Denmark. *Appl. Energy* **2013**, *110*, 147–162. [CrossRef]
22. Foteinaki, K.; Li, R.; Rode, C.; Korsholm, R. Andersen Modelling household electricity load profiles based on Danish time-use survey data. *Energy Build.* **2019**, *202*, 109355. [CrossRef]
23. Grünewald, P.; Diakonova, M. The specific contributions of activities to household electricity demand. *Energy Build.* **2019**, *204*, 109498. [CrossRef]
24. Mazzola, S.; Vergara, C.; Astolfi, M.; Li, V.; Perez-Arriaga, I.; Macchi, E. Assessing the value of forecast-based dispatch in the operation of off-grid rural microgrids. *Renew. Energy* **2017**, *108*, 116–125. [CrossRef]
25. Kools, L.; Phillipson, F. Data granularity and the optimal planning of distributed generation. *Energy* **2016**, *112*, 342–352. [CrossRef]
26. Good, N.; Zhang, L.; Navarro-Espinosa, A.; Mancarella, P. High resolution modelling of multi-energy domestic demand profiles. *Appl. Energy* **2015**, *137*, 193–210. [CrossRef]
27. Wright, A.; Firth, S. The nature of domestic electricity-loads and effects of time averaging on statistics and on-site generation calculations. *Appl. Energy* **2007**, *84*, 389–403. [CrossRef]
28. Widen, J.; Molin, A.; Ellegard, K. Models of domestic occupancy, activity and energy use based on time use data: Deterministic and stochastic approaches with application to building-related simulations. *J. Build. Perform. Simul.* **2011**, *5*, 1–18. [CrossRef]

29. Zheng, Z.; Chen, H.; Luo, X. A Kalman filter-based bottom-up approach for household short-term load forecast. *Appl. Energy* **2019**, *250*, 882–894. [CrossRef]
30. He, Y.; Xu, Q.; Wan, J.; Yang, S. Short-term power load probability density forecasting based on quantile regression neural network and triangle kernel function. *Energy* **2016**, *114*, 498–512. [CrossRef]
31. Causonea, F.; Carlucci, S.; Ferrando, M.; Marchenko, A.; Erba, S. A data-driven procedure to model occupancy and occupant-related electric load profiles in residential buildings for energy simulation. *Energy Build.* **2019**, *202*, 109342. [CrossRef]
32. Gouveia, J.P.; Seixas, J. Unraveling electricity consumption profiles in households through clusters: Combining smart meters and door-to-door surveys. *Energy Build.* **2016**, *116*, 666–676. [CrossRef]
33. Wang, J.; Wang, J.; Liu, C.; Ruiz, J.P. Stochastic unit commitment with sub-hourly dispatch constraints. *Appl. Energy* **2013**, *105*, 418–422. [CrossRef]
34. Lusis, P.; Rajab Khalilpour, K.; Andrew, L.; Liebman, A. Short-term residential load forecasting: Impact of calendar effects and forecast granularity. *Appl. Energy* **2017**, *205*, 654–669. [CrossRef]
35. Taylor, J.W. An evaluation of methods for very short-term load forecasting using minute-by-minute British data. *Int. J. Forecast.* **2008**, *24*, 645–658. [CrossRef]
36. Jain, R.K.; Smith, K.M.; Culligan, P.J.; Taylor, J.E. Forecasting energy consumption of multi-family residential buildings using support vector regression: Investigating the impact of temporal and spatial monitoring granularity on performance accuracy. *Appl. Energy* **2014**, *123*, 168–178. [CrossRef]
37. Ryu, S.; Noh, J.; Kim, H. Deep neural network based demand side short term load forecasting. *Energies* **2017**, *10*, 3. [CrossRef]
38. Murray, D.; Stankovic, L.; Stankovic, V. An electrical load measurements dataset of United Kingdom households from a two-year longitudinal study. *Sci. Data* **2017**, *4*, 1–12. [CrossRef]
39. Naspolini, H.F.; Rüther, R. The effect of measurement time resolution on the peak time power demand reduction potential of domestic solar hot water systems. *Renew. Energy* **2016**, *88*, 325–332. [CrossRef]
40. Bucher, C.; Betcke, J.; Andersson, G. Effects of variation of temporal resolution on domestic power and solar irradiance measurements. In Proceedings of the IEEE Grenoble Conference, Grenoble, France, 16–20 June 2013; pp. 1–6. [CrossRef]
41. Shi, X.; Ming, H.; Shakkottai, S.; Xie, L.; Yao, J. Nonintrusive load monitoring in residential households with low-resolution data. *Appl. Energy* **2019**, *252*, 113283. [CrossRef]
42. Van der Meer, D.W.; Munkhammar, J.; Widén, J. Probabilistic forecasting of solar power, electricity consumption and net load: Investigating the effect of seasons, aggregation and penetration on prediction intervals. *Sol. Energy* **2018**, *171*, 397–413. [CrossRef]
43. Pfenninger, S. Dealing with multiple decades of hourly wind and PV time series in energy models: A comparison of methods to reduce time resolution and the planning implications of inter-annual variability. *Appl. Energy* **2017**, *197*, 1–13. [CrossRef]
44. Sanchez-Sutil, F.; Cano-Ortega, A.; Hernandez, J.C.; Rus-Casas, C. Development and calibration of an open source, low-cost power smart meter prototype for PV household-prosumers. *Electronics* **2019**, *8*, 878. [CrossRef]
45. Cao, S.; Siren, K. Impact of simulation time-resolution on the matching of PV production and household electric demand. *Appl. Energy* **2014**, *128*, 192–208. [CrossRef]
46. Bracale, A.; Carpinelli, G.; De Falco, P. A Bayesian-based approach for the short-term forecasting of electrical loads in smart grids. Part I: Theoretical aspects. In Proceedings of the International Symposium on Power Electronics, Electrical Drives, Automation and Motion (SPEEDAM), Anacapri, Italy, 22–24 June 2016; pp. 121–128. [CrossRef]
47. Bracale, A.; Carpinelli, G.; De Falco, P. A Bayesian-based approach for the short-term forecasting of electrical loads in smart grids. Part II: Numerical application. In Proceedings of the International Symposium on Power Electronics, Electrical Drives, Automation and Motion (SPEEDAM), Anacapri, Italy, 22–24 June 2016; pp. 129–136. [CrossRef]
48. Widen, J.; Wackelgard, E.; Paatero, J.; Lund, P. Impacts of different data averaging times on statistical analysis of distributed domestic photovoltaic systems. *Sol. Energy* **2010**, *84*, 492–500. [CrossRef]
49. Hernandez, J.C.; Sanchez-Sutil, F.; Muñoz-Rodriguez, F.J. Design criteria for the optimal sizing of a hybrid energy storage system in PV household-prosumers to maximize self-consumption and self-sufficiency. *Energy* **2019**, *186*, 115827. [CrossRef]

50. Romero Rodríguez, L.; Brennenstuhl, M.; Yadack, M.; Boch, P.; Eicker, U. Heuristic optimization of clusters of heat pumps: A simulation and case study of residential frequency reserve. *Appl. Energy* **2019**, *233*, 943–958. [CrossRef]
51. Papavasiliou, A.; Mou, Y.; Cambier, L.; Scieur, D. Application of stochastic dual dynamic programming to the real-time dispatch of storage under renewable supply uncertainty. *IEEE Trans. Sustain. Energy* **2018**, *9*, 547–558. [CrossRef]
52. IEC. IEC Standard 61000-4-30. Electromagnetic Compatibility (EMC): Testing and Measurement Techniques—Power Quality Measurement Methods. In Proceedings of the International Electrotechnical Commission, Geneva, Switzerland, May 2015.
53. Wilcox, T.; Jin, N.; Flach, P.; Thumim, J. A Big Data platform for SM data analytics. *Comput. Ind.* **2019**, *105*, 250–259. [CrossRef]
54. Wang, S.; Chen, H.; Wu, L.; Wang, J. A novel smart meter data compression method via stacked convolutional sparse auto-encoder. *Int. J. Electr. Power Energy Syst.* **2020**, *118*, 105761. [CrossRef]
55. Pecan Street. Available online: https://www.pecanstreet.org (accessed on 10 September 2020).
56. Mack, B.; Tampe-Mai, K. An action theory-based electricity saving web portal for households with an interface to SMs. *Util. Policy* **2016**, *42*, 51–63. [CrossRef]
57. Kelly, J. UK-DALE. Available online: https://jack-kelly.com/data (accessed on 10 September 2020).
58. ACS. Available online: https://icosys.ch/acs-f2 (accessed on 10 September 2020).
59. DRED. Available online: http://www.st.ewi.tudelft.nl/akshay/dred (accessed on 10 September 2020).
60. ECO. Available online: http://vs.inf.ethz.ch/res/show.html?what=eco-data (accessed on 10 September 2020).
61. GREEN. Available online: https://sourceforge.net/projects/greend (accessed on 10 September 2020).
62. SMART. Available online: http://traces.cs.umass.edu/index.php/smart/smart (accessed on 10 September 2020).
63. NZERlF. Available online: https://catalog.data.gov/dataset/the-net-zero-energy-residential-test-facility-nzertf-42ce6 (accessed on 10 September 2020).
64. Hoevenaars, E.J.; Crawford Curran, A. Implications of temporal resolution for modeling renewables-based power systems. *Renew. Energy* **2012**, *41*, 285–293. [CrossRef]
65. Haghshenas, S.A.; Razavi, A.A.; Haghighi, A.; Ghader, S. AGP-based approach for improving wind-wavesimulations over the persian gulf. In Proceedings of the International Conference on Coasts, Ports and Marine Structures (Icopmas) Ports & Maritime Organization, Tehran, Iran, 31 October–2 November 2016; pp. 1–6.
66. Broersen, P.M.T. *Automatic Autocorrelation and Spectral Analysis*; Springer: Delft, The Netherlands, 2006.
67. Box, G.E.P.; Jenkins, G.M.; Reinsel, G.C. *Time Series Analysis*, 4th ed.; Wiley: Oxford, UK, 2008.
68. Bartos, S. Prediction of Energy Load Profiles. Master's Thesis, Charles University, Prague, Czech Republic, 2017.
69. Arvamtis, S. A note on the limit theory of a Dickey—Fuller unit root test with heavy tailed innovations. *Stat. Probab. Lett.* **2017**, *126*, 198–204. [CrossRef]
70. Kwiatkowski, D.; Phillips, P.C.; Schmidt, P.; Shin, Y. Testing the null hypothesis of stationarity against the alternative of a unit root: How sure are we that economic time series have a unit root? *J. Econom.* **1992**, *54*, 159–178. [CrossRef]
71. Shin, Y.; Schmidt, P. The KPSS stationarity test as a unit root test. *Econ. Lett.* **1992**, *38*, 387–392. [CrossRef]
72. Eroglu, B.A. Wavelet variance ratio cointegration test and wavestrapping. *J. Multivar. Anal.* **2019**, *171*, 298–319. [CrossRef]
73. Caner, M.; Kilian, L. Size distortions of tests of the null hypothesis of stationarity: Evidence and implications for the PPP debate. *J. Int. Money Finance* **2001**, *28*, 639–657. [CrossRef]
74. Escobari, D.; García, S.; Mellado, C. Identifying bubbles in Latin American equity markets: Phillips-Perron-based tests and linkages. *Emerg. Mark. Rev.* **2017**, *33*, 90–101. [CrossRef]
75. McCullagh, P. *Tensor Methods in Statistics*; Chapman and Hall: London, UK, 1987.
76. Kendall, M.G.; Stuart, A. *The Advanced Theory of Statistics*; Charles Grin and Company Limited: London, UK, 1963; Volume I.
77. Strickland, J. *Predictive Modeling and Analytics*; lulu.com: Morrisville, NC, USA, 2014.
78. Box, G.E.P.; Jenkins, G.M.; Reinsel, G.C.; Ljung, G.M. *Time Series Analysis, Forecasting and Control*, 5th ed.; Wiley & Sons, Inc.: Philadelphia, PA, USA, 2016.
79. Khintchine, A. Korrelationstheorie der Stationären Stochastischen Prozessen. *Math. Ann.* **1934**, *109*, 604–615. [CrossRef]

80. Wiener, N. Generalised harmonic analysis. *Acta Math.* **1930**, *35*, 117–258. [CrossRef]
81. Priestley, M.B. *Spectral Analysis and Time Series*; Academic Press: London, UK, 1981.
82. Das, P.; Mathur, J.; Bhakar, R.; Amit Kanudi, A. Implications of short-term renewable energy resource intermittency in long-term power system planning. *Energy Strategy Rev.* **2018**, *22*, 1–15. [CrossRef]
83. Lopez, A.; Ogayar, B.; Hernandez, J.C.; Sutil, F.S. Survey and assessment of technical and economic features for the provision of frequency control services by household-prosumers. *Energy Policy* **2020**, *146*, 111739. [CrossRef]
84. Hernandez, J.C.; Sanchez-Sutil, F.; Muñoz-Rodríguez, F.J.; Baier, C.R. Optimal sizing and management strategy for PV household-prosumers with self-consumption/sufficiency enhancement and provision of frequency containment reserve. *Appl. Energy* **2020**, *277*, 115529. [CrossRef]
85. Arduino Uno Rev3. Available online: https://store.arduino.cc/arduino-uno-rev3 (accessed on 10 September 2020).
86. WEMOS Electronics. Available online: https://wiki.wemos.cc/products:d1:d1 (accessed on 10 September 2020).
87. STC013 Dechang Electronics Co. Ltd. Available online: http://en.yhdc.com/product/SCT013-401.html (accessed on 10 September 2020).
88. Texas Instruments. Available online: http://www.ti.com/lit/ds/symlink/ads1114.pdf (accessed on 10 September 2020).
89. Interplus Industry Co. Ltd. Available online: http://www.interplus-industry.fr/index.php?option=com_content&view=article&id=52&Itemid=173&lang=en (accessed on 10 September 2020).
90. Firebase. Available online: https://firebase.google.com/?hl=es (accessed on 10 September 2020).
91. Najim, K.; Ikonen, E.; Daoud, A.-K. *Stochastic Processes: Estimation, Optimisation and Analysis*, 1st ed.; Butterworth-Heinemann: Oxford, UK, 2004.

Publisher's Note: MDPI stays neutral with regard to jurisdictional claims in published maps and institutional affiliations.

sensors

Technical Note

Configurable IoT Open-Source Hardware and Software I-V Curve Tracer for Photovoltaic Generators

Isaías González *, José María Portalo and Antonio José Calderón

Department of Electrical Engineering, Electronics and Automation, University of Extremadura, Avenida de Elvas, s/n, 06006 Badajoz, Spain; jportalo@alumnos.unex.es (J.M.P.); ajcalde@unex.es (A.J.C.)
* Correspondence: igonzp@unex.es; Tel.: +34-924-289-600

Abstract: Photovoltaic (PV) energy is a renewable energy resource which is being widely integrated in intelligent power grids, smart grids, and microgrids. To characterize and monitor the behavior of PV modules, current-voltage (I-V) curves are essential. In this regard, Internet of Things (IoT) technologies provide versatile and powerful tools, constituting a modern trend in the design of sensing and data acquisition systems for I-V curve tracing. This paper presents a novel I-V curve tracer based on IoT open-source hardware and software. Namely, a Raspberry Pi microcomputer composes the hardware level, whilst the applied software comprises mariaDB, Python, and Grafana. All the tasks required for curve tracing are automated: load sweep, data acquisition, data storage, communications, and real-time visualization. Modern and legacy communication protocols are handled for seamless data exchange with a programmable logic controller and a programmable load. The development of the system is expounded, and experimental results are reported to prove the suitability and validity of the proposal. In particular, I-V curve tracing of a monocrystalline PV generator under real operating conditions is successfully conducted.

Keywords: IoT; renewable energy sources; photovoltaic energy; I-V curve; monitoring and data acquisition; microgrid; open-source; communication protocols

Citation: González, I.; Portalo, J.M.; Calderón, A.J. Configurable IoT Open-Source Hardware and Software I-V Curve Tracer for Photovoltaic Generators. *Sensors* **2021**, *21*, 7650. https://doi.org/10.3390/s21227650

Academic Editors: Antonio Cano-Ortega and Juan M. Corchado

Received: 25 September 2021
Accepted: 16 November 2021
Published: 18 November 2021

1. Introduction

Photovoltaic (PV) technology is one of the most widespread renewable energy sources (RES) [1,2] and contributes to reducing greenhouse gas emissions and fighting against climate change [3]. In intelligent energy facilities conceived under the paradigm of smart grids and microgrids, PV generators are commonly the main source of renewable energy [4]. In these facilities, PV can be combined with other equipment for energy production and consumption such as wind turbines, batteries, and hydrogen-related devices (fuel cells and electrolyzers).

In PV-based grids, it is required to monitor the state and operation of the PV devices. In this regard, the efficiency of PV cells under natural conditions is measured using current-voltage (I-V) characteristic curves [5]. Consequently, to evaluate the performance of PV modules, it is necessary to measure their I-V output characteristics [6].

I-V curves are obtained by performing a voltage sweep on the PV module, while measuring the output current which is delivered to a connected load [5]. Such curves display maximum voltage and current values of a module in a given setting [5]. This way, the analysis of the curves provides direct information on the electrical state of the module, allowing the researcher to obtain data on the expected performance under different conditions of irradiance and load [7]. The measurement system used to acquire data of the PV modules and to visualize the I-V curve is commonly known as the I-V curve tracer.

In the context of RES, monitoring and data acquisition are essential to recognize the resources available on-site, evaluate electrical conversion efficiency, detect failures, and optimize electrical production [8]. In particular, the characterization of PV modules through I-V curves is required for different purposes, and the applicability of such curves has been widely reported in the literature. For instance, I-V curve tracing is the most commonly applied technique for the electrical characterization of PV modules [9]. An in situ measurement system of PV characteristics can provide valuable information for optimized power generation [10]. I-V curves are widely used to evaluate power generation performance and detect fault conditions of PV generators [11].

Aging effects of PV cells affect the I-V curve [10], and the consequent degradation is usually assessed by means of such I-V characteristics [6]. It must be noted that failures in PV modules may be caused by several reasons such as corrosion failures, cell cracks, hot-spots, encapsulation failures, electrical or mechanical connection failures, potential induced degradation, accumulation of dust or soiling, or partial shading, among others [12].

In the context of degradation and failure analyses, detailed parameter fitting can be carried out using the I-V characteristics of the strings or modules for a deeper understanding of degradation mechanisms prior to failure [13]. According to [14], the I-V curve measurement method is time-consuming, but it is very reliable and considered a paramount step in fault detection. For diagnosis, electrical parameters are extracted from the measured I-V curves such as the short circuit current, the open-circuit voltage, the maximum power-point, and the fill factor [6,14]. Indeed, decisions on the replacement of faulty or degraded devices are better taken based on a direct measurement than based on estimation [15].

In addition, another application of I-V curves is related to the modeling of PV modules, which requires estimating or measuring certain parameters. In this regard, a data-acquisition system is essential to collect and store I-V curves so simulated I-V curves can be plotted based on different models [16]. Namely, the well-known single diode model (SDM) requires estimating the values of series and parallel resistances which can be calculated from the I-V curves [17].

However, the manufacturer provides curves measured at laboratory conditions, obtained under standard conditions of temperature and irradiance (standard test conditions, STC), which do not correspond to real operation in physical facilities. These differences affect the I-V characteristic and, thus, the module production, so it is crucial to use an I-V curve tracer [6].

The relevance of I-V tracing can be witnessed in the literature; the instrumentation and monitoring equipment for such a task have received important research efforts. For example, in-depth reviews of curve tracers according to their topologies can be found in [7,11]. Additionally, diverse equipment is available in the market and in the scientific literature. On the one hand, commercial curve tracers can be found in the market, such as in [18–20]. Their main advantage is the reliable measurements that are performed, guaranteed by the manufacturer of the device. On the contrary, the most noticeable drawbacks are that commercial systems are generally expensive and closed for modifications [8]. Furthermore, another important disadvantage is that the control software of commercial tracers is not prepared for an automatic experimental campaign of measurements [21].

On the other hand, custom-designed I-V curve tracers constitute an important trend in recent years. Diverse works have designed curve tracers using general purpose electronic boards such as microcontrollers and digital signal processors (DSP). For example, in [22,23], a portable I-V curve tracer is based on a DSP, being connected to a personal computer (PC) through serial communication. Vega et al. [6] combine a peripheral interface controller (PIC) and an electronic load (MOSFET) to implement an I-V curve tracer. Acquired data are stored on a PC or a smartphone. A low-cost PIC is used by Ortega et al. [12] to develop a prototype of an I-V curve tracer for individual modules in large photovoltaic systems. A curve tracer developed around a commercial low-cost embedded microcontroller (TivaC) is presented in [24]. Additionally, a low-cost microcontroller is the core of the curve tracer presented in [15], where measurements are stored in local memory and downloaded through a universal synchronous asynchronous receiver transmitter (USART) connected to a Bluetooth device.

Moreover, within custom-designed curve tracers, new developments are progressively including Internet of Things (IoT) open-source technology. Namely, open-source hardware platforms such as Arduino and Raspberry Pi (RPi) are being introduced in research projects and facilities. In the scope of PV energy, these devices are applied for data acquisition and monitoring tasks. For example, Arduino is used in [25,26] to sense the temperature of PV modules. With higher computation capabilities, the RPi microprocessor is used to implement monitoring systems for PV-based microgrids in [4,27,28] and for PV plants in [5,29,30].

In particular, to deploy I-V curve tracers, there are still scarce developments involving such IoT open-source technology, so the most recent research works will now be discussed. Within the open-source community, there is information publicly available about I-V curve tracers using Python and the Arduino microcontroller [31]. In the curve tracer proposed in [21], Arduino is responsible for managing a capacitive load, while data storage and visualization are performed by a PC. Arduino is used together with a commercial data logger in [32] to handle a MOSFET load in order to trace I-V curves of PV modules. An Arduino together with a PC is used in [33] to deploy an I-V curve tracer, the PC acting as a storage means for the measured data. The work reported in [34] applies an Arduino board with data storage on an SD card to collect the data of PV modules under shading conditions. Papageorgas et al. [10] develop a low-cost curve tracer involving an open-source platform with an embedded microcontroller called Polytropon and message queueing telemetry transport (MQTT) as a communication protocol.

Regarding the use of RPi, the following works are found. In [35], an IoT-based remote I-V tracing system is developed using an RPi and a cloud-based server aimed at analyzing soiling losses in distributed solar facilities. An RPi is used in [36] to implement a plug and play I-V curve tracer oriented toward the diagnosis of PV modules. A power MOSFET transistor is used as the electronic load during characterization, the data being recorded on the RPi and on an intermediate file transfer protocol (FTP) server. In [37], an RPi is used as the main component in a so-called outdoor test facility (OTF) with IoT capabilities employed to capture I-V and P-V curves of PV modules. Python scripts are used, and experimental validation is reported.

Some relevant requirements and trends of curve tracers have been identified in the previous literature. For example, in [33], it is pointed out that the measure of the entire I-V curve in a short time requires a suitable data acquisition device. Reference [7] identifies various trends in the advancement of curve tracers, among which low-cost measurement systems and low-cost communications are highlighted. In the same sense, the important role that reliable low-cost communications play is emphasized in [15].

The utilization of open-source and IoT technologies for curve tracing and monitoring constitutes another new trend [16,38]. Furthermore, such technologies encourage the previously mentioned trends of low-cost measurements and communication systems. These technologies provide rapid development and cost-effective solutions for smart monitoring systems [16]. Related to costs, as pointed out in [24], the low-cost characteristic of open-source platforms provides greater accessibility to I-V curve tracing equipment for any research or academic center.

An issue of the existing literature is signaled in [38]: there are papers that do not offer information either about the measurement performance or about the equipment used, showing only the results. Moreover, as it is asserted in [7], most of the curve tracers found in the literature are complex and difficult to integrate in real scenarios.

Aiming at overcoming the aforesaid drawbacks and integrating the identified trends, this paper presents the development of a novel IoT open-source hardware and software I-V curve tracer to characterize PV generators. The RPi microcomputer composes the hardware level; concerning software, Python, MariaDB, and Grafana are applied for data acquisition, storage, and visualization. Open communication protocols, such as Modbus TCP, enable seamless data exchange with proprietary equipment. The program coded in Python is responsible for automating the curve tracing of the PV modules through the modification of the current demanded by an electronic programmable load. The developed system is oriented towards the I-V curve tracing for already existent functioning facilities that require diagnostics, analyses, and/or modeling. In such a situation, this curve tracer is coupled to the facility by means of open protocols and shares data without altering the installation.

For the sake of clarity, a table summarizing the aforementioned literature as well as the present proposal has been elaborated (Table 1). The considered categories are the following: Device, referred to as the equipment used to collect data from PV modules; Load, to indicate the type of load; Data storage means, to discriminate if local (within the Device) or remote accumulation is applied; Language, for a clear identification of the programming language used to gather data; and Communication, in order to specify the protocols for data sharing. In the cases where the information has not been found, Unspecified has been written.

Table 1. Table comparative of previous literature dealing with custom-made curve tracers.

References	Device	Load	Data Storage	Language	Communication
[6]	PIC	MOSFET	External (PC or smart device)	Unspecified	Bluetooth/USB
[10]	Microcontroller	MOSFET	Local database (NoSQL)	Unspecified	MQTT
[12]	PIC	Capacitive	Unspecified	Unspecified	Power line communication
[15]	Microcontroller	MOSFET	Local memory	Unspecified	USART/Bluetooth
[22,23]	DSP	Capacitive	External (PC)	Unspecified	Serial
[24]	Microcontroller	Capacitive	Unspecified	Unspecified	Unspecified
[21]	Arduino	Capacitive	External (PC)	Arduino sketch	USB
[32]	Arduino	MOSFET	External (datalogger)	Arduino sketch	No
[33]	Arduino	MOSFET	External (PC)	Arduino sketch	USB
[34]	Arduino	Resistive	Local SD card	Arduino sketch	USB
[35]	RPi	Capacitive	External cloud server	Unspecified	MQTT
[36]	RPi	MOSFET	FTP server or local SD card	Python	Ethernet
[37]	RPi	Capacitive	Unspecified	Python	Ethernet/Wi-Fi
Present work	RPi	Programmable electronic load	Local database (mariaDB)	Python	Modbus TCP

In view of the previous table, it can be derived that the presented curve tracer is a novelty due to the fact that it is a unique proposal which combines open-source components (hardware and software) and open communication protocols, all being managed by the RPi. In other words, in most of the literature, an IoT open-source device is used to gather data from sensors and handle the load, data storage and visualization being performed by external equipment or services such as PCs or cloud servers. The present work is the only proposal where the RPi is responsible for automating all the tasks involved in the I-V tracing: load sweep, data acquisition, data storage, communications, and visualization in real-time.

Moreover, as can be observed in Table 1, some works do not report information about certain aspects such as the programming language or data storage means (software or hardware), as it has been previously indicated in [38]. Moreover, none of the surveyed references apply a programmable electronic load to perform the I-V tracing. Resistive, capacitive, or electronic loads (power MOSFET) are among the common methods [7], but they require designing specific electronic circuitry for the curve tracing process and only serve for such a purpose. On the contrary, programmable loads are commonly used in microgrids and PV facilities [4,20,39–41] to emulate the behavior of DC or AC loads in order to test control algorithms and energy management strategies under different load profiles. In this regard, the target groups of this paper are scientists and practitioners in the scope of PV-based microgrids and facilities involved in research and development (R&D) activities.

In addition, the validation of the proposal is performed with a medium-scale PV generator under real conditions, which constitutes a requirement to demonstrate the suitability of open-source technologies [4].

It must be remarked that the presented curve tracer is used to characterize and diagnose a PV generator integrated in a smart microgrid (SMG) which combines renewable sources with hydrogen. Such a facility is framed in an R&D project envisioned to develop a digital replica of the subsystems of the microgrid.

The main contributions of the work are now summarized:

- Open-source hardware and software is applied for data storage and visualization. As a consequence, easy and low-cost deployment and replication are feasible;
- Open communication protocols are used to provide a seamless data exchange;
- The curve tracer can be coupled to an already existent PV generator and gather operational data;
- Experimental results achieved on a medium-scale PV generator, not only for a laboratory scale, are reported to prove the suitability of the developed curve tracer;
- Capability of configuration for facilities with a larger number of sensors to manage as well as different communication protocols;
- Utilization of proprietary medium-scale programmable electronic load providing accurate and reliable measurements;
- IoT-enabled remote monitoring of real-time data in the form of time-series;
- Automated data acquisition under programmed conditions without requiring operator intervention.

The structure of the rest of the paper is as follows. The Section 2 describes the developed I-V curve tracer concerning hardware, software, and communications. Section 3 deals with the achieved results from a PV generator of 1100 W, whereas the associated discussion is carried out in Section 4. Finally, the main conclusions of the reported work and further research guidelines are addressed.

2. Developed I-V Curve Tracer

The developed curve tracer is solved by a software application made in Python and executed on an RPi, as well as a database and a data visualization interface. The version of the microcomputer is the RPi 3 model B+. As commented in the previous section, the proposed curve tracer is applied to an existing SMG equipped with an automation system based on a Programmable Logic Controller (PLC) model S7_1516, which is in charge of the energy management of the SMG. Figure 1 shows the interconnection of all the devices used in the I-V curve tracer. This figure shows the sensors involved (irradiance, voltage, current) together with the PLC, the RPi, and a programmable electronic load. The proposed platform takes advantage of an electronic programmable load model Prodigit 32612A (New Taipei City, Taiwan). This legacy device accepts communication through an RS232 interface in order to exchange commands and is used to configure the current profiles demanded by the photovoltaic panels.

The communications diagram of the deployed system can be seen in Figure 2. The RPi acts as a Grafana server, so the user/operator can visualize and download the data processed by the curve tracer through a web browser running on a computer or smartphone connected to the Internet. Namely, the Grafana software provides a user-friendly graphical user interface (GUI) for real-time access to numerical and graphical information about the measurements of the PV system during the tracing.

Among other elements, the RPi includes serial communication through universal serial bus (USB) ports, so a protocol converter from USB to RS232 has been required to establish communication between the load and the RPi.

Concerning sensors, Table 2 summarizes the magnitudes that are measured and the corresponding sensor. It must be noted that the required sensors could be connected to the RPi in a direct manner or through proper electronic boards. In this sense, the presented solution is applicable to already existing automation and monitoring systems or for new facilities without such systems. Using open protocols, such as Modbus TCP, enables easy communication given the widespread availability of this protocol in automation and energy-related equipment [28]. In addition, Modbus TCP has been pointed out as an industrial IoT communication protocol [42] and is supported by both open-source and proprietary equipment.

Figure 1. Interconnection of the I-V curve tracer components together with equipment of the existing microgrid.

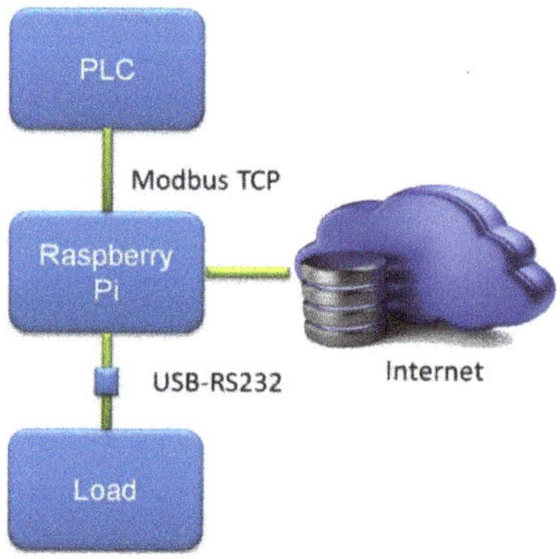

Figure 2. Communications diagram of the I-V curve tracer.

Table 2. Magnitudes and sensors used in the curve tracer.

Magnitude	Sensor
Current	Hall-effect sensor
Voltage	Potentiometric voltage divider
Irradiance	Pyranometer
Temperature	Pt-100

A block diagram with the functionalities associated with each component is illustrated in Figure 3. From a functional viewpoint, it is interesting to note that all the tasks required for the process of I-V curve tracing rely on the RPi. On the one hand, this microprocessor acquires data from the PLC and stores them in a database (mariaDB). Data backup is also handled by the RPi together with other system tasks. On the other hand, the programmable load is managed through Python commands. Lastly, the Grafana server is hosted to establish Internet-enabled data visualization in real-time.

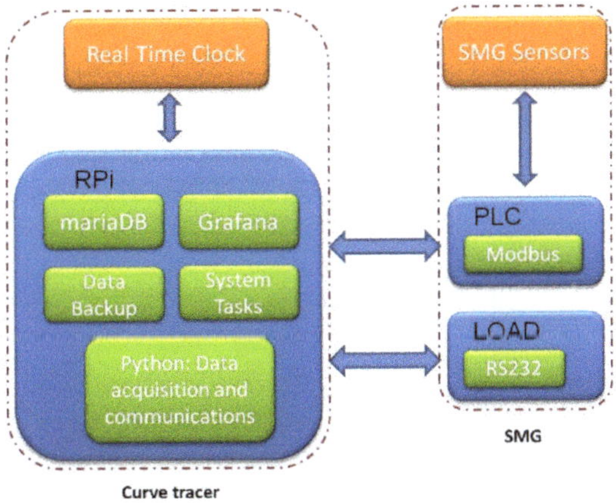

Figure 3. Functionalities implemented by the curve tracer.

Figure 4 shows the flow diagram of the algorithm that implements the I-V tracer. First of all, the irradiance existing at a given time is measured. It is verified whether the measured irradiance exceeds the preset threshold of 100 W/m^2 to initiate the current profile generation and data acquisition processes. Once the minimum irradiance condition is met, a profile of the current demanded by the electronic load is created, which progressively increases from zero to the maximum possible current that the PV module(s) can deliver for the sensed irradiance. The maximum value is calculated based on the existing modules configuration and the irradiance at each moment. This ensures that the PV generator will not be required to provide currents that cannot be achieved for this irradiance value. In particular, the following equation has been used to determine the maximum current, Imax:

$$\text{Imax} = \text{np} \times (0.0055 \times G + 0.1), \quad (1)$$

where np is the number of paired modules, and G is the incident irradiance on the plane of the modules.

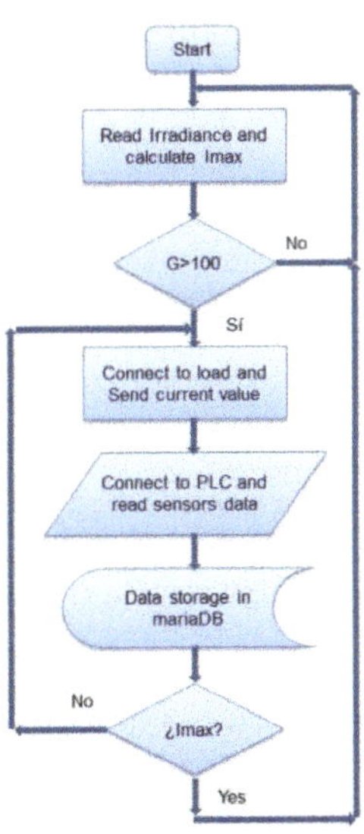

Figure 4. Flowchart of operations performed by the curve tracer.

The following step consists of establishing communication via RS232 from the RPi to the programmable electronic load, so the current value corresponding to each instant is sent. Next, the RPi takes the sensor data obtained by the PLC through a Modbus TCP channel, as well as from the electronic load itself through the RS232 connection. After this, the retrieved data are stored in a mariaDB database specifically designed for this purpose (Figure 5). This process is carried out continuously for each PV module current until the maximum current set is reached.

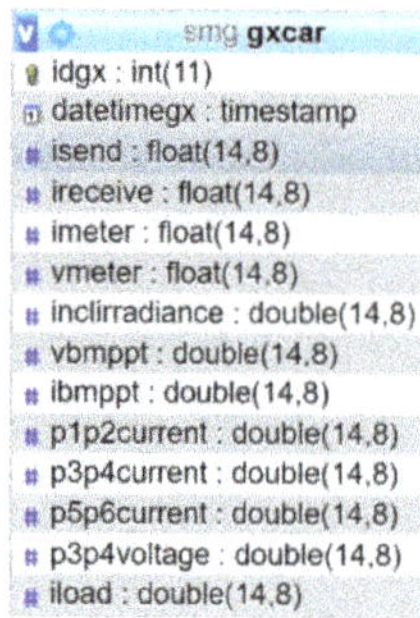

Figure 5. Database created with mariaDB for characterization of PV modules.

In this way, all the necessary data for the characterization of the photovoltaic panels are acquired and stored. As a sample, Figure 6 shows the data taken for the characterization of the PV modules.

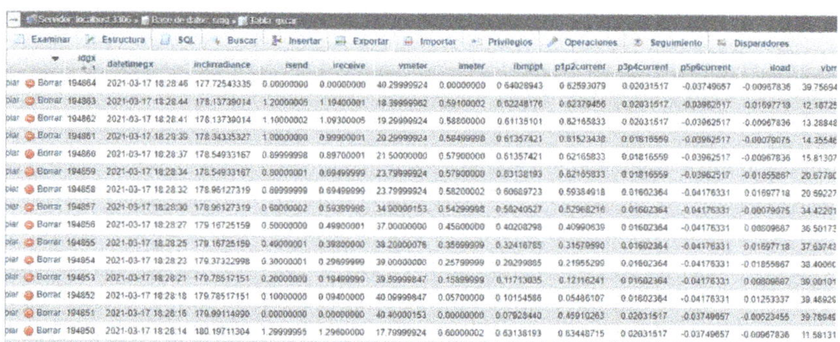

Examinar | Estructura | SQL | Buscar | Insertar | Exportar | Importar | Privilegios | Operaciones | Seguimiento | Disparadores

idgx	datetimegx	inclirradiance	isend	ireceive	vmeter	imeter	ibmppt	p1p2current	p3p4current	p5p6current	iload	vbm
194864	2021-03-17 18:28:46	177.72543335	0.00000000	0.00000000	40.29999924	0.00000000	0.64028943	0.62593079	0.02031517	-0.03749657	-0.00967836	39.75694
194863	2021-03-17 18:28:44	178.13739014	1.20000005	1.19400001	18.39999962	0.59100002	0.62248176	0.62379456	0.02031517	-0.03962517	0.01697718	12.18725
194862	2021-03-17 18:28:41	178.13739014	1.10000002	1.09300005	19.29999924	0.58800000	0.61135101	0.62165833	0.02031517	-0.03962517	-0.00967836	13.28848
194861	2021-03-17 18:28:39	178.34335327	1.00000000	0.99900001	20.29999924	0.58499998	0.61357421	0.61523438	0.01816559	-0.03962517	-0.00079075	14.35546
194860	2021-03-17 18:28:37	178.54933167	0.89999998	0.89700001	21.50000000	0.57900000	0.61357421	0.62165833	0.01816559	-0.03962517	-0.00967836	15.81307
194859	2021-03-17 18:28:34	178.54933167	0.80000001	0.69499999	23.79999924	0.57900000	0.63138193	0.62165833	0.01816559	-0.03962517	-0.01855867	20.67780
194858	2021-03-17 18:28:32	178.96127319	0.69999999	0.69499999	23.79999924	0.58200002	0.60689723	0.59384918	0.01602364	-0.04176331	0.01697718	20.69227
194857	2021-03-17 18:28:30	178.96127319	0.60000002	0.59399998	34.90000153	0.54299998	0.58240527	0.52968216	0.01602364	-0.04176331	-0.00079075	34.42201
194856	2021-03-17 18:28:27	179.16725159	0.50000000	0.49900001	37.00000000	0.45600000	0.40208298	0.40990639	0.01602364	-0.04176331	0.00809687	36.50173
194855	2021-03-17 18:28:25	179.16725159	0.40000001	0.39800000	38.20000076	0.35699999	0.32416785	0.31579590	0.01602364	-0.04176331	0.01697718	37.63743
194854	2021-03-17 18:28:23	179.37322998	0.30000001	0.29699999	39.00000000	0.25799999	0.29299885	0.21955299	0.01602364	-0.04176331	-0.01855867	38.40060
194853	2021-03-17 18:28:21	179.78517151	0.20000000	0.19499999	39.59999847	0.15899999	0.11713035	0.12116241	0.01602364	-0.04176331	0.00809687	39.00101
194852	2021-03-17 18:28:18	179.78517151	0.10000000	0.09400000	40.09999847	0.05700000	0.10154586	0.05486107	0.01602364	-0.04176331	0.01253337	39.46926
194851	2021-03-17 18:28:16	179.99114990	0.00000000	0.00000000	40.40000153	0.00000000	0.07928440	0.45910263	0.02031517	-0.03749657	-0.00523455	39.78945
194850	2021-03-17 18:28:14	180.19711304	1.29999995	1.29600000	17.79999924	0.60000002	0.63138193	0.63448715	0.02031517	-0.03749657	-0.00967836	11.58131
194849	[illegible]	[illegible]	[illegible]	[illegible]	[illegible]	[illegible]	[illegible]	[illegible]	[illegible]	[illegible]	[illegible]	[illegible]

Figure 6. Data taken during characterization of PV modules.

To achieve representative data, the irradiance should change as little as possible during the characterization. Each acquisition cycle can vary from 24 s, for an irradiance of 200 W/m^2 and a single panel configuration (12 samples), to 320 s, considering an irradiance of 1000 W/m^2 for the whole group of panels (320 samples), keeping the step sampling rate at 0.1 A. During these short intervals, irradiance is scarcely altered.

On the other hand, Figure 7 shows a screenshot of part of the Python code running to automate the labeled stages and, hence, the I-V curve tracing.

Aiming to illustrate the described sequence, the main code of Python concerning the load management is shown in Algorithm 1. To begin with, an instance for communication is created, specifying parameters such as port, transmission bit rate, parity bit, etc. After that, the connection is open, and the commands to determine the load current are sent. Namely, the load is activated, the operation mode is selected, and the current demanded to the PV module(s) is established. Moreover, the reached voltage and current values are retrieved. Finally, the connection is closed.

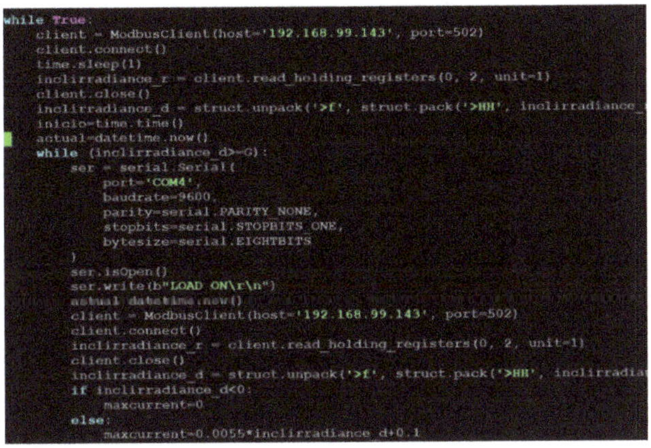

Figure 7. Python code for I-V curve tracing.

Algorithm 1 RS232 Communication through Python

```
while (inclirradiance_d>=G):
  ser = serial.Serial(
    port = 'COM4',
    baudrate = 9600,
    parity = serial.PARITY_NONE,
    stopbits = serial.STOPBITS_ONE,
    bytesize = serial.EIGHTBITS
  )
  ser.isOpen()
   ser.write(b"LOAD ON\r\n")
...
     cadena = "LIN:A "+str(i/10)[0:5]+";\r\n"
     ser.write(cadena.encode())
     ser.write(b"LIN:A?\r\n")
     out = ser.readline()
     ser.write(b"MEAS:CURR?\r\n")
     out1 = ser.readline()
     ser.write(b"MEAS:VOLT?\r\n")
     out2 = ser.readline()
...
   ser.close()
```

3. Results

In this section, the experimental results of applying the I-V curve tracer are reported to demonstrate its successful operation. Namely, a PV generator hybridized with hydrogen in a stand-alone SMG, placed at the University of Extremadura (Spain), is fully characterized.

3.1. Experimental Setup

The PV generator (Figure 8) consists of six monocrystalline modules, each one with maximum output of 185 W, providing a total power of 1110 W. These modules have a fixed inclination angle, the irradiance measured being in the same plane. The main parameters of the PV modules are listed in Table 3. Note that electric characteristics are given for STC by the manufacturer.

Figure 8. PV generator for experimentation.

Table 3. Main parameters of PV modules.

Parameter	Value
Model	LDK Solar 185D-24S
Maximum power (Pmax)	185 W
Voltage at max power (Vmp)	36.9 V
Current at max power (Imp)	5.02 A
Open circuit voltage (Voc)	45.1 V
Short-circuit current (Isc)	5.48 A
Nominal voltage	24 V
Number of solar cells	72
Cell efficiency	17.77%
Module efficiency	14.49%

The curve tracer is coupled to the PLC and the load of the SMG in the laboratory as can be observed in Figure 9a. Note that an Ethernet switch allows data exchange between the RPi and the PLC. The programmable load can be seen in Figure 9b, also placed in the laboratory setup. On the other hand, the block diagram of the SMG is depicted in Figure 10. As can be observed, the PV array is linked to a DC voltage bus through a solar charger. A battery acts as electrochemical energy storage whilst the programmable load conducts the energy consumer role. Regarding hydrogen generation and consumption, an electrolyzer (EL) produces hydrogen, harnessing the surplus of PV energy, and a fuel cell (FC) performs the opposite process, converting hydrogen into electricity when there is no renewable energy availability. A more detailed description of the SMG components can be found in [4,28].

3.2. Data Visualization

The measurement process was carried out during different days due to the variability of weather conditions (cloudy and rainy days, etc.). More than 194,000 samples were recorded during the whole measurement campaign. The stored data are represented through the GUI created in Grafana, which displays the involved magnitudes in the form of time-series.

As a proof of the visualization capabilities, Figure 11 depicts the aspect of the GUI showing the measurements during a day of the PV generator characterization, in particular, 16 March 2021 from 8:00 to 19:00. The typical curve of solar irradiance along the day can be appreciated in the lower graph, reaching a maximum value of 1031.48 W/m^2 at 13:17. During this day, the procedure begins at 8:18 once the irradiance threshold of 100 W/m^2 is exceeded and lasts until 18:47. The top chart represents the current delivered by the PV generator, which fluctuates according to the management performed by the Python program of the curve tracer.

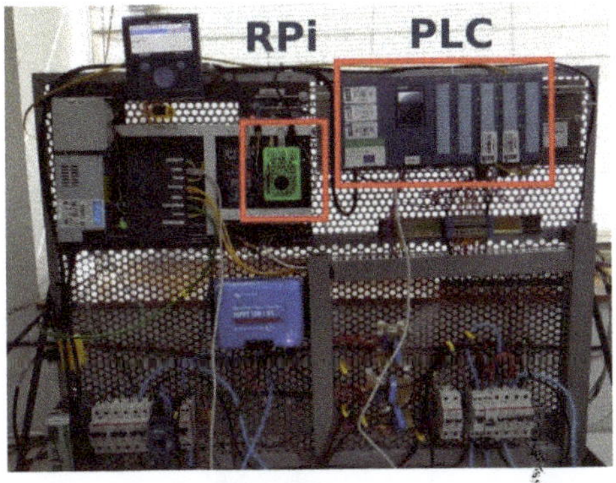

(**a**)

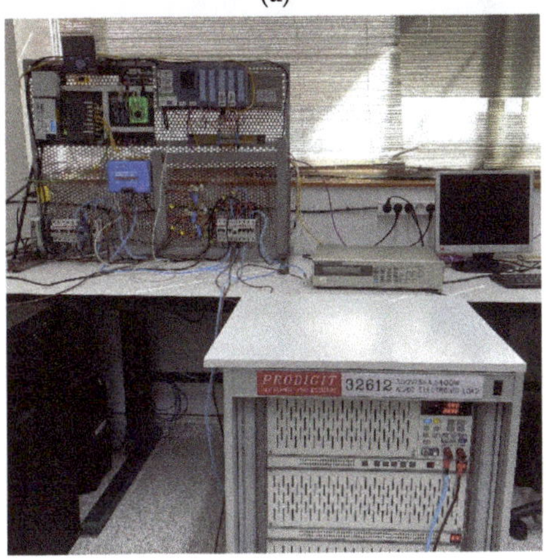

(**b**)

Figure 9. Setup in laboratory: (**a**) detailed view of curve tracer; (**b**) entire view including the programmable load.

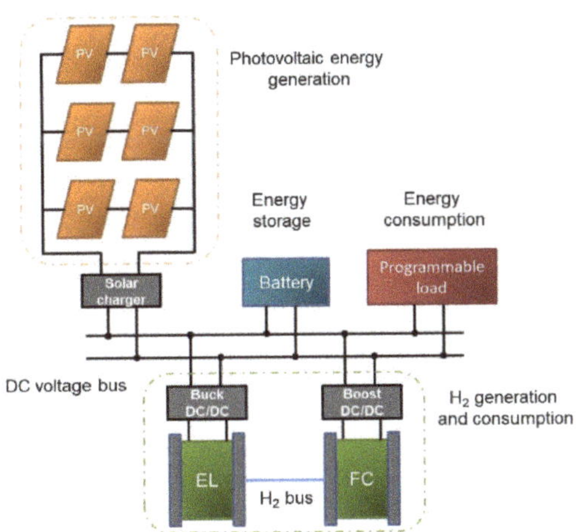

Figure 10. Block diagram of SMG where the PV modules are installed.

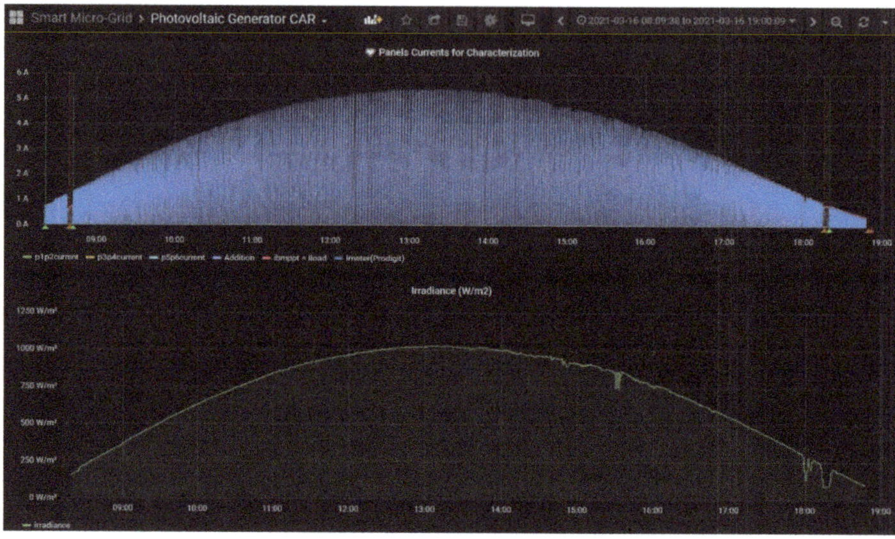

Figure 11. Grafana GUI displaying time-series of PV current and irradiance during a day of characterization.

Figure 12 contains a detailed view of the GUI during the same day for a better observation of the magnitude evolution. In the top graph, it can be seen that the current delivered by the PV generator (blue color) and the load current (red color) are coincident and both exhibit a saw tooth-shaped evolution, coherent with the implemented algorithm. The sensed irradiance during the viewed interval is 1027 W/m^2.

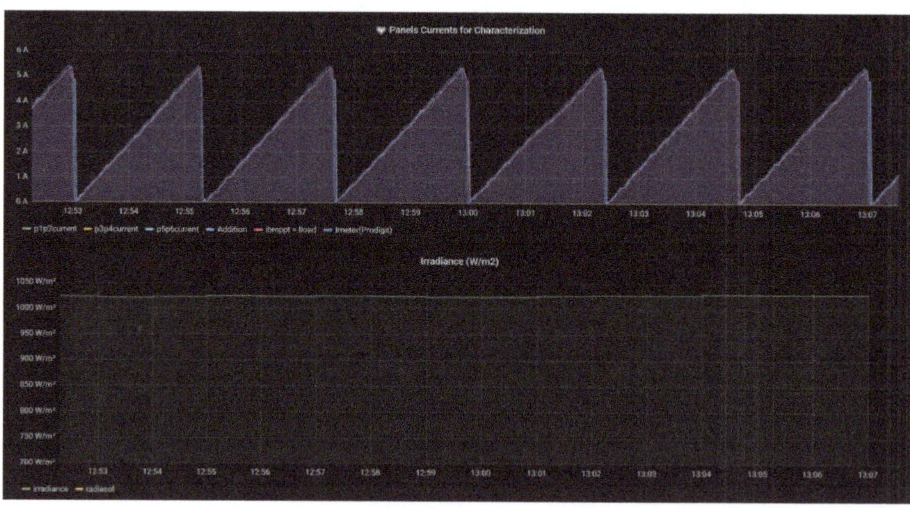

Figure 12. Detailed view of Grafana GUI to observe PV current and irradiance during characterization.

In order to verify the capabilities of the curve tracer, the computational resources of the RPi are also monitored by means of Grafana. To this aim, the GUI includes a dashboard based on Telegraf [43] to visualize the central processing unit (CPU) temperature and load, memory usage, and network statistics. Figure 13 shows this dashboard during the characterization experiments on the same day shown in the previous figures. There are some relevant aspects to discuss in this sense. The usage of CPU is observable in the graph placed in the top left position, and its nominal value is around 4%. There are certain intervals during which the usage rises up to 17% for the system (yellow color) and up to 54% for the user (green color), respectively. These increments are due to Grafana operations, e.g., access for online monitoring and requests to the database. Another parameter is the memory usage (graph in the low and left part) where less than 1 GB is used (yellow line) and around 1 GB is cached (blue line), leaving 2 GB free (orange line), showing a stable behavior. Concerning the CPU temperature, it has a stable value around 35 °C, being an appropriate level to avoid overheating issues.

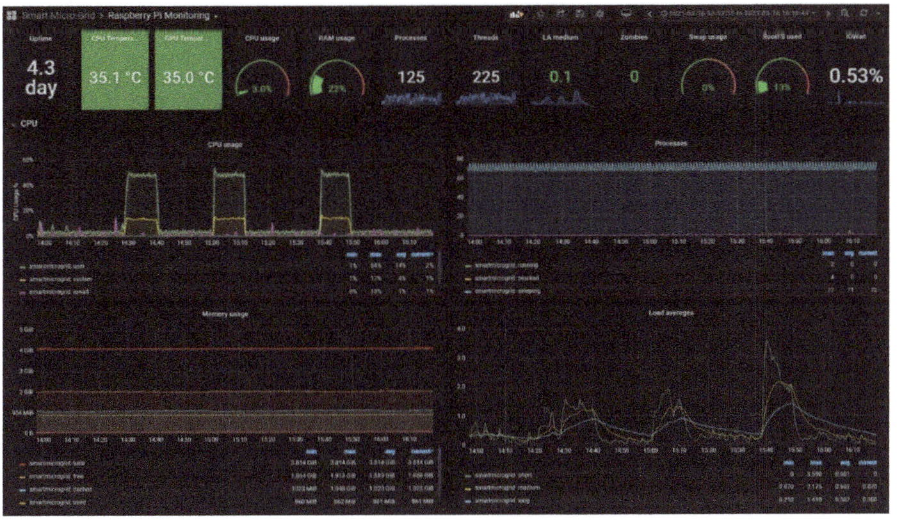

Figure 13. Dashboard devoted to monitoring the resources of RPi during experiments.

3.3. I-V Curves of PV Generator

To achieve a proper validation, I-V curves have been obtained under real operating conditions for the PV generator. In addition, three configurations of the modules have been applied: a single module, a pair of modules connected in series, and the whole generator, consisting of the parallel connection of three pairs.

For the curve tracing, it has been required to select the data for the different irradiances close to the values commonly provided by manufacturers and reported in the literature, namely 200 W/m^2, 400 W/m^2, 600 W/m^2, 800 W/m^2, and 1000 W/m^2. Due to the short duration of the data acquisition intervals, the initial and final values of irradiance are averaged. Table 4 shows the measurements of the incident irradiance and the temperature of the modules during the characterization campaign for each one of the described electrical configurations. Moreover, electrical parameters of the generator can be measured in the I-V curves, such as short circuit current, open-circuit voltage, fill factor, etc.; hence, in Table 4, such parameters are also included.

Figure 14 shows the I-V curves obtained for a single module. The shape and trend of the curves correspond to those expected, matching the information provided by the manufacturer. As can be observed, the open circuit voltage (Voc) decreases whilst the irradiation increases. This effect is due to the associated temperature increase, which causes the curves to move to the left. In particular, the open circuit voltage strongly depends on temperature, while its dependence on irradiance has a modest effect [17]. This relationship can be expressed through Equation (2) [17]:

$$\mathrm{Voc(T)} = \mathrm{Voc}_{,\mathrm{STC}} + \mu_{\mathrm{Voc}}\,(\mathrm{T} - \mathrm{T}_{\mathrm{STC}}), \tag{2}$$

where $Voc_{,STC}$ is the open circuit voltage for STC, T_{STC} corresponds to the STC temperature, and μ_{Voc} is the voltage temperature coefficient, found in the PV module datasheet. For the LDK Solar 185D-24S, such a coefficient has a value of $-0.34\%/^{\circ}C$, so it is easy to check that temperature increments give place to decrements of Voc.

Table 4. Irradiance, temperature, and electrical parameters measured during characterization for different configurations of the PV modules.

Configuration	G (W/m^2)	T (°C)	Voc (V)	Isc (A)	Vmp (V)	Imp (A)	Pmax (W)
Single module	195	11.46	42.23	0.84	37.38	0.74	27.07
	401	19.70	42.49	1.91	35.50	1.79	63.72
	603	26.06	42.39	3.12	34.75	2.83	98.41
	871	41.39	41.01	5.17	29.08	4.78	139.03
	1019	45.1	40.93	6.20	29.20	5.68	165.89
Two modules in series	198.34	6.44	85.49	0.86	77.55	0.74	57.23
	396.59	14.71	85.71	1.92	74.87	1.74	130.66
	608.45	28.75	83.91	3.46	69.44	3.10	215.27
	799.63	39.05	82.57	5.07	64.50	4.42	283.80
	1001.68	50.96	80.07	7.05	58.02	6.45	374.11
Whole PV generator	190.18	11.46	80.65	2.63	67.69	2.46	166.59
	389.94	19.70	84.17	6.18	69.57	5.82	404.89
	603.68	26.06	82.89	9.35	66.74	8.96	597.85
	776.29	41.39	81.84	12.45	64.06	11.83	757.94
	878.30	45.1	80.82	13.81	62.12	13.16	820.19

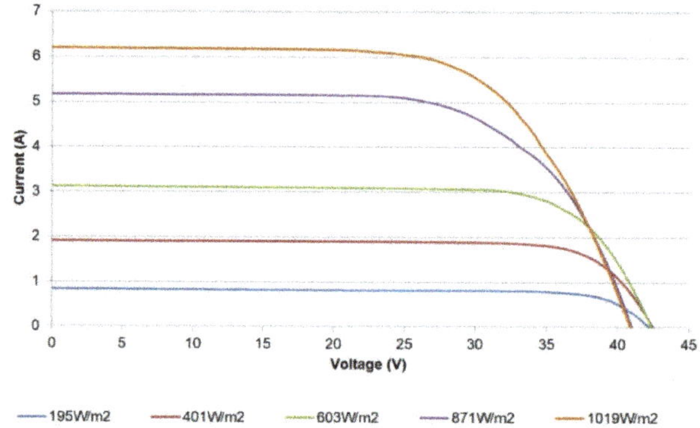

Figure 14. I-V curves for a single PV module.

In a similar sense, the power-voltage (P-V) curve can also be plotted from the acquired data; for instance, Figure 15 shows such a curve for the single PV module. Valuable information such as the maximum power point values (power, current, and voltage) for sensed irradiances can be studied through these curves.

The maximum power produced (165 W) by the module is lower than that reported by the manufacturer (185 W) given the fact that the existing conditions differ from the STC. Moreover, the degradation of the module also contributes to reducing the peak power that can be delivered.

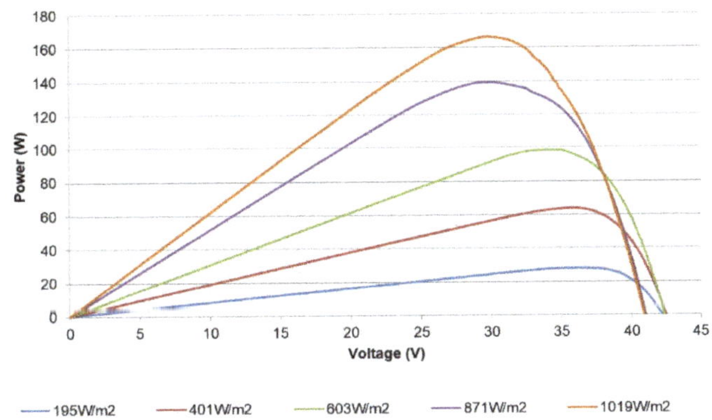

Figure 15. P-V curves for a single PV module.

Following the validation procedure reported in [23,33,34,36,37], the experimental measurements are reproduced by means of a simulator of PV modules based on the SDM. This model is based on the equivalent circuit and is the most widely used method to provide an estimation of the current generated by a PV cell. The circuit consists of a single diode connected in parallel with a photo-generated current source (I_{PH}), a series resistance (R_S) to represent voltage drops and internal losses, and a shunt resistance (R_{SH}) to take into account the leakage currents. Equation (3) describes the model for a module of N_S cells in series:

$$I = I_{PH} - I_o\left[exp\left(\frac{V + IR_S}{nN_S V_{TH}}\right) - 1\right] - \frac{V + IR_S}{R_{SH}} \tag{3}$$

where I_o is the saturation current of the diode, V is the output voltage, and V_{TH} is the thermal equivalent voltage. The last variable is given in terms of the electron charge, q; the Boltzmann constant, K; the cell temperature, T; and the diode ideality factor, n, according to Equation (4):

$$V_{TH} = KT/q, \tag{4}$$

The I-V curve experimentally measured with the curve tracer at an irradiance of 1019 W/m^2 and temperature of 45.1 °C is plotted in Figure 16 (black color) together with the curve provided by the SDM simulator (orange color). As can be observed, the curves show the same trend with very scarce differences. Namely, the ideality factor of the SDM explains the difference appreciated in the knee of the curve [36].

For a better appreciation, the difference between the currents (simulated and measured) can be used to illustrate the achieved fitting [44,45]. In this regard, Figure 17 shows the difference of currents for the characterized module versus the voltage at the reported irradiance levels. The errors are small, reaching a maximum value of 0.21 A for 870 W/m^2. In Figures 14 and 17, it can be seen that the maximum values of these differences are located in a reduced range between the maximum power point (voltage higher than 29 V) and the Voc. These results exhibit proper agreement with the well-known SDM.

The traced I-V curves for a pair of modules connected in series are depicted in Figure 18. The shape observed in the traced curves allows diagnosing or detecting diverse effects in the PV modules, as pointed out in previous works [7,36]. In this case, the curves show a certain alteration in the inflection point and slopes, which indicate the degradation of one of the modules. Therefore, these curves serve for fault detection and diagnostics; namely, aging effects, cell cracking, hot spots, potential induced degradation, and other deterioration situations can be detected. In fact, the modules have been working for 10 years, so aging effects can be expected. Nonetheless, in-depth diagnosis and fault analyses are out of the scope of this paper. On the other hand, Figure 19 contains the traced P-V curves for the pair of modules.

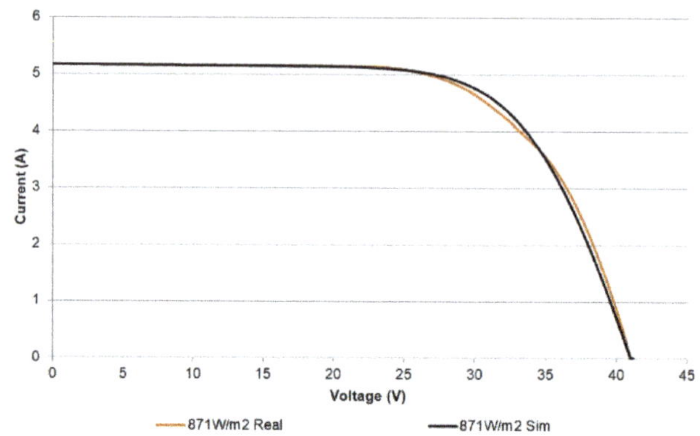

Figure 16. Experimental and simulated I-V curves for a single PV module at 871 W/m^2.

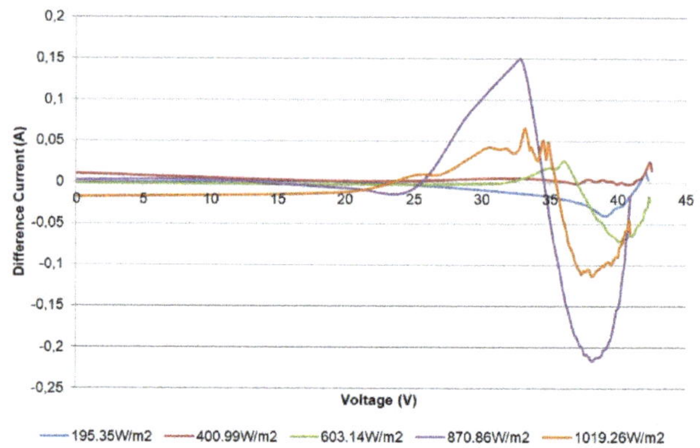

Figure 17. Difference of measured and simulated currents for the single PV module.

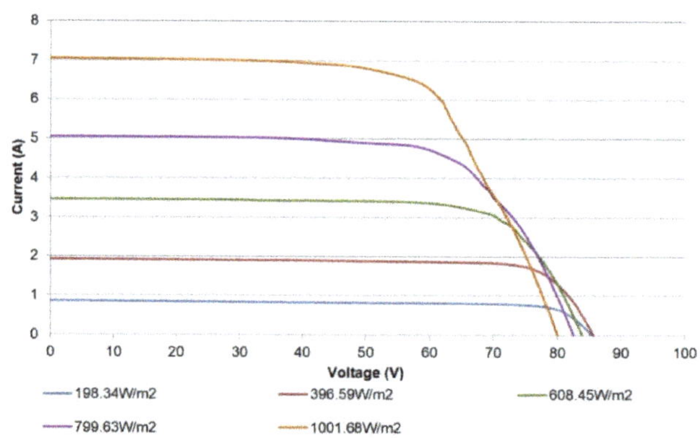

Figure 18. I-V curves for two PV modules connected in series.

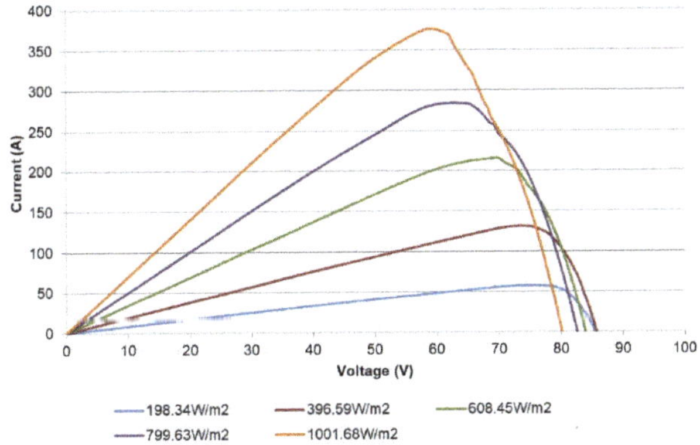

Figure 19. P-V curves of a pair of PV modules.

Finally, the I-V curves captured for the whole PV generator are shown in Figure 20. The corresponding P-V curves are depicted in Figure 21. As in the previous figures, the curves display the expected operation of the generator and can be applied for diagnostics purposes.

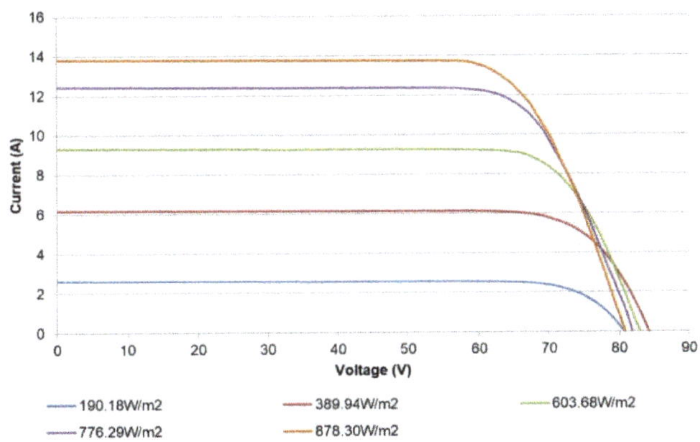

Figure 20. I-V curves of PV generator.

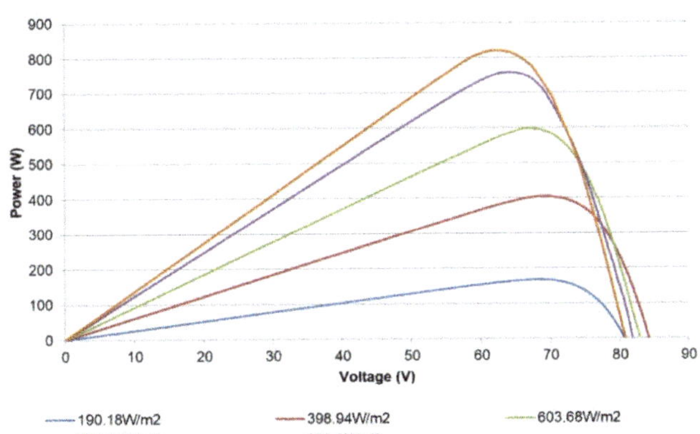

Figure 21. P-V curves of PV generator.

4. Discussion

Experimental results provide I-V curves for different electrical configurations and environmental conditions, emphasizing the suitability of the designed curve tracer.

The main strength is that the developed system is not limited to data acquisition of PV modules for I-V curves, but data recording and visualization in real-time during the characterization are also entirely approached. Indeed, once the desired conditions are programmed, the fully autonomous operation of the curve tracer is achieved without requiring the intervention of the operator.

The deployed curve tracer consists of the RPi and the associated software, whilst a PLC and a programmable load of an experimental SMG are used to validate the operation of such curve tracer.

The computational capabilities of the microprocessor are proven to be adequate to resolve for data acquisition, storage, and visualization. It must be emphasized that none of the previous literature provides information in this regard.

Using an in-house database (mariaDB) and a web-enabled user interface (Grafana) avoids dependencies on external servers and the associated hosting or licensing costs. Hosting on one's own databases even implies a total control of administration aspects [4].

As a proof of concept, in the reported application case, Modbus TCP and RS232 have been used. However, the curve tracer can manage virtually any communication protocol given the wide availability of libraries on the Internet. Furthermore, this ability to support many other protocols provides features such as configurability and modularity, facilitating interoperability [46].

In particular, the use of open communication protocols such as Modbus TCP together with the ability of the open-source equipment allows for the establishment of seamless data exchange. This way, proprietary equipment (PLC) is combined with the curve tracer without interoperability issues. In fact, logical connections through communication protocols enable measurement information sharing and facilitate integration in real scenarios, which constitutes a disadvantage in most of the curve tracers in the literature [5]. In this regard, the deployed system is focused on PV facilities already existent, so the coupling is made through the aforementioned open protocol. The curve tracer even makes use of already existing sensors, which is a benefit since the PV generator can be re-characterized when required without essential alterations in the electrical and communications schemes.

Instead of using a variable resistor, capacitive load, or a power MOSFET, the proposal employs a real electronic programmable load to perform the I-V tracing. In addition, the used load is legacy equipment which does not support modern communication interfaces, so being able to manage such valuable equipment is an important advantage. In fact, IoT technologies must contribute to solve compatibility and interoperability issues with legacy devices [47,48].

Regarding economic assessments, the cost of the curve tracer is very low given the inexpensive nature of the IoT open-source equipment, which constitutes an advantage of scientific equipment based on this type of technology [49]. Namely, taking into account that all the software is free (Python, mariaDB, Grafana), only the RPi involves expenses; the overall cost is around EUR 70. Auxiliary elements such as a memory card, heatsink and fan for cooling, and power adapter are also included.

Analyses of the retrieved I-V curves allow decision making with respect to operation and maintenance of the PV modules as well as implementing accurate models. Moreover, further experiments will include partial shading of the PV modules in order to obtain and analyze the measured I-V curves.

Thanks to the flexibility and availability of open-source equipment, the system can be customized to fulfill particular requirements in research or academic contexts. The RPi provides a large number of analogue and digital inputs, allowing the connection of additional sensors or instruments. Indeed, advances in IoT technology, both hardware and software, can be integrated in the presented system.

Despite the obtained results, the presented system has some limitations which are now briefly described. To begin with, managing open-source technology does not imply ease of configuration when advanced functions are required. For example, programming skills and a certain expertise in communication protocols and networks are needed. In addition, the proposal does not allow online measurements of the PV modules; it is only devoted to offline characterization. For a proper data exchange, it is necessary that the automation unit (PLC or similar device) and the programmable load provide communication interfaces that the RPi can handle. It is not a probable boundary in modern devices, but for legacy equipment, it must be carefully tackled. Finally, the representation of the I-V curves requires manual data extraction from the files that Grafana stores and provides. This can be a time-consuming task when a large number of measurements have been conducted.

5. Conclusions

RES are key enablers for the evolution towards a more sustainable energetic global scenario, PV technology being one of the most applied RES in microgrids. In order to characterize and study the behavior of PV modules, an I-V curve tracer based on IoT open-source technologies has been presented. Namely, software such as Python, MariaDB, and Grafana run on an RPi are responsible for automating all the required tasks: load sweep, data acquisition, data storage, communications, and visualization in real-time. An open communication protocol (Modbus TCP) has been applied to exchange information with a PLC, whilst an RS232 allows for managing a legacy programmable load. Both proprietary devices belong to a research-oriented microgrid facility and serve as proof of concept to prove the suitability of the curve tracer.

It must be emphasized that this development is a novelty in the existing literature, addresses trends, and overcomes limitations identified in previous works, among which short-time measurements, low-cost measurement systems, low-cost communications, and IoT open-source technology can be highlighted.

Experimental results under real operating conditions are used to validate the proposal. Namely, a PV generator of 1110 W integrated into an SMG is characterized by means of the developed curve tracer.

Future research includes diagnostics and fault detection of the PV modules. Furthermore, another interesting topic deals with the development of an on-line characterization procedure using the presented system.

Author Contributions: Conceptualization, J.M.P., I.G. and A.J.C.; Methodology, I.G. and A.J.C.; Validation, J.M.P. and A.J.C.; Investigation, I.G. and A.J.C.; Data Curation, J.M.P. and A.J.C.; Writing—Original Draft Preparation, J.M.P. and I.G.; Writing—Review and Editing, I.G. and A.J.C.; Supervision, A.J.C. All authors have read and agreed to the published version of the manuscript.

Funding: This project was co-financed by European Regional Development Funds FEDER and by the Junta de Extremadura (IB18041).

Institutional Review Board Statement: Not applicable.

Informed Consent Statement: Not applicable.

Data Availability Statement: Not applicable.

Conflicts of Interest: The authors declare no conflict of interest.

Abbreviations

The following abbreviations are used in this manuscript:

CPU	Central Processing Unit
DSP	Digital Signal Processor
EL	Electrolyzer
FC	Fuel cell
FTP	File Transfer Protocol

GUI	Graphical User Interface
I-V	Current-Voltage
IoT	Internet of Things
MQTT	Message Queue Telemetry Transport
OTF	Outdoor Test Facility
P-V	Power-Voltage
PC	Personal Computer
PLC	Programmable Logic Controller
PV	Photovoltaic
R&D	Research and Development
RES	Renewable Energy Sources
RPi	Raspberry Pi
SMG	Smart MicroGrid
STC	Standard Test Conditions
TCP/IP	Transmission Control Protocol/Internet Protocol
USART	Universal Synchronous Asynchronous Receiver Transmitter
USB	Universal Serial Bus

References

1. Gimeno-Sales, F.J.; Orts-Grau, S.; Escribá-Aparisi, A.; González-Altozano, P.; Balbastre-Peralta, I.; Martínez-Márquez, C.I.; Gasque, M.; Seguí-Chilet, S. PV Monitoring System for a Water Pumping Scheme with a Lithium-Ion Battery Using Free Open-Source Software and IoT Technologies. *Sustainability* **2020**, *12*, 10651. [CrossRef]
2. Ansari, S.; Ayob, A.; Lipu, M.S.H.; Saad, M.H.M.; Hussain, A. A Review of Monitoring Technologies for Solar PV Systems Using Data Processing Modules and Transmission Protocols: Progress, Challenges and Prospects. *Sustainability* **2021**, *13*, 8120. [CrossRef]
3. Gielen, D.; Boshell, F.; Saygin, D.; Bazilian, M.D.; Wagner, N.; Gorini, R. The role of renewable energy in the global energy transformation. *Energy Strateg. Rev.* **2019**, *24*, 38–50. [CrossRef]
4. Portalo, J.M.; González, I.; Calderón, A.J. Monitoring System for Tracking a PV Generator in an Experimental Smart Microgrid: An Open-Source Solution. *Sustainability* **2021**, *13*, 8182. [CrossRef]
5. Dupont, I.M.; Carvalho, P.C.M.; Jucá, S.C.S.; Neto, J.S.P. Novel methodology for detecting non-ideal operating conditions for grid-connected photovoltaic plants using Internet of Things architecture. *Energy Convers. Manag.* **2019**, *200*, 112078. [CrossRef]
6. Vega, A.; Valiño, V.; Conde, E.; Ramos, A.; Reina, P. Double sweep tracer for I-V curves characterization and continuous monitoring of photovoltaic facilities. *Sol. Energy* **2019**, *190*, 622–629. [CrossRef]
7. Morales-Aragonés, J.I.; Dávila-Sacoto, M.; González, L.G.; Alonso-Gómez, V.; Gallardo-Saavedra, S.; Hernández-Callejo, L. A Review of I–V Tracers for Photovoltaic Modules: Topologies and Challenges. *Electronics* **2021**, *10*, 1283. [CrossRef]
8. Melo, G.C.G.D.; Torres, I.C.; Araújo, Í.B.Q.D.; Brito, D.B.; Barboza, E.D.A. A Low-Cost IoT System for Real-Time Monitoring of Climatic Variables and Photovoltaic Generation for Smart Grid Application. *Sensors* **2021**, *21*, 3293. [CrossRef]
9. Quansah, D.A.; Adaramola, M.S.; Takyi, G.; Edwin, I.A. Reliability and Degradation of Solar PV Modules—Case Study of 19-Year-Old Polycrystalline Modules in Ghana. *Technologies* **2017**, *5*, 22. [CrossRef]
10. Papageorgas, P.; Piromalis, D.; Valavanis, T.; Kambasis, S.; Iliopoulou, T.; Vokas, G. A low-cost and fast PV I-V curve tracer based on an open source platform with M2M communication capabilities for preventive monitoring. *Energy Procedia* **2015**, *74*, 423–438. [CrossRef]
11. Zhu, Y.; Xiao, W. A comprehensive review of topologies for photovoltaic I–V curve tracer. *Sol. Energy* **2020**, *196*, 346–357. [CrossRef]
12. Ortega, E.; Aranguren, G.; Jimeno, J.C. New monitoring method to characterize individual modules in large photovoltaic systems. *Sol. Energy* **2019**, *193*, 906–914. [CrossRef]
13. Toledo, C.; Serrano-Lujan, L.; Abad, J.; Lampitelli, A.; Urbina, A. Measurement of Thermal and Electrical Parameters in Photovoltaic Systems for Predictive and Cross-Correlated Monitorization. *Energies* **2019**, *12*, 668. [CrossRef]
14. Sarikh, S.; Raoufi, M.; Bennouna, A.; Ikken, B. Characteristic curve diagnosis based on fuzzy classification for a reliable photovoltaic fault monitoring. *Sustain. Energy Technol. Assess.* **2021**, *43*, 100958. [CrossRef]
15. Morales-Aragonés, J.I.; Gallardo-Saavedra, S.; Alonso-Gómez, V.; Sánchez-Pacheco, F.J.; González, M.A.; Martínez, O.; Muñoz-García, M.A.; Alonso-García, M.d.C.; Hernández-Callejo, L. Low-Cost Electronics for Online I-V Tracing at Photovoltaic Module Level: Development of Two Strategies and Comparison between Them. *Electronics* **2021**, *10*, 671. [CrossRef]
16. Mellit, A.; Kalogirou, S. Artificial intelligence and internet of things to improve efficacy of diagnosis and remote sensing of solar photovoltaic systems: Challenges, recommendations and future directions. *Renew. Sustain. Energy Rev.* **2021**, *143*, 110889. [CrossRef]
17. Anani, N.; Ibrahim, H. Adjusting the Single-Diode Model Parameters of a Photovoltaic Module with Irradiance and Temperature. *Energies* **2020**, *13*, 3226. [CrossRef]

18. Web of Commercial Curve Tracer of Manufacturer Atecorp. Available online: https://www.atecorp.com/products/tritec/tri-ka-iv (accessed on 15 September 2021).
19. Web of Commercial Curve Tracer of Manufacturer LabX. Available online: https://www.labx.com/item/daystar-energy-engineering-ds-100c-photovoltaic-iv/LV41115631 (accessed on 16 September 2021).
20. Web of Commercial Curve Tracer of Manufacturer Ht-Instruments. Available online: https://www.ht-instruments.com/es-es/productos/instrumentacion-fotovoltaica/medidores-curva-i-v/i-v500w/ (accessed on 17 September 2021).
21. Montes-Romero, J.; Piliougine, M.; Muñoz, J.V.; Fernández, E.F.; De la Casa, J. Photovoltaic Device Performance Evaluation Using an Open-Hardware System and Standard Calibrated Laboratory Instruments. *Energies* **2017**, *10*, 1869. [CrossRef]
22. Chen, Z.; Lin, W.; Wu, L.; Long, C.; Lin, P.; Cheng, S. A capacitor based fast I-V characteristics tester for photovoltaic arrays. *Energy Procedia* **2017**, *145*, 381–387. [CrossRef]
23. Chen, Z.; Lin, Y.; Wu, L.; Cheng, S.; Lin, P. Development of a capacitor charging based quick I-V curve tracer with automatic parameter extraction for photovoltaic arrays. *Energy Convers. Manag.* **2020**, *226*, 113521. [CrossRef]
24. Cáceres, M.; Firman, A.; Montes-Romero, J.; González Mayans, A.R.; Vera, L.H.; Fernández, F.E.; de la Casa Higueras, J. Low-Cost I–V Tracer for PV Modules under Real Operating Conditions. *Energies* **2020**, *13*, 4320. [CrossRef]
25. González, I.; Calderón, A.J. Integration of open source hardware Arduino platform in automation systems applied to Smart Grids/Micro-Grids. *Sustain. Energy Technol. Assess.* **2019**, *36*, 100557. [CrossRef]
26. Fuentes, M.; Vivar, M.; Burgos, J.M.; Aguilera, J.; Vacas, J.A. Design of an accurate, low-cost autonomous data logger for PV system monitoring using Arduino™ that complies with IEC standards. *Sol. Energy Mater. Sol. Cells* **2014**, *130*, 529–543. [CrossRef]
27. Vargas-Salgado, C.; Aguila-Leon, J.; Chiñas-Palacios, C.; Hurtado-Pérez, E. Low-cost web-based Supervisory Control and Data Acquisition system for a microgrid testbed: A case study in design and implementation for academic and research applications. *Heliyon* **2019**, *5*, e02474. [CrossRef] [PubMed]
28. González, I.; Calderón, A.J.; Portalo, J.M. Innovative Multi-Layered Architecture for Heterogeneous Automation and Monitoring Systems: Application Case of a Photovoltaic Smart Microgrid. *Sustainability* **2021**, *13*, 2234. [CrossRef]
29. Paredes-Parra, J.M.; Mateo-Aroca, A.; Silvente-Niñirola, G.; Bueso, M.C.; Molina-García, Á. PV module monitoring system based on low-cost solutions: Wireless raspberry application and assessment. *Energies* **2018**, *11*, 3051. [CrossRef]
30. Pereira, R.I.S.; Dupont, I.M.; Carvalho, P.C.M.; Jucá, S.C.S. IoT embedded linux system based on Raspberry Pi applied to real-time cloud monitoring of a decentralized photovoltaic plant. *Measurement* **2018**, *114*, 286–297. [CrossRef]
31. IV Swinger Documentation. Available online: https://github.com/csatt/IV_Swinger (accessed on 18 September 2021).
32. Willoughby, A.A.; Osinowo, M.O. Development of an electronic load I-V curve tracer to investigate the impact of Harmattan aerosol loading on PV module pern2tkformance in southwest Nigeria. *Sol. Energy* **2018**, *166*, 171–180. [CrossRef]
33. Amiry, H.; Benhmida, M.; Bendaoud, R.; Hajjaj, C.; Bounouar, S.; Yadir, S.; Rais, K.; Sidki, M. Design and implementation of a photovoltaic I-V curve tracer: Solar modules characterization under real operating conditions. *Energy Convers. Manag.* **2018**, *169*, 206–216. [CrossRef]
34. Pachauri, R.K.; Mahela, O.P.; Khan, B.; Kumar, A.; Agarwal, S.; Alhelou, H.H.; Bai, J. Development of arduino assisted data acquisition system for solar photovoltaic array characterization under partial shading conditions. *Comput. Electr. Eng.* **2021**, *92*, 107175. [CrossRef]
35. Shapsough, S.; Takrouri, M.; Dhaouadi, R.; Zualkernan, I. An IoT-based remote IV tracing system for analysis of city-wide solar power facilities. *Sustain. Cities Soc.* **2020**, *57*, 102041. [CrossRef]
36. Sarikh, S.; Raoufi, M.; Bennouna, A.; Benlarabi, A.; Ikken, B. Implementation of a plug and play I-V curve tracer dedicated to characterization and diagnosis of PV modules under real operating conditions. *Energy Convers. Manag.* **2020**, *209*, 112613. [CrossRef]
37. Sayyad, J.; Nasikkar, P.; Singh, A.P.; Ozana, S. Capacitive Load-Based Smart OTF for High Power Rated SPV Module. *Energies* **2021**, *14*, 788. [CrossRef]
38. Cotfas, P.A.; Cotfas, D.T. Comprehensive Review of Methods and Instruments for Photovoltaic–Thermoelectric Generator Hybrid System Characterization. *Energies* **2020**, *13*, 6045. [CrossRef]
39. Torres-Moreno, J.L.; Gimenez-Fernandez, A.; Perez-Garcia, M.; Rodriguez, F. Energy Management Strategy for Micro-Grids with PV-Battery Systems and Electric Vehicles. *Energies* **2018**, *11*, 522. [CrossRef]
40. Elkazaz, M.; Sumner, M.; Pholboon, S.; Davies, R.; Thomas, D. Performance Assessment of an Energy Management System for a Home Microgrid with PV Generation. *Energies* **2020**, *13*, 3436. [CrossRef]
41. Yerasimou, Y.; Kynigos, M.; Efthymiou, V.; Georghiou, G.E. Design of a Smart Nanogrid for Increasing Energy Efficiency of Buildings. *Energies* **2021**, *14*, 3683. [CrossRef]
42. Jaloudi, S. Communication Protocols of an Industrial Internet of Things Environment: A Comparative Study. *Future Internet* **2019**, *11*, 66. [CrossRef]
43. Grafana Dashboard for Monitoring System Metrics Based on Telegraf. Available online: https://grafana.com/grafana/dashboards/928 (accessed on 20 September 2021).
44. Pindado, S.; Cubas, J.; Roibás-Millán, E.; Bugallo-Siegel, F.; Sorribes-Palmer, F. Assessment of Explicit Models for Different Photovoltaic Technologies. *Energies* **2018**, *11*, 1353. [CrossRef]

45. Álvarez, J.M.; Alfonso-Corcuera, D.; Roibás-Millán, E.; Cubas, J.; Cubero-Estalrrich, J.; Gonzalez-Estrada, A.; Jado-Puente, R.; Sanabria-Pinzón, M.; Pindado, S. Analytical Modeling of Current-Voltage Photovoltaic Performance: An Easy Approach to Solar Panel Behavior. *Appl. Sci.* **2021**, *11*, 4250. [CrossRef]
46. Hosseinzadeh, N.; Al Maashri, A.; Tarhuni, N.; Elhaffar, A.; Al-Hinai, A. A Real-Time Monitoring Platform for Distributed Energy Resources in a Microgrid—Pilot Study in Oman. *Electronics* **2021**, *10*, 1803. [CrossRef]
47. Akerman, M.; Fast-Berglund, A.; Ekered, S. Interoperability for a dynamic assembly system. *Procedia CIRP* **2016**, *44*, 407–411. [CrossRef]
48. Nițulescu, I.-V.; Korodi, A. Supervisory Control and Data Acquisition Approach in Node-RED: Application and Discussions. *IoT* **2020**, *1*, 76–91. [CrossRef]
49. Pearce, J.M. Economic savings for scientific free and open source technology: A review. *HardwareX* **2020**, *8*, e00139. [CrossRef] [PubMed]

sensors

MDPI

Communication

A Customized Energy Management System for Distributed PV, Energy Storage Units, and Charging Stations on Kinmen Island of Taiwan

Hsi-Chieh Lee [1,*], Hua-Yueh Liu [2], Tsung-Chieh Lin [1] and Chih-Ying Lee [3]

[1] Department of Computer Science and Information Engineering, National Quemoy University, Kinmen County 892, Taiwan
[2] Department of Architecture, National Quemoy University, Kinmen County 892, Taiwan
[3] Department of Horticulture and Landscape Architecture, National Taiwan University, Taipei 106, Taiwan
* Correspondence: cjlee@email.nqu.edu.tw; Tel.: +886-82-313978

Abstract: Kinmen, the famous Cold War island also known as Quemoy, is a typical island with isolated power grids. It considers the promotion of renewable energy and electric charging vehicles to be two essential strategies to achieve the goal of a low-carbon island and smart grid. With this motivation in mind, the main objective of this study is to design and deploy an energy management system for hundreds of current PV sites distributed on the island, energy storage systems, and charging stations on the island. In addition, the real-time acquisition of the data for power generation, power storage, and power consumption systems will be used for future demand and response analysis. Moreover, the accumulated dataset will also be utilized for the forecast or prediction of renewable energy generated by the PV systems or power consumed by the battery units or charging stations. The results of this study are promising since a practical, robust, and workable system and database are developed and implemented with a variety of Internet of Things (IoT), data transmission technologies, and the hybrid of on-premises and cloud servers. Users of the proposed system can remotely access the visualized data through the user-friendly web-based and Line bot interfaces seamlessly.

Keywords: distributed PV; energy management system; energy storage units; charging piles; smart grid; redundancy; IoT; Home Assistant; low-carbon island; Kinmen

Citation: Lee, H.-C.; Liu, H.-Y.; Lin, T.-C.; Lee, C.-Y. A Customized Energy Management System for Distributed PV, Energy Storage Units, and Charging Stations on Kinmen Island of Taiwan. *Sensors* **2023**, *23*, 5286. https://doi.org/10.3390/s23115286

Academic Editors: Antonio Cano-Ortega and Francisco Sánchez-Sutil

Received: 13 March 2023
Revised: 29 May 2023
Accepted: 30 May 2023
Published: 2 June 2023

1. Introduction

1.1. Climate Responsibility and Energy Generation of Kinmen, Taiwan

From Taiwan's perspective, three forces firmly push the renewable energy strategy forward. First, as a member of a global society, Taiwan provided its Intended Nationally Determined Contribution (INDC) on the 17 September 2015, including targets to achieve a 50% reduction below the BUA GHG emission level by 2030 [1]. Furthermore, Taiwan has demonstrated its commitment to achieving net zero by 2050 [2] through concrete actions, including implementing the Climate Change Response Act [3] in response to the 2021 26th Session of the Conference of the Parties (COP26), the U.N. climate conference held in Glasgow. Second, from an energy source viewpoint, Taiwan's dependency on imported energy was 97.5% in 2020 [4] and even higher over the past 10 years. Looking into the composition of net power generated and purchased energy in 2021 [5], thermal energy dominated at 81.6%, as shown in Figure 1. It is apparently a risk regarding energy dependency and diversity. Third, Taiwan is still ambitious to strive for the vision of a nuclear-free homeland in 2025 with a clear energy target: 50% by natural gas, 30% by coal, and 20% by renewable energy. For the aforementioned goals, it is clear that promoting low-carbon renewable energy plays an essential role in achieving INDC and the nuclear-free vision and further balancing energy generation dependency and diversity.

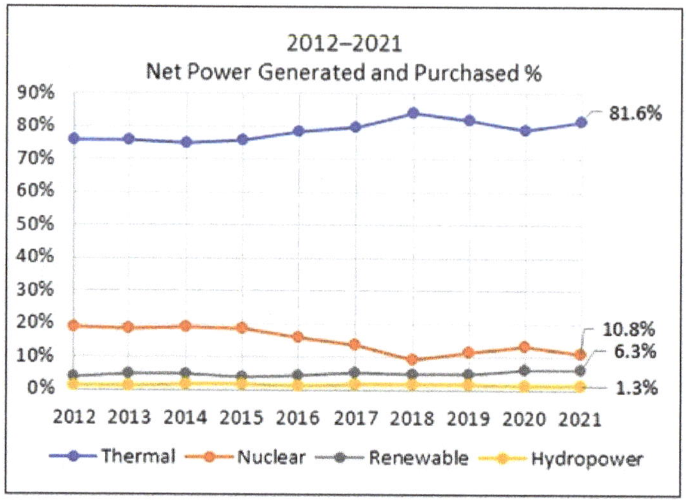

Figure 1. The net power generated and purchased by Taiwan's Taipower Company, a state-owned exclusive enterprise.

Kinmen is an outlying island of Taiwan with a 150 km^2 area, which has an isolated power grid for its electricity supply due to a distance of 248 km from western Taiwan. In Kinmen's history, it had 43 years of being front-line against Communists until the abolishment of the military administration in 1992 [6]. With the gradually improved relationship between Taiwan and China and direct transportation across the national border, more and more tourists came to the Kinmen islands, resulting in higher and higher energy demand. Since 2013, Kinmen has been selected as a demonstrated low-carbon island by Taiwan Executive Yuan, and visions and strategies were set to reach zero carbon by 2030 [7]. The installation of renewable energy stations and low-carbon transportation have become two of the main strategies to achieve these goals.

In 43 years of military administration, this front-line constructed many distributed military facilities across the whole Kinmen islands, which were gradually released or abandoned since the troops left. From the analysis of [8–11], a distributed renewable energy power grid integrated with those facilities is a suggested and suitable strategy to fulfill Kinmen's low-carbon vision.

Compared with traditional power networks, the smart grid is an advanced electricity platform that emphasizes two-way communication based on digital information technology. Key elements of an advanced smart grid include bulk electricity generation, demand response, distribution, utility companies, customers, transmission, service providers, and renewable energies [12]. Among these, sustainable analysis and management of data and information generated along with all activities is one of the most critical and valuable measures to get system efficiency and constant improvement.

In September 2021, the total installed PV capacity in Kinmen reached about 10.7 MW. While a distributed PV and energy storage system has become an essential approach for the Kinmen local government to move the low-carbon island vision forward, a reliable monitoring and data acquisition system that can constantly work for future data analysis of energy generation and efficiency under different circumstances is needed for this goal of a low-carbon island.

1.2. Remote Real-Time Monitoring and Controlling System for Distributed PV and Energy Storage Stations

Rahman et al. [13] conducted a very detailed review of different monitoring systems for PV since 1994, including RTAI, ZigBee, DAQ, SCXI, PIC, PLC, etc., in terms of their fundamental features, architecture, performance, and budget. Some of these remote systems further embrace the Internet of Things, web applications, and cloud platforms. The value of gathered data includes the sustainable status of the PV system, failure or error detection, and warning notification.

Al-Fuqaha et al. [14] reviewed the overview of the most relevant architectures and protocol standards for IoT. This study summarized the five-layer IoT model as the most functional architecture to develop an IoT system: Objects, Object abstraction, Service Management, the Application Layer, and the Business Layer. The Application layer relies on high computational machine resources. IoT functionality includes Identification, Sensing, Communication, Computation, Service, and Semantics. Challenges of Availability, Reliability, Mobility, Performance, Management, Scalability, Interoperability, and Security Privacy should be carefully considered in developing an IoT system.

In [15], a comparison of IoT sensor modules among Arduino, Raspberry Pi, PLC, and BeagleBone shows perspectives of data handling, cost and module size, and coding language. Ansari et al. concluded that Raspberry Pi is the most recommended due to its extension capability.

Plenty of research has explored this field regarding local IoT networks and remote system connections. The popular wireless communication technologies used by IoT are shown in Figure 2.

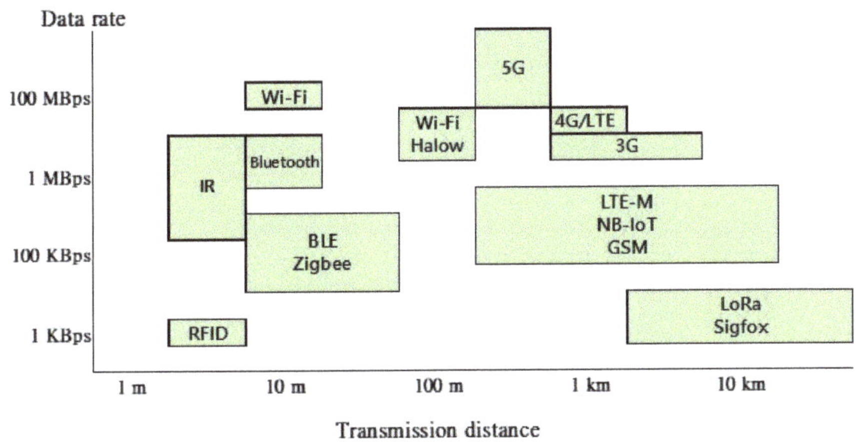

Figure 2. Comparison of different wireless technologies in terms of data rate and transmission distance.

Belghith et al. [16] designed a remote monitoring system that features star architecture of sensors, GSM communication, and a human–machine interface. Zego et al. [17] developed a wireless network to send sensed data to a local Raspberry Pi server via Zigbee. Li et al. [18] proposed a local ZigBee network and GSM connection for PV monitoring and fault diagnosis. It consists of data acquisition, data gateways, and a monitoring website based on the PHP Laravel framework. In Low-Power Wide-Area Network (LPWAN) applications, LTE-M, Sigfox, LoRa, and NB-IoT were developed. Among them, LoRa and NB-IoT are the most promising. LoRa is used in [19] for long-range and low power consumption requirements. In another implemented study [20], an Arduino-based data logger was designed to integrate 3G communication to serve stand-alone PV sites. Ascensión et al. described detailed designed data logger specifications corresponding to the IEC61724 standard. In [21], ZigBee was used as a local sensor network. After that, a 4G gateway was used to connect the local network to the internet for remote real-time monitoring. Melo et al. [22] proposed LoRa and Wi-Fi as local wireless networks. The structure comprises three key parts: data loggers, a local IoT system, and a Web application for monitoring.

Key research and comparisons are summarized in Table 1.

Table 1. Comparisons with cited research.

Title	Published Year	Data Log System	Monitoring System or SCADA	Data Transmission (LAN)	Data Transmission (to Internet)
Remote GSM module monitoring and photovoltaic system control [16]	2014	PIC18F4550 MCU	LabVIEW	Wired	GSM
A low-cost solar generation monitoring system suitable for Internet of Things [17]	2017	Raspberry Pi	Web APP	ZigBee	NA
Online monitoring system of PV array based on Internet of Things technology [18]	2017	DSP-TMS320F28335/ Raspberry Pi	Web APP	ZigBee/ Wi-Fi	Wired
An Alternative Internet-of-Things Solution Based on LoRa for PV Power Plants [19]	2019	Arduino/ Raspberry Pi	NA	LoRa	Wired
IoT Application for Real-Time Monitoring of Solar Home Systems Based on ArduinoTM With 3G Connectivity [20]	2019	Arduino UNO	ThingSpeak	Wired	3G
A Real-time Monitoring System Based on ZigBee and 4G Communications for Photovoltaic Generation [21]	2020	Cloud server	Web APP	ZigBee	4G
A Low-Cost IoT System for Real-Time Monitoring of Climatic Variables and Photovoltaic Generation for Smart Grid Application [22]	2021	Heltec Wi-Fi LoRa 32	Web APP	LoRa/Wi-Fi	Wired
Proposed system	--	Central/Cloud servers	Web APP	Wired/Wi-Fi/IR/BLE	Wired/ 4G LTE

Redundancy refers to the backup of the system to prevent service disruption due to single-point failure. Namely, redundancy is the measure to achieve a robust and reliable service system. In order to ensure system redundancy, extra replicated servers are created with the same functions, applications, and other important service components. Failover means seamlessly and automatically switching to prepared backup servers while the primary system is down. The purpose of failover is to reduce the impact when a system failure happens. To the best of the authors' knowledge, no previous studies have ever explored this mechanism in the renewable energy field or established reliable systems with this approach.

Moniruzzaman et al. [23] proposed a reliable web system supporting continuous service even if a system component fails. This high-availability system features computer cluster and loading balancing deployment via a three-tier architecture consisting of a Linux virtual server, virtualization, and shared storage.

Nguyen et al. [24] analyzed a hospital MIS system and suggested integrating different load balance and failover strategies to sustain hospital services under heavy system workloads. This edge/fog-based system design evaluated three load balance techniques: probability, random, and shortest queue-based approaches with or without failover function at different layers.

The main objective of this study is to develop and deploy a robust, reliable, workable, and suitable IoT-based PV monitoring system specific to Kinmen as a significant approach to achieving zero-carbon and smart-grid visions. This monitoring system is implemented in Kinmen with coverage of more than 40 sites, which is about half of the whole PV installed capacity in Kinmen. It is capable of collecting and archiving real-time data into on-premises and cloud database servers with IoT subsystem support that leverages Home Assistant, an open source of IoT Hub, to monitor the status and electricity usage of appliances, power generation of PV panels, and charging stations for electric motorcycles.

The main contributions and novelty of this study are as follows.

- A reliable and workable system: the relevant solar power generation facilities covered in this study started as early as 2015, including self-generation and self-use, Feed-In Tariff (FIT) wholesale sales, and grid connection. The total number of monitoring sites in Kinmen reached 50 by early 2023, proving that this system is a long-term effective practical information system and a crucial demonstration of island-level independent power grids.
- Leveraged technologies: from the data acquisition perspective, this study covers several technologies corresponding to different facility environments and data sources, including wired networks, wireless networks, TCP/IP, HTTP, crawler, IoT technology, and cloud technology. It demonstrates that IoT and cloud technology can significantly facilitate and manage large-scale renewable energy facilities.
- Established dataset: data established and collected by this research are an essential dataset for future power generation and consumption research in the Kinmen area.
- Integration of distributed PV, energy storage, and charging stations: this research includes integrating electric vehicle charging stations, solar power generation, and energy storage, which is vital as leading pre-research on demand response and smart grid research in the future.
- Redundancy and failover design with a hybrid of on-premises and cloud systems: no previous studies have ever explored this field or established reliable systems with this approach.

2. System and Methods

This study designed and deployed a set of information systems for data acquisition and monitoring, which was applied to many distributed energy storage and renewable energy sites on a medium-sized island with an independent power grid as a basis for system security, performance, maintenance, and data technology development.

2.1. System Overview and General Description

The overview of the proposed system, which aims to contribute a smart grid in Kinmen, is composed of five layers of critical functions, as illustrated in Figure 3. The first layer is distributed facility sites, including PV, battery, and charging stations, and it is the core of the green energy facilities of the whole project. The second layer is IoT, the front tier of the proposed monitoring information system, which is deployed to sense real-time data of the daily running facilities. The third layer is data acquisition, which is designed to get all real-time data back to on-premises servers. The fourth layer is the hybrid of cloud and on-premises deployment, which is capable of handling ample data information flow and designed with the perspective of redundancy and failover. The last layer is a custom SCADA system designed by this project with various friendly user interfaces.

The most valuable element of this proposed system is the data. These facilities operate daily and generate real-time big data, which could be further analyzed and transformed into periodical reports, or as critical datasets for predictions in future uses. The four-layer architecture of data processing is explained in Figure 4, namely, data sensing, data transmission, data storage and process, and data display and access. For the long-term study of the following plan, demand response analysis, this plan is deployed mainly to collect three types of data, power generation, power consumption, and power storage, depicted in Figure 5.

In addition to the overview mentioned above, Figure 6 presents low-level intra-system interactions, dataflow, network, interface, and user GUI, which is depicted based on on-premises deployment and will be explained in the following sections.

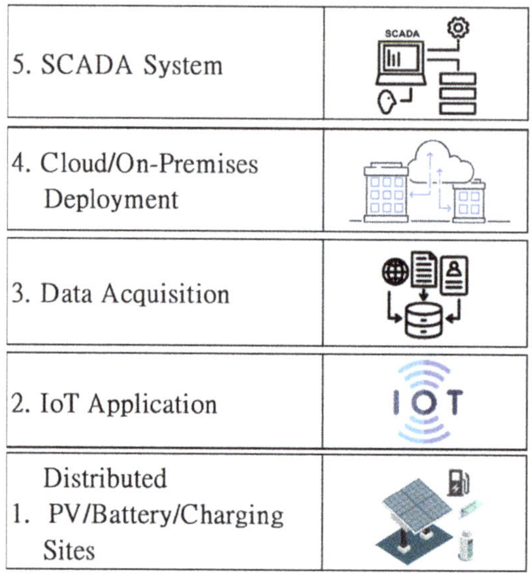

Figure 3. The high-level and five-layer architecture of the proposed system.

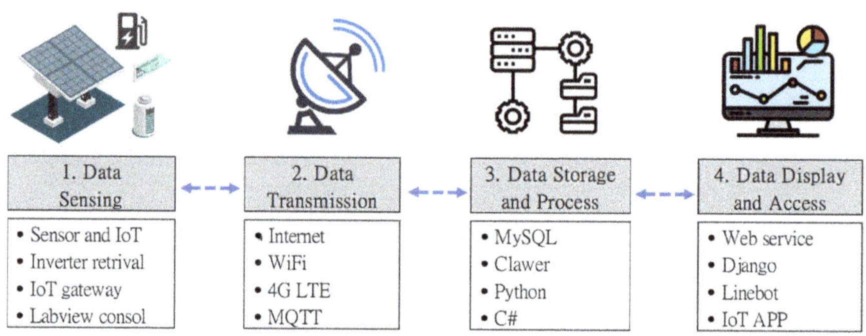

Figure 4. The four-layer architecture of the data processing flow of the proposed system.

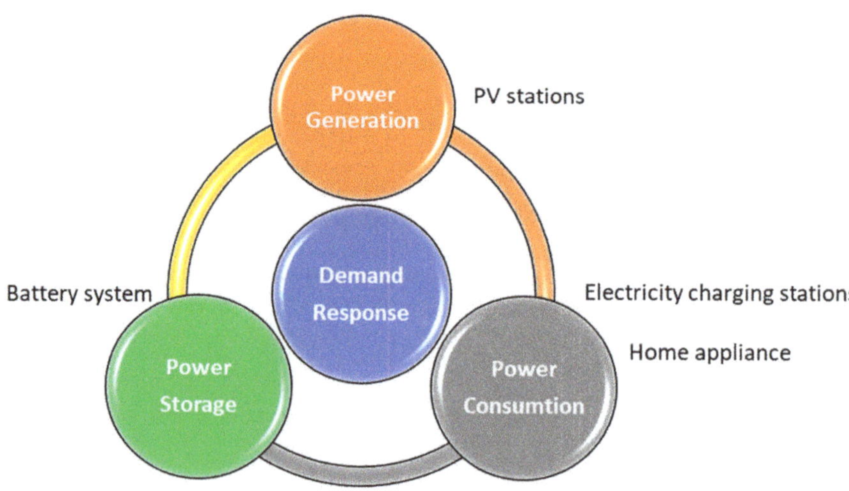

Figure 5. Purpose of dataset collection.

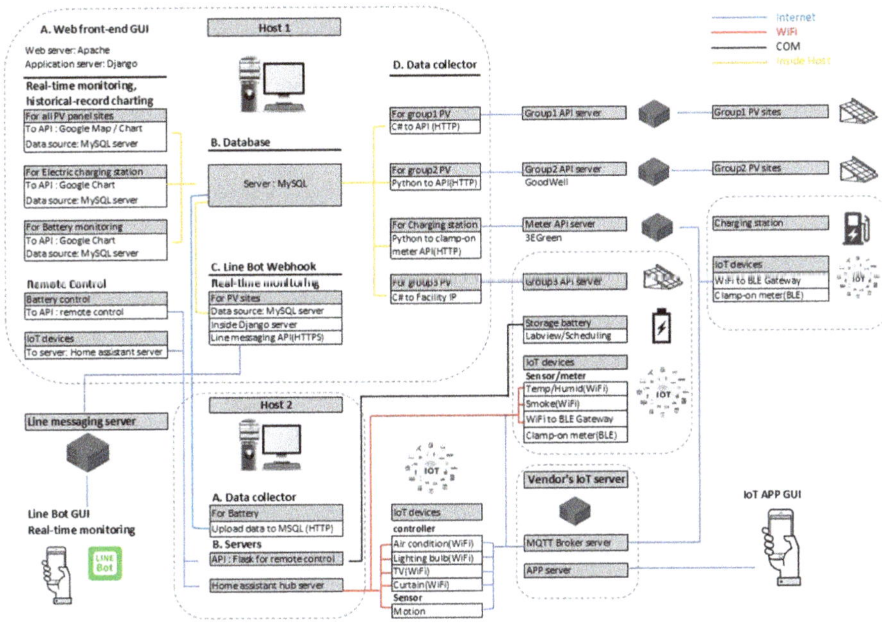

Figure 6. Low-level block diagram of on-premises deployment of the proposed system.

Redundancy and failover design are basic requirements for a sustainable and robust system. AWS cloud service is leveraged in the redundancy plan in this study. Figure 7 depicts the cooperation and backup among servers belonging to on-premises or cloud.

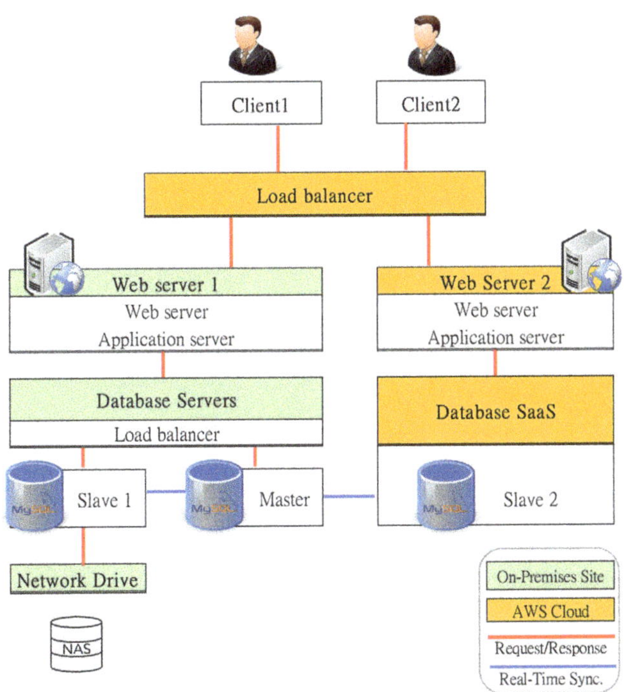

Figure 7. Cluster redundancy design based on a hybrid of cloud and on-premises deployment.

Figure 8 shows key software and hardware technologies that serve the system per the site's conditions and connection flexibility. Open-source software is leveraged as much as possible for better coding extensionality while hardware and facility are developed and deployed.

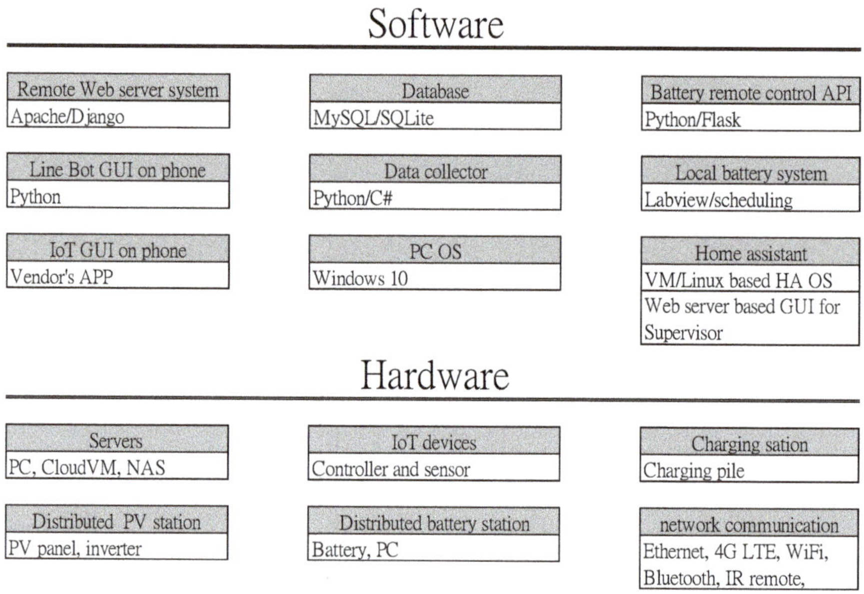

Figure 8. Key technologies of the software and hardware used.

2.2. Detailed Design

As described in Figures 3 and 4, the system operates within five-layer system architecture in which each subsystem interacts and four-layer data processing architecture in which data are generated, transmitted, stored, processed, and displayed. This section explains low-level activities and critical designs.

2.2.1. On-Premises Remote Central Monitoring and Archiving Database System

- Web server

The SCADA, a custom web application as the monitoring and controlling core, can be remotely accessed from anywhere and at any time. It is designed with Python-based Django architecture and mainly leveraged with Google Maps and Google Chart APIs for site localization and statistics visualization. Regarding remote controlling, Python-based APIs were developed for front-end requests through HTTP. This server lives in Windows OS with an Apache web server in the production phase.

- MySQL master server

The MySQL server supports the back-end data archive and retrieval. The master is installed with the web server in one host for better transmission speed. The database application GUI example is shown in Figure 9. The earliest data were established in 2015.

- Line bot GUI

The Apache server uses an SSL certificate for the HTTPS channel. Line bot web-hook lives in Django with HTTPS support. Users can actively query from a smart device or passively receive daily reports via this automatic bot publication functionality.

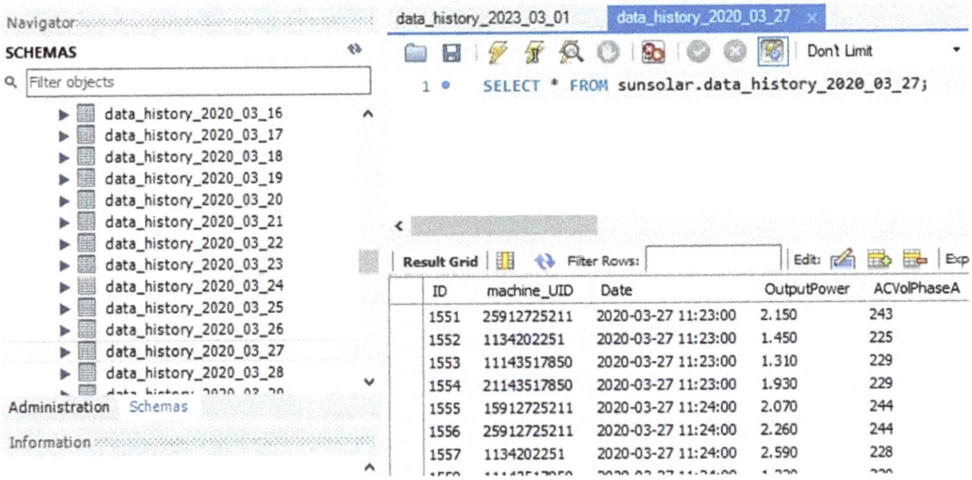

Figure 9. MySQL Workbench.

- Data Collector

In Figure 6, group 1 PV is mainly state-owned facilities. A C# API was designed as a data collector for retrieving the data servers of this group. Meanwhile, a C# TCP/IP application was designed for direct connection to group 3's PV inverters, which were designed without middle data servers. As for group 2's PV and charging station, the crawler is used to collect from a third-party's API from middle servers. These groups were built for different purposes at different times, so different data acquisition approaches are used to retrieve and observe their real-time data. Nevertheless, all data are finally archived in the same database with the same data format.

2.2.2. IoT Hub, Local Database, and IoT Network

- IoT Hub

Home Assistant (HA), a popular hub tool for most IoT devices, was introduced as the IoT Hub, a Python-based open-source platform specific to smart-home applications. Dataflow between HA and IoT devices could be direct and local via LAN or indirect via external third-party API. The former is preferred because of privacy considerations. For the GUI of HA, users can access it via a web browser or smart device APP. In a LAN case, it may need VLAN to get a HA link to a different subnet, while VPN is required in order to be through the internet. This GUI is mainly for developers or system administrator access, not for regular users.

- Local Database

SQLite is used locally to work with HA. It also works as a data logger for IoT devices and local backup for the central MySQL database in case the internet is out of the connection.

- IoT Network

The wireless IoT controllers and sensors are connected to LAN via Wi-Fi, BLE, IR, or sub-1G. In case the facility site condition is complex for mentioned wireless or wired internet access, 4G LTE is used for internet connection, such as for stations in rural areas.

2.2.3. PV, Battery, and Charging Stations

- PV station

The ongoing project continues to increase data collection of newly built PV stations in Kinmen. So far, the relevant data includes information from state-owned stations, privately owned resident stations, Taipower project stations, and Lab stations, among others. Some sites are based on FIT contracts, and some are for private use or research. The earliest sites have been running since 2015. The total installed and monitored capacity in this system is about 5 MW. More than 5 years of data from state-owned sites are incorporated. Figure 10a shows all monitored sites in the system via Google Maps, and Figure 10b shows one case with clear PV panels on the roof in satellite picture mode.

Figure 10. (**a**) All monitored PV sites in Kinmen are shown on Google Maps; (**b**) a clear PV-panel image on the roof of the monitored site.

- Battery station

Distributed battery stations were added to this project in 2021, mainly for demand response research. Until now, one site stably runs for over a year with a 10 kWh storage capacity. The key components are a Windows PC, inverters, meters, and batteries inside this facility.

- Charging stations

Kinmen has 65 state-owned free charging sites for electric motorcycles. It started monitoring the charging data from some newly built charging piles for vehicles and motorcycles in 2021. Figure 11 shows one newly built site in Kinmen National Park.

(a) (b)

Figure 11. (**a**) Field picture with charging piles, (**b**) power distribution box with an IoT clamp-meter marked in red.

2.2.4. Redundancy and Failover

- Cloud redundancy

In Figure 6, a cloud AWS VM is used for the replicated web server and load balancers. As to the Mysql database, SaaS database service is used as well. A load balancer contributes workload balancing and automatic web service failover functions.

- On-premises redundancy

The load balancer is used for database failover with a shared NAS drive. All Mysql servers get real-time synchronization by setting one master and two slaves.

2.2.5. Software and Language

- Python 3.9

Python and open-source Python-based applications were mainly used for better integration, extensions, and sustainability, such as Django, Flask, HA, HTTP API, Line bot, data collector, and battery charging scheduling application.

- C# 10

For the site group 1,2, a C# application was developed to work as the data collector. This application GUI is shown in Figure 12.

- Vendors' APP for IoT device

This is a backup alternative to web GUI and HA GUI of deployed IoT devices. However, the disadvantage is privacy concerns due to data uploaded to third-party servers.

- Labview for battery module

The battery control console was designed by Labview.

- HA

A VM of Linux-based HA OS is used inside the Windows server. GUI for HA is used via web service. It is easy to access from anywhere with the internet.

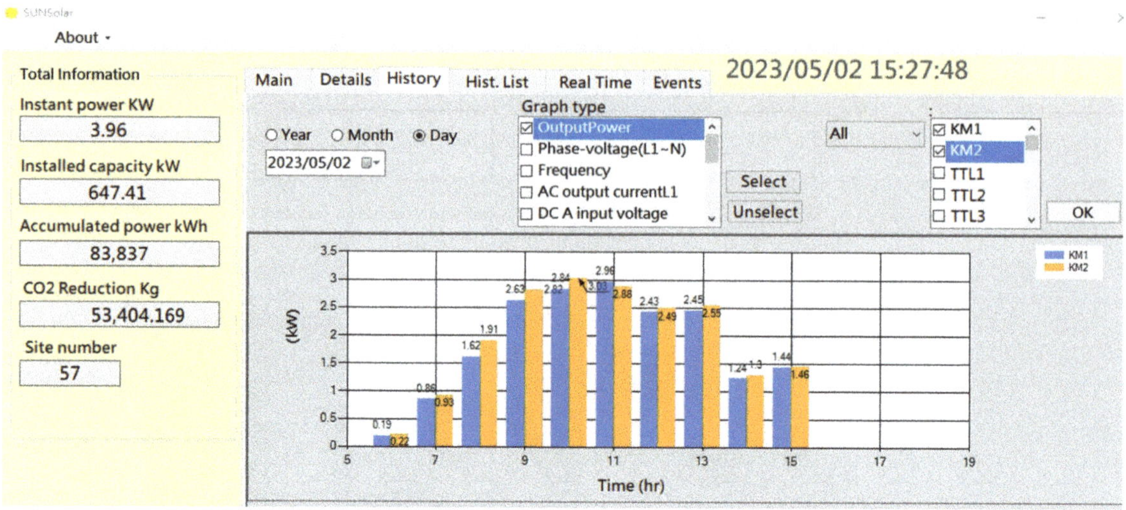

Figure 12. C# GUI for PV stations.

2.2.6. Hardware

- PC

  Regular PCs are used with Windows OS for servers.

- IoT devices and network devices

  Devices including smoke sensors, temperature and humidity sensors, motion sensors, clamp-on meters for electricity measurement, switches, curtain controllers, air conditioning controllers, and smart lighting bulbs were installed, as well as network devices including routers, Wi-Fi AP, Wi-Fi/BLE gateways, and infrared (IR) remote controllers.

- Facility Stations

  The facility station mainly includes PV panels, inverters, batteries, and charging piles.

3. Results

3.1. PV Stations

For PV site real-time monitoring and historical data review, users can accomplish this via web GUI on a desktop or line bot on smart devices, as shown in Figure 13. Users can actively or passively receive detailed daily data from the Line bot application (Figure 13b).

3.2. IoT Devices

For IoT hub monitoring and controlling, users can accomplish this via HA GUI or web GUI as in Figures 14 and 15. Due to the higher risk of battery operation, a temperature/humidity sensor and a smoke sensor were put inside the battery cabinet to monitor environmental security.

3.3. Battery Station

Battery monitoring and schedule control can be accomplished via HA GUI or web GUI as in Figure 16.

3.4. Redundancy and Failover Based on a Hybrid of On-Premises and Cloud Servers

The on-premises central hosts are located in a lab of Quemoy University. Typically, several power failures or internet disconnections happen each year. In the initial stage of the project, these events would lead to the web service going down or discontinuity of collected real-time data in database servers. Since the introduction of redundancy and failover mechanisms, the supporting servers are globally deployed on AWS with much less chance of being down in the meantime.

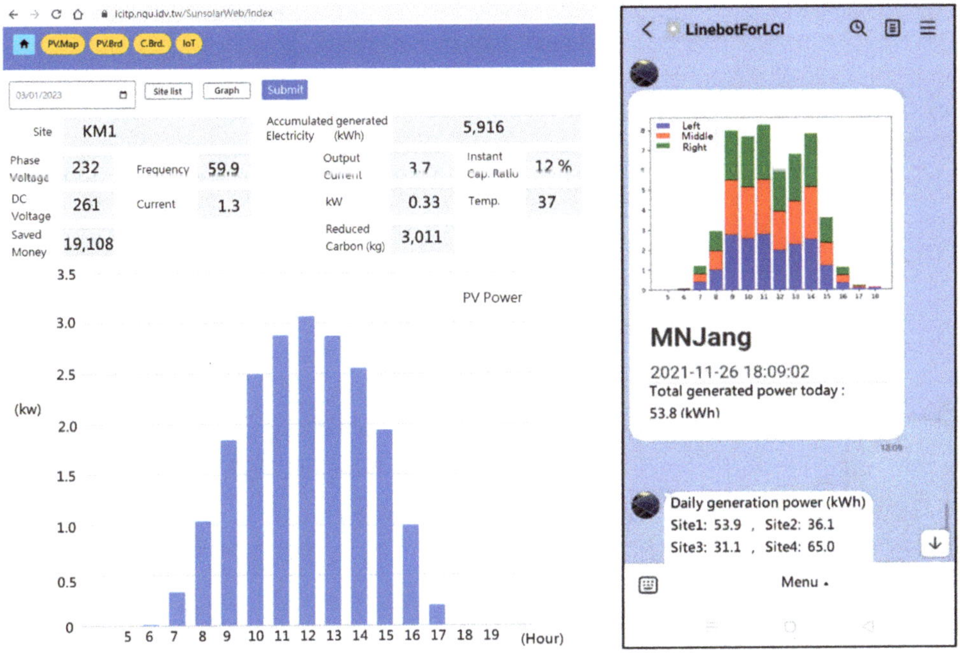

(**a**) (**b**)

Figure 13. (**a**) Desktop web GUI, (**b**) smartphone line bot GUI. Both show PV daily data.

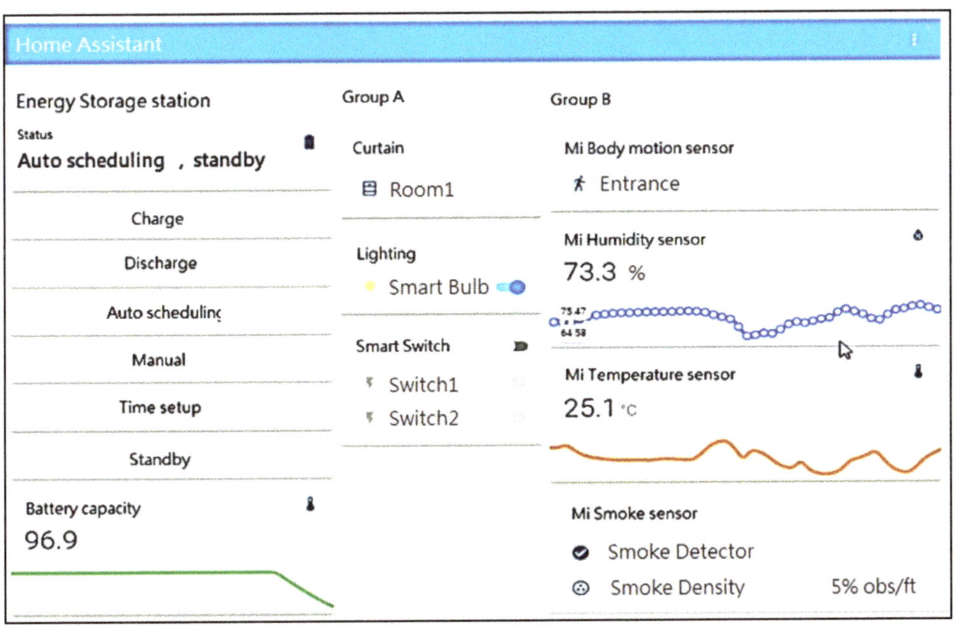

Figure 14. HA browser GUI for IoT device controlling and monitoring.

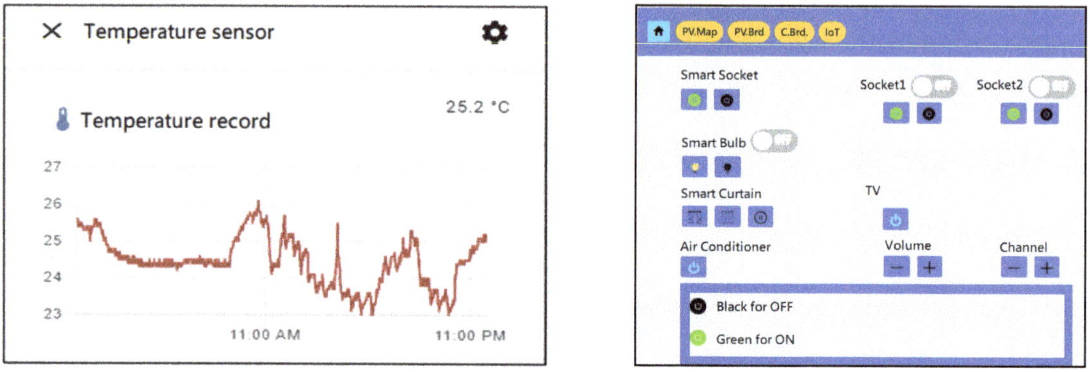

(a) (b)

Figure 15. (**a**) HA temperature daily charting of IoT sensor, (**b**) Web browser GUI for IoT device control.

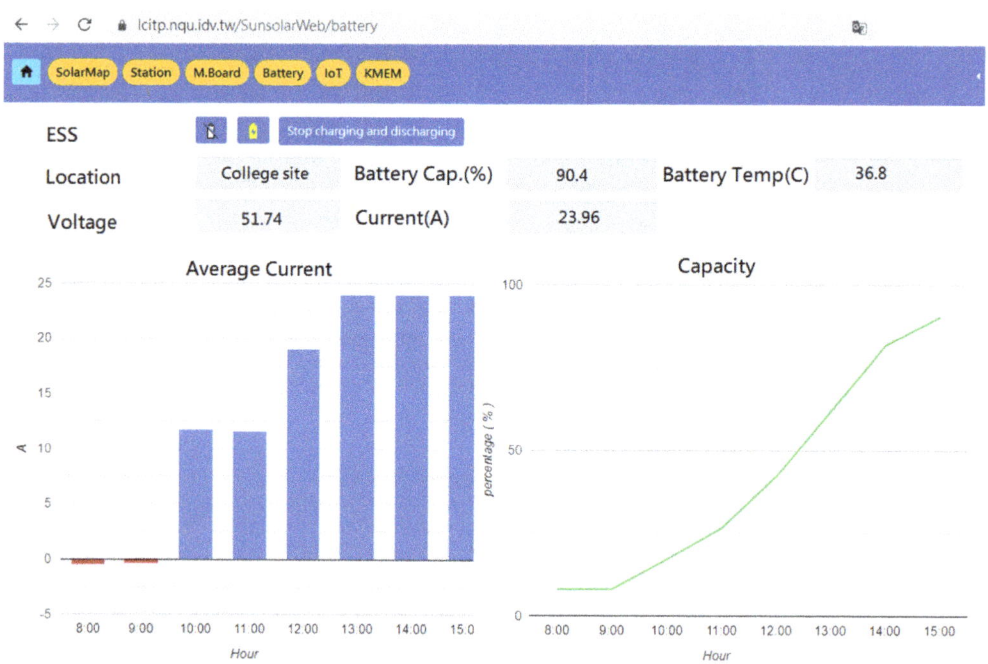

Figure 16. Web GUI for battery.

4. Discussion

SCADA mainly uses the Django web application and Home Assistant to monitor and control the facility and IoT devices. They are browser-based, so users can easily access them anywhere on any computer or smart device. It also provides Line Bot, which has "reply message" and "push message" functions as monitoring alternatives. The front-end service servers are deployed both on-premises and in the cloud as a redundancy design, and back-end database servers are deployed similarly. From the perspective of service accessibility, reliability, flexibility, and availability, this proposed system is much more comprehensive and functional than the cited research.

Facility stations in this study spread across the whole main island of Kinmen. Some of the stations downtown can be connected to the internet via wired or wireless methods, but some rural areas must use 4G LTE for wireless internet access. 4G LTE has higher quality and transmission rate than the other technologies shown in Figure 2 in case more extensive data transmission is needed, such as video surveillance.

Via scheduling setup, the energy storage system is beneficial to balance PV power generating fluctuation due to sunlight intensity and time-of-use rate mechanism. Electrical transportation is a sure trend for low-carbon policy. A good understanding of vehicle user charging behavior could contribute to stabilizing the power grid. None of the cited research has worked on integrating power generation, power storage, and power consumption.

The coverage of this work is more versatile than the cited research. Moreover, all collected data, system facilities, and approaches are beneficial for future demand response plans based on distributed virtual power plants. Good utilization of accumulated raw big data would make this system valuable and in line with the future smart grid vision.

Current limitations and future work:

- This study selected available methods for the IoT transmission approach but only explored some relevant technologies. It is believed there is room for optimization.
- Due to budget limitations and the availability of data sources, the quantity of energy storage facilities and wind energy generation stations is insufficient, and the monitoring data for power consumption is insufficient to produce an informative dataset for demand response analysis.
- Data value is based on good extraction and transformation. Although the current system can collect raw data and visualize it well, it needs to upgrade system capability further and integrate artificial intelligence models to make meaningfully data-driven predictions and optimize future demand response design to the automatic level of the machine-to-machine (M2M) by machine learning.

5. Conclusions

Kinmen is a resource-limited island with good solar and wind energy potential. The low-carbon trend is a must-do item to fulfill responsibility as a world member. Smart grid and low-carbon requirements could simultaneously move forward well with the support of a well-designed information system. Technically, only a system that can dynamically adjust demand response balance could make a smart grid possible.

In this work, a comprehensive monitoring and data collection system is well developed and deployed with versatile technologies corresponding to different environments and service requirements. With the redundancy deployment on a hybrid of on-premises and cloud systems, this robust, reliable, workable, and suitable IoT-based PV monitoring system specific to Kinmen is a practical approach to achieving zero-carbon and smart grid visions. Users can remotely access visualized data through the developed user-friendly web browser and Line bot. This implemented system collected and archived real-time data in terms of power generation, power storage, and power consumption since 2015 with IoT subsystem support to monitor site status and electricity usage of each site. The established dataset is essential for future power generation and consumption research in the Kinmen area.

The proposed system is on the way to integrating dataflow of distributed energy generation and storage, charging stations, and home electricity usage via IoT to make Kinmen a benchmark city with the smart grid.

Author Contributions: Conceptualization, H.-C.L.; Software, T.-C.L. and C.-Y.L.; Formal analysis, H.-C.L.; Investigation, H.-Y.L.; Resources, H.-Y.L.; Writing—original draft, H.-C.L. and T.-C.L.; Writing—review & editing, H.-C.L. and C.-Y.L.; Visualization, C.-Y.L.; Supervision, H.-C.L.; Project administration, H.-C.L. All authors have read and agreed to the published version of the manuscript.

Funding: This research received no external funding.

Institutional Review Board Statement: Not applicable.

Informed Consent Statement: Not applicable.

Data Availability Statement: The data presented in this study are available on request from the corresponding author.

Acknowledgments: The authors sincerely acknowledge the significant support of HA system setup and C#, Python, and Web programs by Ming-Chih Liao, Tzu-Hao Lin, and Chung-En Chang from the Department of Computer Science and Information Engineering, National Quemoy University.

Conflicts of Interest: The authors declare no conflict of interest.

Abbreviations

The following abbreviations are used in this paper:

AP	Access Point
API	Application Programming Interface
APP	Application
AWS	Amazon Web Services
BLE	Bluetooth Low Energy
DAQ	Data Acquisition
FIT	Feed-In Tariff
GUI	Graphical User Interface
HA	Home Assistance
HTTP	Hyper Text Transfer Protocol
HTTPS	Hyper Text Transfer Protocol Secure
IoT	Internet of Things
LoRa	Long Range
LPWAN	Low-Power Wide-Area Network
LTE	Long-Term Evolution
MCU	Microcontroller Unit
MQTT	Message Queuing Telemetry Transport
NAS	Network Attached Storage
TCP/IP	Transmission Control Protocol/Internet Protocol
PIC	Peripheral Interface Controller
PLC	Programmable Logic Controller
PV	Photovoltaic
RTAI	Real-Time Application Interface
SCADA	Supervisory Control and Data Acquisition
SQL	Standard Query Language
SaaS	Software as a Service
Wi-Fi	Wireless Fidelity

References

1. Hu, W.-C.; Lin, J.-C.; Fan, C.-T.; Lien, C.-A.; Chung, S.-M. A Booming Green Business for Taiwan's Climate Perspective. *Renew. Sustain. Energy Rev.* **2016**, *59*, 876–886. [CrossRef]
2. President Tsai Addresses COP26 Taiwan Day Event. Available online: https://english.president.gov.tw/NEWS/6186 (accessed on 30 April 2023).
3. Greenhouse Gas Reduction and Management Act Preannounced to Change Name to Climate Change Response Act. Available online: https://www.epa.gov.tw/ENG/F7AB26007B8FE8DF/74023638-a2ee-4d1f-ad5a-8941a2415ba2 (accessed on 30 April 2023).
4. Energy Statistical Annual Reports, Energy Indicators. Available online: https://www.moeaboe.gov.tw/ECW/english/content/ContentLink.aspx?menu_id=1540 (accessed on 30 April 2023).
5. Net Power Generated and Purchased in Past 10 Years. Available online: https://www.taipower.com.tw/en/news_noclassify_info.aspx?id=4212&chk=64ebb1d1-2448-47ba-94a9-a495e72b8609&mid=4440&param=pn%3d1%26mid%3d4440%26key%3d (accessed on 30 April 2023).
6. Liu, H.-Y. From Cold War Island to Low Carbon Island. *Int. J. Technol. Hum. Interact.* **2012**, *8*, 63–74. [CrossRef]
7. CEPD Approves Kinmen Low-Carbon Project. Available online: https://taiwantoday.tw/news.php?unit=10,23,45,10&post=20035 (accessed on 30 April 2023).

8. Liu, H.-Y. Regeneration of Kinmen Military Installations through the Integration of Renewable Energy Technologies and Battlefield Resources. *Renew. Energy* **2012**, *43*, 165–171. [CrossRef]
9. Liu, H.-Y. Sustainable Reuse of Military Facilities with a Carbon Inventory: Kinmen, Taiwan. *Sustainability* **2019**, *11*, 1810. [CrossRef]
10. Liu, H.-Y.; Lee, H.-C. The Carbon Inventory of the Reuse Phase's Life Cycle: The Example of the Reconstruction of a Zero-Carbon Campus on an Unused Military Camp. *Sustainability* **2022**, *14*, 1064. [CrossRef]
11. Lee, H.-C.; Liu, H.-Y.; Teng, S.-Y. Distributed Energy Strategy Using Renewable Energy Transformation in Kinmen Island: Virtual Power Plants That Take the Military Camps as the Mainstay. *Energy Strategy Rev.* **2022**, *44*, 100993. [CrossRef]
12. Shakeri, M.; Pasupuleti, J.; Amin, N.; Rokonuzzaman, M.; Low, F.W.; Yaw, C.T.; Asim, N.; Samsudin, N.A.; Tiong, S.K.; Hen, C.K.; et al. An Overview of the Building Energy Management System Considering the Demand Response Programs, Smart Strategies and Smart Grid. *Energies* **2020**, *13*, 3299. [CrossRef]
13. Rahman, M.M.; Selvaraj, J.; Rahim, N.A.; Hasanuzzaman, M. Global Modern Monitoring Systems for PV Based Power Generation: A Review. *Renew. Sustain. Energy Rev.* **2018**, *82*, 4142–4158. [CrossRef]
14. Al-Fuqaha, A.; Guizani, M.; Mohammadi, M.; Aledhari, M.; Ayyash, M. Internet of Things: A Survey on Enabling Technologies, Protocols, and Applications. *IEEE Commun. Surv. Tutor.* **2015**, *17*, 2347–2376. [CrossRef]
15. Ansari, S.; Afida, A.; Lipu, M.H.; Saad, M.H.M.; Hussain, A. Comparison of the IoT Based Modules for Solar PV Environment: A Review. In Proceedings of the Student Conference on Research and Development, Batu Pahat, Malaysia, 27–29 September 2020. [CrossRef]
16. Ben Belghith, O.; Sbita, L. Remote GSM module monitoring and Photovoltaic system control. In Proceedings of the First International Conference on Green Energy ICGE, Sfax, Tunisia, 25–27 March 2014. [CrossRef]
17. Zago, R.M.; Fruett, F. A Low-Cost Solar Generation Monitoring System Suitable for Internet of Things. In Proceedings of the 2nd International Symposium on Instrumentation Systems, Circuits and Transducers (INSCIT), Fortaleza, Brazil, 28 August–1 September 2017. [CrossRef]
18. Li, Y.; Lin, P.S.D.; Zhou, H.; Chen, Z.; Wu, L.H.; Cheng, S.; Fengping, S. On-Line Monitoring System of PV Array Based on Internet of Things Technology. In Proceedings of the IOP Conference Series: Earth and Environmental Science, International Conference on New Energy and Future Energy System (NEFES 2017), Kunming, China, 22–25 September 2017; Volume 93, p. 012078. [CrossRef]
19. Paredes-Parra, J.M.; García-Sánchez, A.J.; Mateo-Aroca, A.; Molina-Garcia, A. An Alternative Internet-of-Things Solution Based on LoRa for PV Power Plants: Data Monitoring and Management. *Energies* **2019**, *12*, 881. [CrossRef]
20. López-Vargas, A.; Fuentes, M.; Vivar, M. IoT Application for Real-Time Monitoring of Solar Home Systems Based on ArduinoTM with 3G Connectivity. *IEEE Sens. J.* **2019**, *19*, 679–691. [CrossRef]
21. Xia, K.; Ni, J.; Ye, Y.; Xu, P.; Wang, Y. A Real-Time Monitoring System Based on ZigBee and 4G Communications for Photovoltaic Generation. *CSEE J. Power Energy Syst.* **2020**, *6*, 52–63. [CrossRef]
22. Melo, G.C.G.d.; Torres, I.C.; Araújo, Í.B.Q.d.; Brito, D.B.; Barboza, E.d.A. A Low-Cost IoT System for Real-Time Monitoring of Climatic Variables and Photovoltaic Generation for Smart Grid Application. *Sensors* **2021**, *21*, 3293. [CrossRef] [PubMed]
23. Moniruzzaman, A.B.M.; Waliullah, M.; Rahman, M.S. A High Availability Clusters Model Combined with Load Balancing and Shared Storage Technologies for Web Servers. *Int. J. Grid Distrib. Comput.* **2015**, *8*, 109–120. [CrossRef]
24. Guyen, T.V.; Fe, I.; Brito, C.; Kaliappan, V.K.; Choi, E.; Min, D.; Lee, J.S.; Silva, F. Performability Evaluation of Load Balancing and Fail-over Strategies for Medical Information Systems with Edge/Fog Computing Using Stochastic Reward Nets. *Sensors* **2021**, *21*, 6253. [CrossRef]

sensors

MDPI

Article

High-Performance Breaking and Intelligent of Miniature Circuit Breakers

Jianning Yin [1,2,*], Xiaojian Lang [1], Haotian Xu [1] and Jiandong Duan [1]

1 School of Electrical Engineering, Xi'an University of Technology, Xi'an 710048, China
2 Zhejiang Tengen Electric Co., Ltd., Yueqing 325604, China
* Correspondence: yinjianning@xaut.edu.cn

Abstract: The exploitation and utilization of clean energy such as wind and photovoltaic power plays an important role in the reduction in carbon emissions to achieve the goal of "emission peak and carbon neutral", but such a quantity of clean energy accessing the electric system will foster the transition of the electric power system structure. The intelligentization of power equipment will be an inevitable trend of development. High breaking performance, remote control and a digital detection platform of miniature circuit breaker, a protective equipment of a power distribution system, have also been inevitable requirements of the power Iot system. Based on the above, this paper studies three aspects: high-performance AC and DC general switching technology, remote control technology and operation status' digital monitoring. A new DC non-polar breaking technology is proposed, which improves the short circuit breaking ability. An experimental prototype using the above techniques was fabricated and passed the DC 1000 V/10 kA short-circuit breaking test. On the basis of the above, an intelligent circuit breaker is developed, which contains multiple functions: remote switching, real-time temperature detection, energy metering and fault warning. Moreover, a software for digital condition monitoring and remote control is developed. This work has certain theoretical and practical significance for the development of the power Internet of things.

Keywords: DC interrupting; digitization; remote control; electric energy measurement; miniature circuit breaker

Citation: Yin, J.; Lang, X.; Xu, H.; Duan, J. High-Performance Breaking and Intelligent of Miniature Circuit Breakers. *Sensors* **2022**, *22*, 5990. https://doi.org/10.3390/s22165990

Academic Editors: Antonio Cano-Ortega and Francisco Sánchez-Sutil

Received: 11 July 2022
Accepted: 9 August 2022
Published: 11 August 2022

1. Introduction

Today, the regulation of the global pollution has been urgent, especially massive quantities of carbon dioxide emissions which lead to a more serious greenhouse effect and higher sea level that are threatening the global environment. Therefore, it has been a Global Common Mission to reduce the carbon emissions. China has also put forward the long-term vision of "emission peak and carbon neutral", and the raising of the goal will necessitate the exploitation and utilization of clean energy. Such substantial quantities of distributed clean energy accessing the electric system will foster the transition of the power system framework and equipment. Aiming to adequately absorb clean energy, it has been a trend of electric system reform to construct a new power system taking clean energy as the main body, and the promised power system will be more intelligent, shared and controllable [1,2]. At the same time, smart grid and distributed power technology will also boom. The digitalization and intellectualization of the core power equipment among the new power system will hoist visualization, regulation capacity and promote consumption utilization level of new energy generation, so as to speed up the transformation from power grid to energy internet; electrical equipment's intellectualization and digitalization endow its status with visualization, so as to support a more flexible and moderate energy rationing platform [3,4]. Digitalization of power distribution allows facility managers and maintainers to efficiently solve problems with less energy, reducing operating and maintenance expenses [5].

Acting as the core equipment of the power distribution system's terminal protection and regulation, MCB (miniature circuit breaker)'s intellectual trend promotes the achievement of power distribution digitalization [6]. At the same time, the exploitation and utilization of new energy generation, especially the boom of photovoltaic power, raises higher requirements to the DC current breaking capacity of MCB. With the development of AC–DC hybrid distribution network, high voltage (1000 V) and AC/DC universal MCB faces a great challenges.

The research on the breaking ability and the invention of new products concerning MCB have been going on for many years. A large number of scholars have conducted thorough research on its arc extinguishing ability [7,8]. Most of the research focuses on arc characteristics. There are two main schemes to study the arc characteristics of the circuit breaker: simulation and experiment [9–14]. For simulation, magneto-hydrodynamics (MHD) has become an effective auxiliary mean to study characteristics of arc's motion and vanishing [15–18]; for experiment, the arc's diagnostic means mainly include optical fiber testing, high-speed photography and laser light filling. The dynamic characteristics and breaking performance of arc via recording the arc motion form are studied [19–21]. Most of the DC MCB products on the market are below 1000 V, and the breaking capacity is insufficient (6 kA) and unstable. Regarding remote control and digitalization, as the technology of the Internet of things experiences continuous development in recent years, remote monitoring technology based on WIFI/4G is proposed [22]. Some industry conducted preliminary studies on the intelligent circuit breaker and developed some of those products. However, there are few circuit breakers that can realize electric energy measurement and real-time online monitoring of operating status [23,24]. Therefore, it is of great theoretical and practical significance to study the remote control, digitalization and AC/DC general high-performance breaking technology. Based on this, the DC non-polar breaking technology of MCB is studied firstly, and an arc extinguishing strategy of coordinated control of magnetic blowing and air blowing is proposed. An experimental prototype using the above techniques was fabricated and passed the DC 1000 V/10 kA short-circuit breaking test. Secondly, the intelligent technology of MCB is studied. The integrating electric operation mechanism to realize remote control, visualization system, digital monitoring platform and mobile application (APP) based on a cloud platform are developed. The remote control, temperature monitoring, power management, automatic alarm and real-time monitoring of MCB are realized. The safe and efficient operation of load and terminal network is enhanced. The above research contributes to the development of a new power system via providing theoretical and technical reference for the improvement and optimization of system protection appliances after the high proportion of new energy access.

This paper is organized as follows: Section 2 presents high-performance (1000 V/10 kA) DC/AC breaking technology, and the principle and experimental results are introduced. Section 3 presents intelligent technologies of MCB, and the prototype realization principle and product performance are introduced. Section 4 is the conclusion.

2. High-Performance DC/AC Breaking Technology

For the breaking characteristics of a micro-circuit breaker, whether the circuit breaker can be successfully broken is directly determined by whether the air arc can be extinguished smoothly and quickly. Because of the inexistence of natural zero point, a DC arc is more difficult to put out compared with that of AC, and its breaking ability cannot be improved via increasing the distances and quantities of open gate pieces or any other conventional technologies due to the limitation of small volume of MCB; furthermore, single arc extinguishing measures have been unable to meet the higher voltage (1000 V) of the DC circuit breaker open performance requirements. At the same time, the requirements of power system distribution equipment hope the DC circuit breaker can realize the non-polarity breaking, so the structure design of its arc extinguishing chamber is facing tougher challenges.

In order to enhance the energy dissipation of the arc, this paper puts forward an arc extinguishing strategy coordinated by air blowing and magnetic blowing, so as to realize the purpose of breaking large DC current with a small volume by increasing the effect of air blowing and magnetic blowing in the arc extinguishing chamber.

2.1. Theoretical Analysis and Practical Scheme of DC Non-Polar Arc Extinguishing

The traditional arc extinguishing strategy has been difficult to match the high voltage DC interruption. Therefore, an arc extinguishing scheme with coordinated control of magnetic blowing and air blowing is proposed. Specifically, the permanent magnets and gas-producing materials are added to the arc extinguishing chamber: on the one hand, permanent magnet is used to enhance the magnetic blowing effect; on the other hand, gas generation material is used to enhance the air blowing effect. The overall layout scheme of the arc extinguishing chamber is shown in Figure 1. Permanent magnets are placed on both sides of the contact and arc running area, respectively, and the gas-producing material is wrapped on the outside of the permanent magnet. On the one hand, the air blowing can be enhanced, and on the other hand, the permanent magnet can be prevented from losing magnetism due to direct contact with the high-temperature arc. Due to the polarity of permanent magnets, it is a necessity to arrange the position of permanent magnets reasonably in order to realize the non-polar breaking of DC arc. The schematic diagram of non-polar permanent magnet layout scheme is shown in Figure 2. Figure 2 shows the schematic diagram of the arrangement of two permanent magnets' S poles opposite each other. In order to realize the DC non-polar breaking, the permanent magnets are arranged on both sides of the arc extinguishing chamber wall, respectively, with the same magnetic poles being opposite with each other.

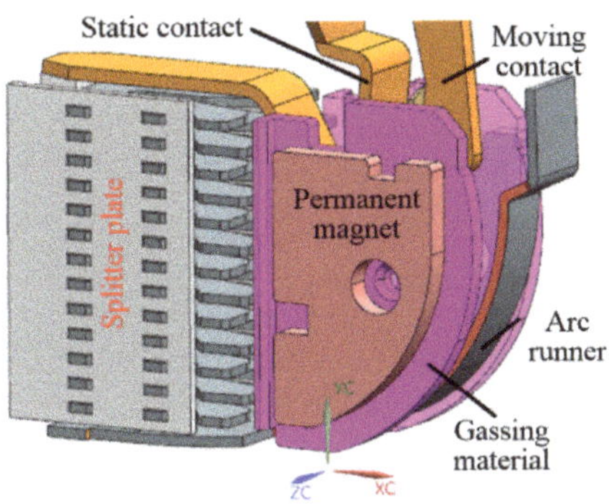

Figure 1. Layout scheme of arc extinguishing chamber.

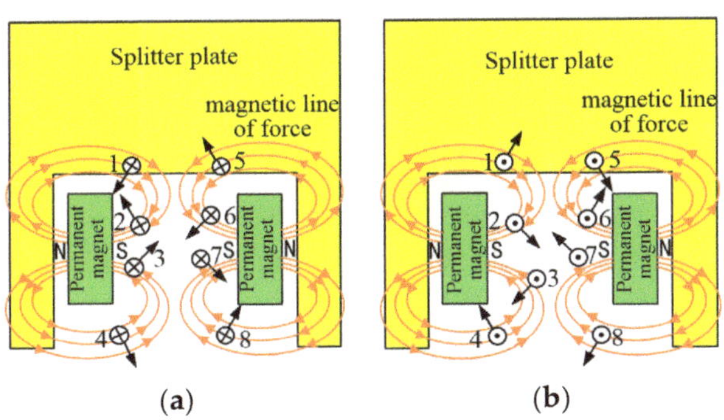

Figure 2. Arrangement schematic diagram of permanent magnet. (**a**) The current goes into the paper. (**b**) The current goes out of the paper.

The extinguishing of the DC arc mainly depends on the current limiting of the arc voltage, which forces the current to cross zero to extinguish. Therefore, increasing the arc voltage is the fundamental measure for arc extinguishing. Increasing the arc voltage mainly depends on the splitter plate cutting the arc, forming multiple near-pole voltage drops. Therefore, the main measures in the arc extinguishing design are to make the arc enter the splitter plate area quickly. For DC non-polar breaking, that is, after the current direction is changed, it does not affect the smooth entry of the arc into the arc extinguishing chamber.

Figure 2a shows the direction of the Loren magnetic force in the arc columns at eight different locations when the current direction is straight into the page. It can be seen from the figure that when arc column is located in the point 2 or 5, the Loren magnetic force moves arc to left and up direction; when arc column is located in the point 3 or 8, the Loren magnetic force moves arc to right and up direction, which is of benefit for the arc being blown into the grid, cooled and cut so as to improve the arc voltage.

When the arc column is located at point 1, it will move to the point 2 under the action of Loren magnetic force, and the changed Loren magnetic force will move arc to right and up direction and push it into grid region; when the arc column is located at point 6, it will move to the point 3 and receive the force to the upper right of the grid; when the arc column is located at point 7, it will move to the point 8, and the Loren magnetic force will move arc to right and up direction and push it into grid region. When the arc column is located at point 4, on the one hand, Loren magnetic force will move the arc toward the contact area; on the other hand, the presence of gas producing materials will force arc toward the grid area, and, due to the complex interaction, the arc will eventually move toward the grid area.

From the above analysis, it can be seen that the arc can enter the grid area and be cut quickly no matter how the direction of current place. At the same time, the permanent magnet arrangement scheme shown in Figure 2 can reduce the pinch force of the magnetic field generated by the arc itself and weaken the hindering effect of the magnetic field on the arc column movement, so as to further hasten the speed of that entering the grid area, promote the rapid rise of the arc voltage and improve the DC arc breaking ability.

2.2. *DC Breaking Test and Result Analysis*

(1) Test prototype and conditions.

Based on the above theory, a MCB test prototype is made, as shown in Figure 3: The permanent magnet is arranged on both sides of the contact and arc running area and wrapped by the gas-producing material. In order to verify the breaking ability of this scheme, the test was carried out under the conditions of 1000 V DC, 10 kA short-circuit current and 5 ms time constant. The test was carried out in the standard circuit breaker test station.

Figure 3. Experimental prototype.

(2) Analysis of test results.

In order to verify the non-polarity breaking capacity of this scheme, a prototype of forward connection and reverse connection was tested in the short-circuit experiment. According to the short-circuit breaking capacity test standard of circuit breakers, an o (open)-co (close-open) standard process needs to be completed under short-circuit current. That is to say, one experiment is closed before power-on and opened directly after power-on, and the second time is closed and then opened after power-on. DC breaking waveform (including the arc current and voltage curves) is shown in Figure 4.

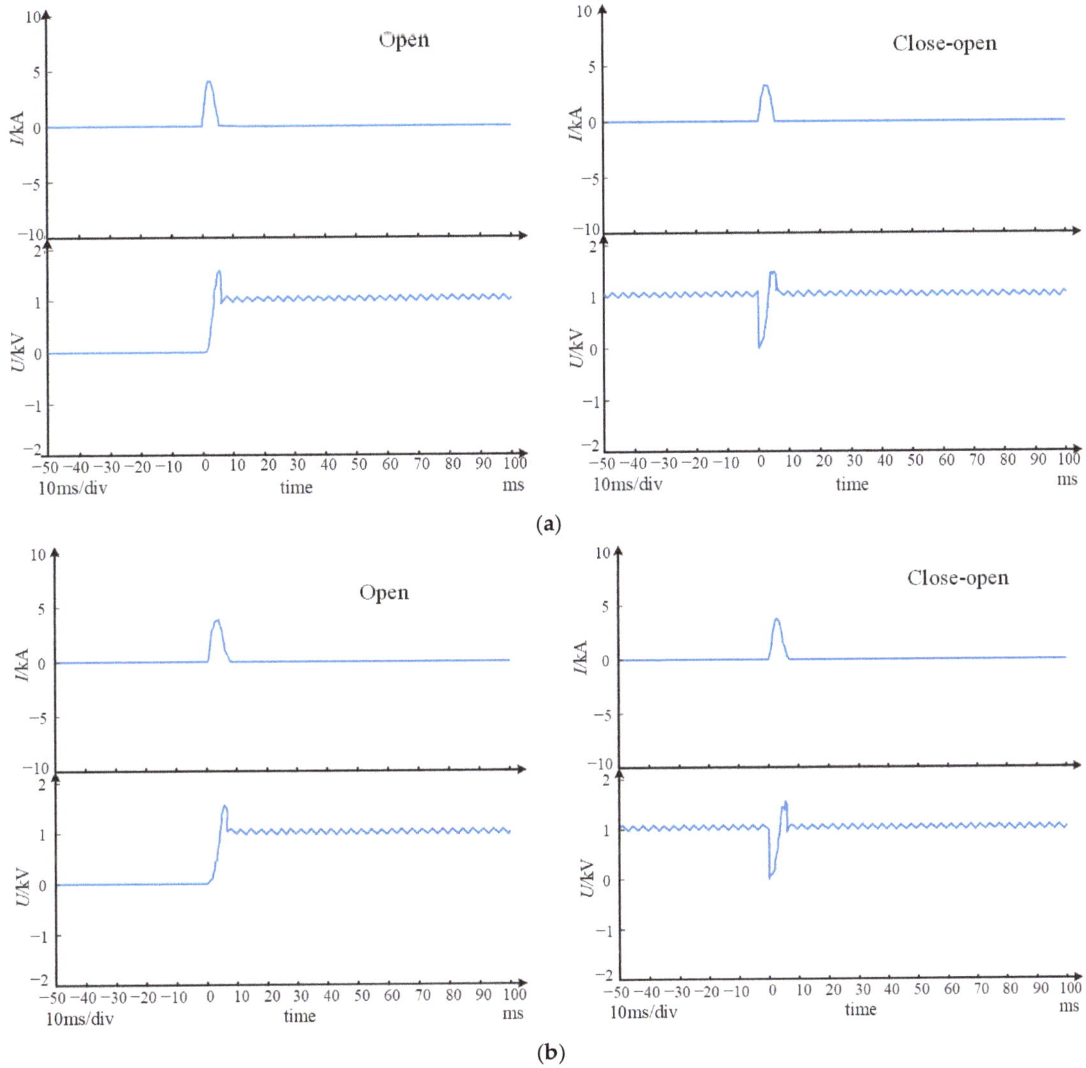

Figure 4. DC breaking test waveform (third party test). (**a**) The waveform of forward connection. (**b**) The waveform of reverse connection.

It can be seen from the arc voltage waveform that the arc voltage rises rapidly, and the highest voltage is more than 1800 V, which exceeds the system voltage 1000 V to a large extent. At the same time, the arcing time is about 5 ms. This is mainly because the arc quickly enters the splitter plate area and is cut by the splitter plate under the combined action of magnetic blowing and air blowing. At the same time the dissipation of arc energy is enhanced. Under the joint action, the arc voltage rises rapidly, and the current limiting effect is obvious. Therefore, the arc current quickly crosses zero and extinguishes, shortening the arcing time.

As it can be seen from the above test waveform, whether it is a forward or reverse connection, the circuit breaker prototype has successfully broken the short circuit current of 10 kA, and shortened the arc burning time, which fully verifies that the above permanent magnet layout scheme can achieve non-polar breaking and improve the breaking capacity compared with that of the market conventional circuit breaker (6 kA). The results show that the magnetic blowing and air blowing coordinated control strategy is effective in improving the breaking capacity.

The coordinated control strategy of arcing above does not need to change the size of the original circuit breaker and the main structure but only needs to place the permanent magnets and gas material on both sides of contact and arc running area; therefore, the scheme can not only be used for the development of high-performance DC circuit breaker but can also be applied to existing AC–DC optimal design of the miniature circuit breaker directly, in order to enhance the breaking capacity of short-circuit current.

3. Intellectualization of MCB

With the development of Internet of things technology, intelligent and digital requirements are put forward for the MCB used in the distribution system terminal. In order to realize the remote opening and closing control and online status monitoring of the MCB, the hardware and software systems are researched and developed.

3.1. Intelligent Platform Architecture

The intelligent platform architecture of MCB is shown in Figure 5. The APP is an abbreviation of "mobile phone application". The remote control is mainly based on the cloud platform, and the communication between the cloud platform and the circuit breaker is realized through the gateway. The gateway is connected with the circuit breaker module by a Type C data cable and can be configured with WIFI and data networks. At the same time, the development of a web version and a portable mobile phone APP realize the digital monitoring of the circuit breaker running state, remote opening and closing through the operation of the APP, power managing, temperature monitoring, overtemperature alarm, automatic trip and other multiple functions.

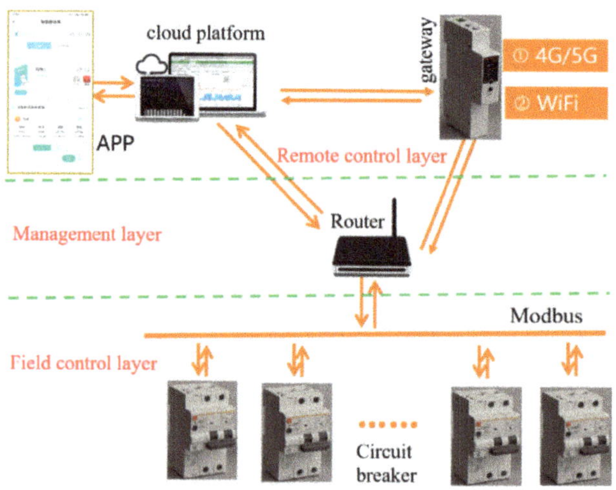

Figure 5. Intelligent overall architecture of circuit breaker.

In order to achieve remote control, it is necessary to configure the circuit breaker hardware as follows: through the operating mechanism to realize the opening and closing operation, through the voltage and current sensor to realize the data acquisition and power management and through the temperature sensor to realize the real-time monitoring of the temperature of the circuit breaker.

3.2. Intelligent and Digital Circuit Breaker System

The whole intelligent circuit breaker is shown in Figure 6, which mainly consists of three modules: power module, gateway and circuit breaker module. The three modules are connected by a Type C data cable. The input of the power module is AC 220 V, and the output is DC 12 V, whose main function is to supply power to the gateway and single chip operating mechanism of the circuit breaker module. The gateway plays the role of network communication; the Type C cable not only provides power but also implements communication for the gateway and the circuit breaker modules.

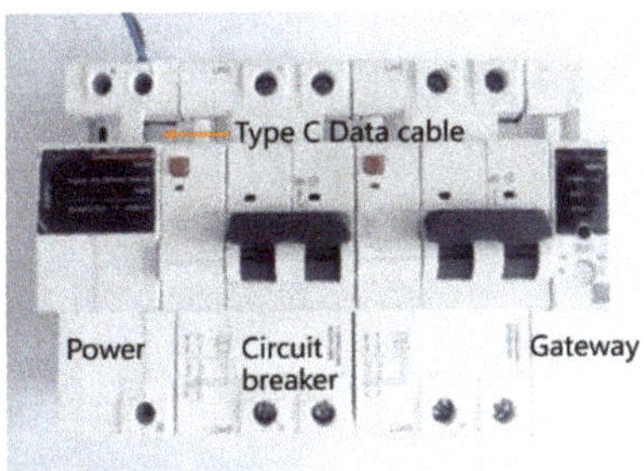

Figure 6. Composition of intelligent MCB.

Compared with the conventional circuit breaker, the new intelligent micro-circuit breaker products share the circuit breaker module plus a pole, used to install operating mechanism, control board, sensors and other devices to build a digital monitoring platform. The hardware system mainly includes a data acquisition system, central processor, actuator, display unit and circuit breaker. The hardware system diagram is shown in Figure 7.

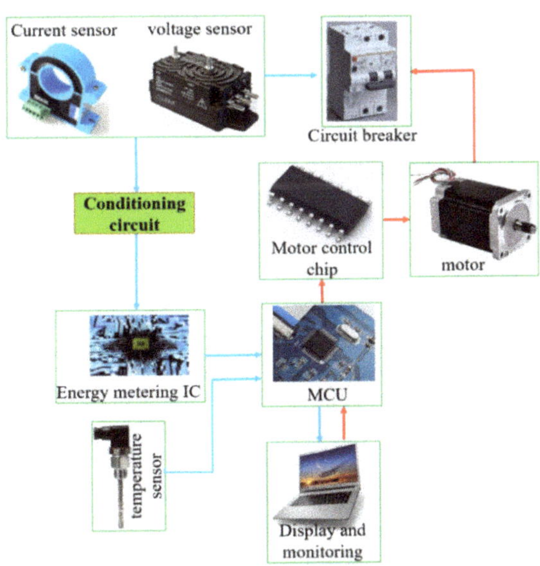

Figure 7. Diagram of hardware system.

(1) Remote control opening and closing technology.

The "conditioning circuit" mainly filters the arc voltage and current signals. The MCU is the Single Chip Microcomputer MKE02Z. The main function is to process and display the collected data and issue opening and closing instructions to the motor control chip.

In order to realize the software remote control circuit breaker opening and closing, the hardware system added a motor with a control chip. When switching remote control points, firstly choose circuit breaker which requires operation in the display interface or the phone APP interface, clicking on the corresponding button; then issue instructions to the chip which control the motor turning forward or reverse by the single chip micro-computer; then drive the gears which link the circuit breaker handle with coaxial connection, so as to realize the remote points and closing operation. The entire operating mechanism and data acquisition hardware layout are shown in Figure 8.

Figure 8. Hardware of operating mechanism and data processing and acquisition.

(2) Data acquisition and digital display interface.

In order to realize the digital monitoring of running status of the circuit breaker, the hardware system uses an NXP single-chip micro-computer as the main control chip, installed the voltage sensor, current transformer and temperature sensor to complete the data collection of voltage, current and temperature. In order to realize the electric energy metering function, the voltage and current signals are sent into the electric energy metering chip through the conditioning circuit to complete the consumption calculation, then are sent into the digital display interface through the MCU main control chip to realize the digital real-time display of voltage, current and electric energy.

Moreover, the single-chip computer stores and processes data. When the circuit breaker's real-time temperature exceeds its rated temperature, the main control chip will emit alarm instructions, on the one hand displaying alarm information on popup windows in the system, on the other hand conveying fault signal to the circuit breaker failure indicator and control its flashing, so as to realize real-time fault alarm, facilitating maintenance personnel.

Based on the above principles, the corresponding monitoring system and APP were developed. The interface of monitoring system is shown in Figure 9. The number of circuit breakers and specific information in different states (include good, alarm, fault and offline) are displayed in the interface. At the same time, the fault recognition rate, push information method, etc. are also displayed on the interface. Opening and closing commands can be issued through the circuit breaker monitoring interface and display the operating status of the circuit breaker. The temperature and power of each circuit breaker can also be viewed in real time through the interface.

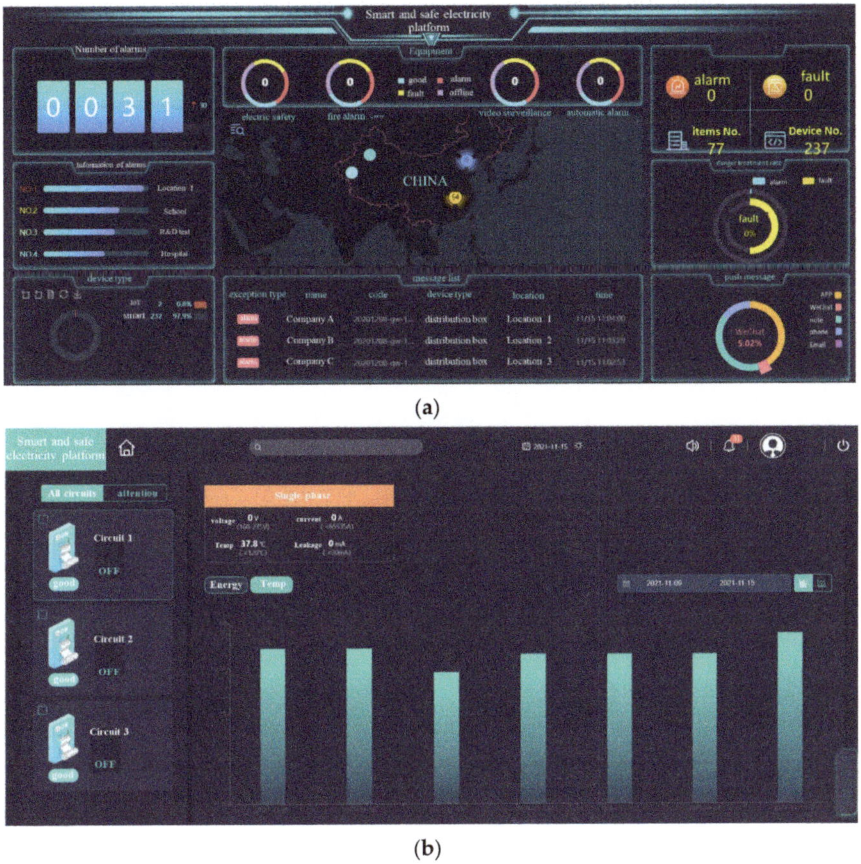

Figure 9. Circuit breaker digital monitoring system. (**a**) The main interface. (**b**) Circuit breaker monitoring interface.

Through this monitoring system, all circuit breaker layout points and operation data in the whole process can be digitally monitored. At the same time, the system can be used for a specific circuit breaker to achieve remote open and close operation, electric energy measurement, current and voltage monitoring, circuit breaker operating temperature display, real-time warning of overtemperature and so on. In the opening and closing operation, one circuit breaker can be operated alone, and multiple channels can be operated at the same time. Voice control is added in the mobile phone APP.

The digital control system above endows the power distribution system terminal protection equipment MCB with intelligent and digital monitoring and endows it with the functions of remote opening and closing, electric energy measurement, over-temperature alarm and so on, which facilitates the operation of users and improves the reliability of power equipment. At the same time, it can be seen that the current digital monitoring of the circuit breaker is still on a computer or mobile phone APP, with limited functions, requiring further research and development aiming digital control pane, so as to realize the integrated design of circuit breaker and digital monitoring platform, providing theoretical foundation for research of new generations of digital circuit breakers.

In combination with the high-performance arc open-circuit technology and intelligent technology mentioned above, the prototype was made and successfully passed through the third party test, providing technical reference for intellectualization and digitalization of power equipment in China, providing reference for the research and development of a high-performance DC intelligent circuit breaker for photovoltaic power generation, intelligent park and energy storage system.

4. Conclusions

(1) The arc extinguishing strategy coordinated by air blowing and magnetic blowing was proposed. The high-voltage DC non-polar breaking capacity of circuit breaker was improved, which increased from 6 kA to 10 kA DC.
(2) Through the reasonable arrangement of permanent magnet, this arc extinguishing scheme can realize DC non-polar breaking, AC and DC universal, while keeping the original circuit breaker structure size unchanged. The arc-extinguishing scheme can be extended to the research and development of high-performance DC molded case circuit breakers and frame circuit breakers.
(3) Based on the cloud platform, the intelligent and digital monitoring system of circuit breaker was developed, which provides a reference for the digitalization of power equipment.
(4) Through the combination of the above high-performance breaking technology and intelligent technology, the prototype was developed, confirming this technology has formed a mature product and has been promoted and transformed.

Author Contributions: Data curation, J.Y.; Formal analysis, X.L.; Funding acquisition, J.Y.; Investigation, X.L. and H.X.; Project administration, J.Y.; Software, X.L. and J.D.; Visualization, J.D.; Writing—original draft, J.Y. and H.X.; Writing—review and editing, H.X. and J.D. All authors have read and agreed to the published version of the manuscript.

Funding: This work is supported by the National Natural Science Foundation of China (NSFC) (52107167), China Postdoctoral Science Foundation (2021M692877), Basic Research Program of Natural Science in Shaanxi Province (2021JQ-473), Scientific Research Projects of Education Department of Shaanxi Provincial Government (21JK0788), Research Fund of Xi'an University and Technology (104–451119032).

Acknowledgments: This work was completed with the support of several funds. Fund information is listed as the following. Besides, many thanks to Ge Shiwei for his support of this paper.

Conflicts of Interest: The authors declare no conflict of interest.

References

1. Yuan, Z.; Guo, P.; Liu, G.; Zhao, Y.; Shi, Z. Review on control and protection for renewable energy integration through VSC-HVDC. *High Volt. Eng.* **2020**, *46*, 1460–1475.
2. Wei, X.; Gao, S.; Zang, T.; Huang, T.; Wang, T.; Li, D. Social energy internet: Concept, architecture and outlook. *Proc. CSEE* **2018**, *38*, 4969–4986.
3. Melo, G.C.G.d.; Torres, I.C.; Araújo, Í.B.Q.d.; Brito, D.B.; Barboza, E.d.A. A Low-Cost IoT System for Real-Time Monitoring of Climatic Variables and Photovoltaic Generation for Smart Grid Application. *Sensors* **2021**, *21*, 3293. [CrossRef]
4. Bedi, G.; Venayagamoorthy, G.K.; Singh, R.; Brooks, R.R.; Wang, K.C. Review of Internet of Things (IoT) in electric power and energy systems. *IEEE Internet Things J.* **2018**, *5*, 847–870. [CrossRef]
5. Khan, M.N.; Rahman, H.U.; Faisal, M.; Khan, F.; Ahmad, S. An IoT-Enabled Information System for Smart Navigation in Museums. *Sensors* **2022**, *22*, 312. [CrossRef]
6. Zhao, S.; Zhao, H.; Shou, P. Discussion on Key Technology and Operation & Maintenance of Intelligent Power Equipment. *Autom. Electr. Power Syst.* **2020**, *44*, 1–10.
7. Kim, H. Aging Characteristics of Contact Electrodes of Low Voltage DC Switches. *Energies* **2021**, *14*, 6838. [CrossRef]
8. Li, X.; Jia, S.; Zhang, B. Research and application on physical farameters calculation and behavior simulation of gas switching arc. *High Volt. Eng.* **2020**, *46*, 757–771.
9. Wang, G.; Wang, Y.; Zhang, L.; Xue, S.; Dong, E.; Zou, J. A Novel Model of Electromechanical Contactors for Predicting Dynamic Characteristics. *Energies* **2021**, *14*, 7466. [CrossRef]
10. Ma, Z.; Peng, Z.; Zhao, C.; Xu, T. Research and Development of 5kV High-speed Medium-voltage DC air circuit breaker. *High Volt. Eng.* **2020**, *46*, 312–318.
11. Li, J.; Liu, K.; Cao, Y.; Hou, C.; Liu, S. Arc root development and its influence on arc reigniting during the breaking process of the DC contactor. *Proc. CSEE* **2019**, *39*, 1241–1250.

12. Yin, J.; Wang, Q.; Li, X.; Xu, H. Numerical study of influence of frequency and eddy currents on arc motion in low-voltage circuit breaker. *IEEE Trans. Compon. Packag. Manuf. Technol.* **2018**, *8*, 1373–1380. [CrossRef]
13. Lindmayer, M.; Marzahn, E.; Mutzke, A.; Ruther, T.; Springstubbe, M. The process of arc splitting between metal plates in low voltage arc chutes. *IEEE Trans. Compon. Packag. Technol.* **2006**, *29*, 310–317. [CrossRef]
14. Dai, R. Simulation and Experimental Research on the Arc Dynamic Characteristics in Low Voltage Arc Quenching Chamber. Ph.D. Thesis, Xi'an Jiaotong University, Xi'an, China, 2009.
15. Yin, J.; Wang, Q.; Zhang, B.; Zhang, P.; Li, X. Effect of frequency on arc motion in multiple parallel contacts' system. *IEEE Trans. Plasma Sci.* **2019**, *47*, 1957–1963. [CrossRef]
16. Iturregi, A.; Barbu, B.; Torres, E.; Berger, F.; Zamora, I. Electric arc in low-voltage circuit breakers: Experiments and simulation. *IEEE Trans. Plasma Sci.* **2017**, *45*, 113–120. [CrossRef]
17. Wu, Y.; Rong, M.Z.; Yang, Q.; Hu, G.X. Simulation on dynamic characteristics of arc in low voltage circuit breaker modelling. *Proc. CSEE* **2005**, *25*, 143–148.
18. Wu, Y.; Rong, M.; Li, J.; Lou, J. Calculation of electric and magnetic fields in simplified chambers of low-voltage circuit breakers. *IEEE Trans. Magn.* **2006**, *42*, 1007–1010. [CrossRef]
19. Qu, J.; Yuan, S.; Wang, Q.; Li, X. Application of laser arc imaging technology to observe arc behavior and contact motion. *Plasma Phys. Technol.* **2015**, *2*, 183–186.
20. Li, X.; Chen, D.; Liu, H.; Chen, Y.; Li, Z. Imaging and spectrum diagnostics of air arc plasma characteristics. *IEEE Trans. Plasma Sci.* **2004**, *32*, 2243–2249. [CrossRef]
21. Smugala, D.; Bonk, M. Study of Arc Parameters of AC Relays Operating under Distorted Supply Voltage Conditions. *Energies* **2020**, *13*, 4785. [CrossRef]
22. Liu, Z.; Duan, X.Y.; Liao, M.F. A model-based measurement method for intelligent circuit breaker with data communication. *Trans. Inst. Meas. Control.* **2018**, *40*, 15–19. [CrossRef]
23. Seničar, B.; Gabrijelčič Tomc, H. User-Centred design and development of an intelligent light switch for sensor systems. *Tehnički Vjesnik* **2019**, *26*, 125–131.
24. Kim, W.; Kim, Y.-J.; Kim, H. Arc Voltage and Current Characteristics in Low-Voltage Direct Current. *Energies* **2018**, *11*, 2511. [CrossRef]

sensors

MDPI

Article

Knowledge-Based Sensors for Controlling A High-Concentration Photovoltaic Tracker

Joaquin Canada-Bago [1,*], Jose-Angel Fernandez-Prieto [1], Manuel-Angel Gadeo-Martos [1] and Pedro Perez-Higueras [2]

[1] Telematic Engineering System Research Group, CEATIC Center of Advanced Studies in Information and Communication Technologies, University of Jaén, Campus Las Lagunillas, C.P. 23071 Jaén, Spain; jan@ujaen.es (J.-A.F.-P.); gadeo@ujaen.es (M.-A.G.-M.)

[2] CEAEMA Center for Advanced Studies in Energy and Environment, Electronic and Automation Department, University of Jaén, Campus Las Lagunillas, C.P. 23071 Jaén, Spain; pjperez@ujaen.es

* Correspondence: jcbago@ujaen.es

Received: 5 February 2020; Accepted: 26 February 2020; Published: 28 February 2020

Abstract: To reduce the cost of generated electrical energy, high-concentration photovoltaic systems have been proposed to reduce the amount of semiconductor material needed by concentrating sunlight using lenses and mirrors. Due to the concentration of energy, the use of tracker or pointing systems is necessary in order to obtain the desired amount of electrical energy. However, a high degree of inaccuracy and imprecision is observed in the real installation of concentration photovoltaic systems. The main objective of this work is to design a knowledge-based controller for a high-concentration photovoltaic system (HCPV) tracker. The methodology proposed consists of using fuzzy rule-based systems (FRBS) and to implement the controller in a real system by means of Internet of Things (IoT) technologies. FRBS have demonstrated correct adaptation to problems having a high degree of inaccuracy and uncertainty, and IoT technology allows use of constrained resource devices, cloud computer architecture, and a platform to store and monitor the data obtained. As a result, two knowledge-based controllers are presented in this paper: the first based on a pointing device and the second based on the measure of the electrical current generated, which showed the best performance in the experiments carried out. New factors that increase imprecision and uncertainty in HCPV solar tracker installations are presented in the experiments carried out in the real installation.

Keywords: knowledge-based sensor; Internet of Things; high-concentration photovoltaic systems; sun tracker

1. Introduction

The European Commission has recently published the photovoltaic (PV) status report [1] in which PV market, electricity costs, and the economics of PV systems are analyzed. Within its conclusions, the following stand out: (a) the new installed capacity of solar PV power and the number and volume of PV markets are increasing; (b) a rapid decarbonization is necessary; (c) a rapid cost reduction exists in PV manufacturing; (d) different studies about subsidies for combustibles, fuels, and electricity have been presented; (e) solar energy will continue to grow at high rates; and (f) electricity from PV systems could be cheaper than residential consumer prices in a wide range of countries.

To analyze the PV system profitability, it is convenient to take into account additional factors such as subsidies and forecasting of PV power generation. According to [1], while fossil fuel subsidies could indirectly increase noxious and greenhouse gases, renewable energies and energy efficient technologies subsidies may help to reduce emissions. A new scheme of subsidies based on the price of CO_2 is presented in the literature [2]. A review of forecasting of PV generation is presented in the literature [3].

High-concentration photovoltaic systems (HCPVs) [4–7] concentrate the sunlight received between 300 and 2000 times onto photovoltaic cells by means of optical concentration devices. The main objective of these systems is to replace semiconductor materials (photovoltaic cells) with more economical optical materials (lenses and mirrors), reducing the cost of power plants and generated energy.

Although HCPV is a young technology, it has already demonstrated a great capacity for growth in recent years. In this sense, the number of companies that develop HCPV systems has grown rapidly, and the installed power has gone from a few kWs in laboratories to several megawatts.

According to [8] concentration photovoltaic (CPV) has potential for reducing the levelized cost of electricity. In this sense, if installations continue growing, CPV could reach a cost ranging between €0.045/kWh and €0.075/kWh. The system prices, including installation for CPV power plants, would then be between €700 and €1100/kWp. On the other hand, HCPV could be competitive in some locations in 2020 [9].

Due to the concentration of energy, tracker or pointing systems [10] are necessary in CPV and HCPV systems, which represents one of the differences with respect to conventional photovoltaic (PV) systems [11,12]. In these systems, power generation decreases dramatically with a sun pointing error greater than 0.5°, becoming practically zero if the error exceeds even a few degrees.

Frequently, a high degree of inaccuracy and uncertainty or imprecision are observed in HCPV tracker installations due to factors such as HCPV module manufacturing errors, module alignment errors, and the precision and accuracy of the tracker control system [13].

Fuzzy rule-based systems (FRBSs) [14] have demonstrated correct adaptation to problems having a high degree of inaccuracy and uncertainty. Based on fuzzy logic (FL) [15], these systems express knowledge by means of a set of linguistic rules grouped in a knowledge base (KB). FRBSs can be used in control systems, e.g., fuzzy logic controllers (FLCs), in which the control algorithm is expressed as a set of actuation linguistic rules.

Currently, there is a persistent trend to integrate knowledge-based systems (e.g., FRBSs) and FLCs into resource-constrained devices and into the paradigm of the Internet of Things (IoT) [16].

The IoT concept was introduced by Kevin Ashton in 1999, in which the physical world is connected to the Internet through ubiquitous sensors [17,18]. The IoT refers to the use of constrained resource devices, data acquisition, actuation, data communication with fog and cloud servers, data storage, and subsequent analysis. The range of applications of the IoT is very wide and includes environmental monitoring systems, fire detection, intelligent buildings, smart cities (traffic, lighting, parking location, garbage containers, etc.), intelligent agriculture, industrial control and monitoring, logistics, health monitoring, etc.

The main objective of this work is to design a knowledge-based controller for an HCPV tracker and to implement it in a real system by means of IoT technology (i.e., constrained resource devices, data communication, and a cloud computing server for data storage and analysis).

The remainder of this paper is organized as follows. The following section shows related work. Section 3 addresses the proposed controller and knowledge-based FRBS sensors. Section 4 presents the real HCPV tracker, the experiment that was carried out, and the results obtained. Finally, conclusions and future work are presented in Section 5.

2. Related Work and Background

To achieve the objective, this work proposes to use FRBS due to the high degree of inaccuracy and uncertainty presented in real installations of PV trackers and IoT technologies to integrate the tracker controller into a resource-constrained device and monitor the data obtained. The following sections show the related work about PV trackers, and an introduction to FRBS and IoT technologies.

2.1. PV Trackers

To point to the sun, PV installations use solar tracking systems or trackers that are composed of a metal structure that may be moved (whether on a dual or single axis) using motors, PV modules, position sensors, and a control system.

A review of different conventional PV tracker systems was presented in the literature [19], in which they are classified as active or passive control systems. The most common are active control systems, which may be differentiated into five types:

(a) Open–closed loop driver systems. The main difference of this type of controller is based on the feedback loop, which broadly uses closed loop systems that use the output variables as inputs to the control system.
(b) Sensor driver systems. These controllers are based on different devices (electronic components, sensors, or probes), such as electro-optical sensors, light-dependent resistors (LDRs) [20], light intensity sensors [21], and pointing devices.
(c) Microprocessor driver systems. The information processing capacity of the microprocessor allows the controller to execute algorithms such as calculating the sun position. The ephemeris algorithm presented in the literature [22] precisely calculates the azimuth and elevation angles of sun position using time, date, and tracker global position.
(d) Intelligent driver systems. These controllers are based on artificial intelligence technologies (using personal computers), such as neural networks and FL [23–27].
(e) Combination of sensors and microprocessors. The last type of controller is based on a hybrid system of sensors and microprocessors. Reference [11] presents a CPV hybrid controller with different strategies and an auto calibration system.

The main objective of controllers is to generate the maximum energy stabilizing the PV system in the maximum power point (MPP) by means of the maximum power point tracking (MPPT) technique [28,29] that is widely used in PV systems. A MPPT tracking based on learning is presented in [30]. To verify the proper operation of PV systems it is necessary to monitor the evolution of the significant magnitudes involved in the system [31–33].

Despite the wide interest in PV tracker controllers, little attention has been paid to tracker controllers for HCPV systems.

When HCPV systems are used, the sun pointing error has a maximum admitted value. The electrical current generation surface presents a maximum if azimuth and elevation errors are zero. If these errors are greater, the electrical current generated by the module decreases dramatically [34]. For example, the HCPV modules used in the experiments in this work require azimuth and elevation errors lower than $\pm$ 0.6°.

Although the algorithms and pointing devices are able to calculate the solar position with sufficient accuracy for HCPV systems, there is a high degree of inaccuracy and uncertainty or imprecision in HCPV tracker installations due to multiple factors. According to [13], there are three factors that cause system mismatches and power losses: manufacture error in HCPV modules, alignment error in the installation of the modules, and imprecision and inaccuracy in the tracker control system.

Due to these errors, the maximum power generation does not coincide with the zero error pointing of the tracker [35] in the installation of HCPV modules; therefore, controllers based on pointing algorithms (e.g., ephemeris) and pointing devices that minimize pointing error may present unacceptable errors in HCPV systems. However, the precision used in CPV installations is lower than that required in HCPV systems.

As a consequence, current trackers are not properly adapted to HCPV systems. These systems require more complex controls in order to obtain the maximum energy. Due to the optical concentration, the complexity of these systems, and the high degree of inaccuracy and uncertainty observed in the installation of these systems, a greater degree of precision and complexity in the control of tracker systems is necessary.

2.2. Fuzzy Rule-Based Systems

A technology that has been demonstrated to adapt correctly to environments with inaccuracy and uncertainty is the FRBS [8], which uses FL and expresses knowledge through IF-THEN-type linguistic rules. These systems (Figure 1) are composed of a fuzzification interface, a KB, an inference engine, and defuzzification interface. The fuzzification interface adapts the actual input values to the fuzzy system. The KB contains the definition of input and output variables, the fuzzy sets defined in the variables, and a set of IF-THEN-type linguistic rules that corelate these variables. The inference engine is responsible for inferring the fuzzy output of the system from the input variables and the KB. Finally, the defuzzification interface adapts the value of the fuzzy output to a real output value.

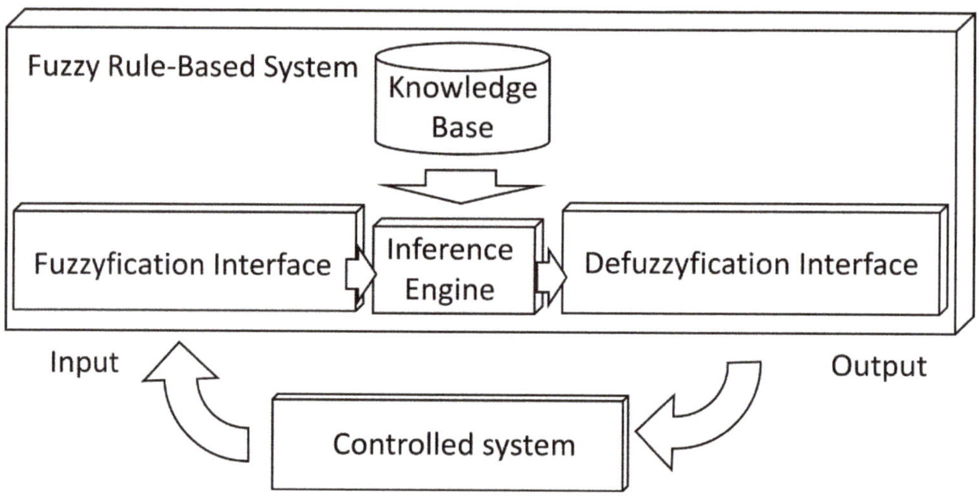

Figure 1. Fuzzy rule-based system.

Two approaches have been proposed within FRBSs: those of Mamdani [36,37] and Takagi–Sugeno–Kang (TSK) [38]. The main difference between the two approaches lies in the consequent knowledge rules. In the Mandani approach, the consequent is expressed as a linguistic variable:

IF X_1 is A_1 and . . . and X_n is A_n THEN Y is B,

where Xi represents input variables, A_i is fuzzy sets associated with input variables, Y is the output variable, and B is a fuzzy set associated with the output variable.

In the TSK approach, the consequent is an analytical function of the input variables:

IF X_1 is A_1 and . . . and X_n is A_n THEN $Y = f(X_1, \ldots, X_n)$.

where X_i represents input variables, A_i is fuzzy sets associated with input variables, Y is the output variable, and $f(X_1, \ldots, X_n)$ is the output function, in most cases, a linear function.

Controllers that use FRBS systems incorporating control knowledge are called FLCs (Figure 1).

In the literature [14], several FRBS applications are presented, such as classification systems, modeling systems, control systems, and robotics. A model of an HCPV module is presented using an FRBS system [39].

Currently, there is a trend to integrate knowledge-based systems into resource-constrained devices. Reference [40] presents a collaborative FRBS system for integration into wireless sensor networks (WSNs). In the literature [41], an optimization for smart spaces is proposed. Mariscal-Ramirez et al. [42] designed a sensor to monitor noise pollution adapted to resource-constrained devices.

2.3. Internet of Things

One of the objectives of this work is the use of IoT technologies to integrate the controller of an HCPV tracker into a resource-constrained device and monitor the data obtained using an IoT cloud platform.

Basically, IoT technologies [17,43] consist of constrained resource devices, data networks, communication protocols, and cloud platforms as follows:

(a) Although IoT devices have constrained resources, they have information processing capacity, sensor capacity of the environment, local information process, action on the environment, and the ability to communicate data with servers on the Internet. A first classification divides them into devices with an operating system (e.g., Raspberry) or without one (e.g., WaspMote, Arduino). These devices are used to obtain environmental data (temperature, pressure, etc.), detect alarms or extreme conditions (fire, gas leaks, etc.), or perform system control.

(b) Although there are data networks commonly used in IoT (IEEE 802.15.4, Long Rang (LORA), etc.), it is also possible to use conventional networks such as local area networks (LAN) (Ethernet, Wi-Fi, etc.) or mobile networks (4G).

(c) Specific IoT application protocols such as Message Queue Telemetry Transport (MQTT) or Constrained Application Protocol (CoAP) have been designed, although the use of other protocols such as HTTP is feasible.

(d) The data generated by IoT devices are sent to platforms for storage, visualization, analysis, and processing. Depending on the location of the servers, the platforms can be divided into cloud computing [44] and fog computing [45,46]. While cloud systems locate servers anywhere on the Internet, fog computing servers are closed to devices, usually on the same local network.

The use of IoT in smart spaces [18] and smart devices is widely referenced in European Commission documents [47,48] concerning the Internet of Things and the Internet of the Future. These documents present devices called smart things in which several algorithms can be executed for intelligent decisions based on real-time measurements of the sensors.

3. HCPV Tracker Knowledge-Based Controller

The main objective of this work is to design a knowledge-based controller for an HCPV tracker. Due to the uncertainty and inaccuracy of the positioning or tracker systems, this work proposes the use of FRBS systems because these knowledge-based systems have demonstrated their effectiveness in these conditions.

In addition, the proposed system will be implemented in a real system using IoT technology with constrained resource devices, data communication, and a platform for storing and analyzing the data obtained. Therefore, all the algorithms used in the control system will be designed to be executed in a low-cost microcontroller with low information processing capacity.

Therefore, the novelty of the proposed controller lies in (a) the design of a knowledge-based controller by means of FRBS, (b) easy to understand control knowledge, and (c) a design to be executed on a resource-constrained device.

The following sections show the structure of the proposed controller and two knowledge-based FRBS sensors: the first uses a positioning device, and the second is based on the electrical current generated by the photovoltaic concentration modules.

3.1. Controller Structure

Figure 2 shows the basic structure of the control system, which is composed of a pointing system, an error inference system, and a solar tracking system. In this way, the positioning of the tracker will be calculated as the sum of the positioning algorithm and the error inferred by the FRBS systems.

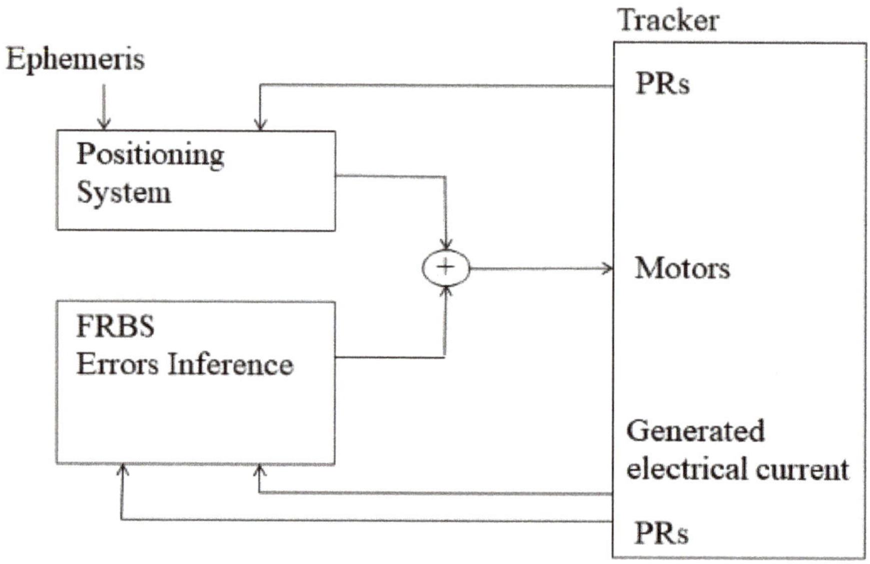

Figure 2. Controller structure.

The calculation of the solar position can be performed by different algorithms (such as ephemeris) or by a solar position sensor. If an ephemeris algorithm is used, it calculates the position of the sun using the date, time, and global position of the solar tracker. After that step, the controller compares the position of the sun with the position to which the tracker is pointing and calculates the azimuth and elevation angles that the solar tracker has to perform in the next movement.

To calculate the pointing error, a knowledge-based FRBS is used. To execute the FRBS in a constrained resource device, we introduced several modifications to the classical structure of Mandani FRBS to minimize computational burden: the device executes a small but complete FRBS; only triangular fuzzy sets are available; fuzzification and defuzzification interfaces only admit linear conversions; a First Infer Then Aggregate (FITA) inference approach is used; the inference engine operates with numerical values instead of linguistic labels; and the number of fuzzy sets defined in each variable and rules in the KB is small.

This work presents two knowledge-based FRBSs to infer the pointing error. The first is a smart sensor that is composed of a pointing device and an FRBS that infers the error. The second one is a different smart sensor composed of a probe that obtains the electric current generated by the HCPV module and another FRBS to infer the error.

3.2. FRBS Sensor Based on a Pointing Device

This smart sensor is composed of a hardware pointing device that allows measuring luminosity by means of four photoresistors (PRs) and an FRBS that infers the pointing error using a specific KB.

The pointing device is composed of an optical light/shadow device, a set of four PRs, and a signal adaptation stage. In this way, the luminosity values measured by the PRs are the inputs to the FRBS that infer the error. If the sensor is pointed correctly to the sun, each of the PRs has the same solar radiation. In the case of a pointing error, the optical device increases the radiation difference that some PRs receive compared to others to detect small pointing errors.

Figures 3 and 4 show the pointing device used by the sensor. The difference between the luminosity measured by PR1 and PR2 allows the sensor to infer the elevation error. Similarly, it can be estimated using PR4 and PR3, which allows the sensor to infer the error with two different systems. On the other hand, the difference in luminosity measured by PR4 and PR1 (as well as between PR3 and PR2) allows the sensor to infer the azimuth error.

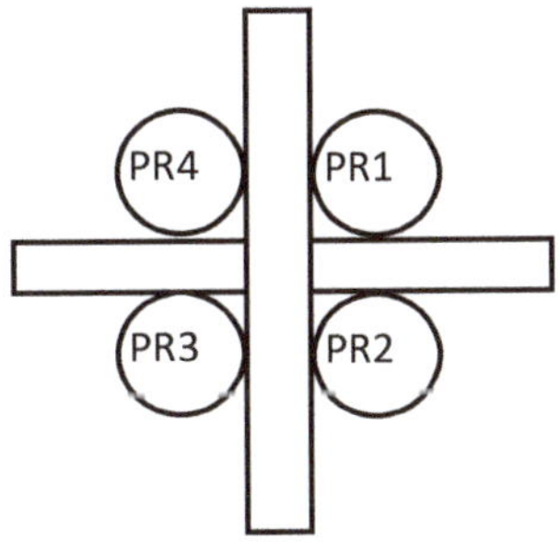

Figure 3. Pointing device. Top view.

Figure 4. Pointing device.

This work proposes to use two KBs, one to infer the elevation error and another to infer the azimuth error. Since KBs will be executed in a constrained resource system, a small number of fuzzy sets defined in variables and action rules will be used to infer the error as quickly as possible.

Elevation error KB consists of two input variables (luminosity measured in two PRs), one output variable (elevation error), and a set of action rules. Figure 5 shows the fuzzy sets defined for all input variables (PRs) and the output variable (elevation error).

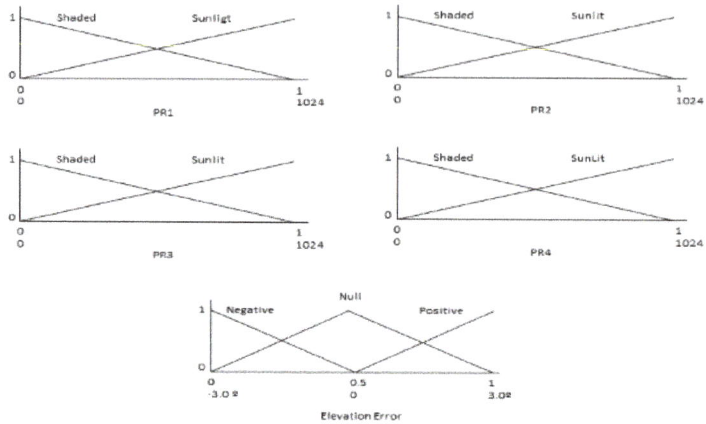

Figure 5. Fuzzy sets defined in input (PRs) and output variables (elevation error).

Table 1 shows the KB elevation error action rules for the sensor using PR1 and PR2.

Table 1. Action rules for the elevation error (PR1 and PR2).

Rule		Antecedents		Consequent
R1	If	PR1 is shaded and PR2 is shaded	then	the elevation error is null
R2	If	PR1 is shaded and PR2 is sunlit	then	the elevation error is positive
R3	If	PR1 is sunlit and PR2 is shaded	then	the elevation error is negative
R4	If	PR1 is sunlit and PR2 is sunlit	then	the elevation error is null

Table 2 shows the KB elevation error action rules for the sensor using PR4 and PR3.

Table 2. Action rules for the elevation error (PR4 and PR3).

Rule		Antecedents		Consequent
R1	If	PR4 is shaded and PR3 is shaded	then	the elevation error is null
R2	If	PR4 is shaded and PR3 is sunlit	then	the elevation error is positive
R3	If	PR4 is sunlit and PR3 is shaded	then	the elevation error is negative
R4	If	PR4 is sunlit and PR3 is sunlit	then	the elevation error is null

The azimuth error KB uses the same definition of input and output variables as the elevation error KB (Figure 5). However, the set of action rules is different. Table 3 shows the KB azimuth error action rules for the sensor using PR4 and PR1.

Table 3. Action rules for the azimuth error (PR4 and PR1).

Rule		Antecedents		Consequent
R1	If	PR4 is shaded and PR1 is shaded	then	the azimuth error is null
R2	If	PR4 is shaded and PR1 is sunlit	then	the azimuth error is positive
R3	If	PR4 is sunlit and PR1 is shaded	then	the azimuth error is negative
R4	If	PR4 is sunlit and PR1 is sunlit	then	the azimuth error is null

Table 4 shows the KB elevation error action rules for the sensor using PR3 and PR2.

Table 4. Action rules for the azimuth error (PR3 and PR2).

Rule		Antecedents		Consequent
R1	If	PR3 is shaded and PR2 is shaded	then	the azimuth error is null
R2	If	PR3 is shaded and PR2 is sunlit	then	the azimuth error is positive
R3	If	PR3 is sunlit and PR2 is shaded	then	the azimuth error is negative
R4	If	PR3 is sunlit and PR2 is sunlit	then	the azimuth error is null

The fuzzification interface linearly converts the range [0, 1024] measured by the PRs to the normalized range [0, 1] and the output range [0, 1] to [−3°, 3°].

3.3. *FRBS Sensor Based on the Electrical Current Generated*

This sensor is composed of a hardware probe that measures the electrical current generated by HCPV modules and an FRBS, which infers the elevation and azimuth errors.

The controller uses the usual sun tracking movements to measure the electrical current generation before and after each movement. After each movement, these two measurements are used to infer the elevation and azimuth errors. The inferred errors are taken into account in the next tracker movement to correct the errors and follow the maximum electrical current generated. Therefore, no extraordinary movements are made.

The KB of this FRBS system is used to infer both the elevation and azimuth errors and is composed of two input variables (electrical current before and after the movement), an output variable (error committed), and a set of action rules. Figure 6 shows the fuzzy sets defined in input variables (I_t and I_{t+1}) and output variable (Error) (with an HCPV module with Imax 6A and a maximum error of ± 3°).

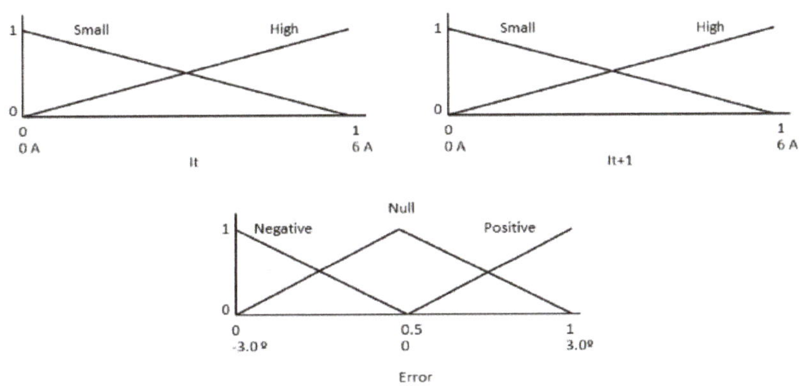

Figure 6. Fuzzy sets defined in input (It and I_{t+1}) and output variables (error).

Table 5 shows the KB error action rules used by the sensor.

Table 5. Action rules in knowledge base (KB) error.

Rule		Antecedents		Consequent
R1	If	I_t is low and I_{t+1} is low	then	the error is null
R2	If	I_t is low and I_{t+1} is high	then	the error is negative
R3	If	It is high and I_{t+1} is low	then	the error is positive
R4	If	I_t is high and I_{t+1} is high	then	the error is null

The fuzzification interface linearly converts the range [0, 6] measured by the electrical current probe to the normalized range [0, 1] and the output range [0, 1] to [−3°, 3°].

4. Experimental Results

To evaluate the controller and FRBS sensors proposed in the previous section, a real two-axis tracker with HCPV modules controlled by a low-cost microcontroller was designed and implemented. In addition, elevation and azimuth errors of tracker pointing with respect to the sun are measured by a precision instrument. On the other hand, data obtained (most significant tracker variables and elevation and azimuth errors) were sent to an IoT platform in order to analyze their evolution and compare the results of the different FRBS systems proposed.

This section describes the HCPV tracker used in the experiments carried out and the results obtained with the following controllers based on (a) an ephemeris algorithm only; (b) an FRBS sensor based on a pointing device; and (c) an FRBS sensor based on the electrical current generated.

4.1. HCPV Tracker

The two-axis solar tracker (Figures 7 and 8) is composed of a metal structure with the possibility of movement in elevation and azimuth by means of gearboxes, various HCVPV modules, a calibrated solar cell, DC azimuth and elevation motors, a measurement system of the angular movement of each motor (encoders), a pointing error sensor, electrical current-generated sensors, and the control system.

Figure 7. The designed high-concentration photovoltaic system (HCPV) tracker.

Figure 8. The designed high-concentration photovoltaic system (HCPV) tracker.

The control system (Figure 9) is based on a low-cost 32-bit microcontroller and several signal adaptation interfaces to the following inputs and outputs:

- Nine analog inputs: four PRs, temperature, solar radiation, direction and speed wind sensors, and electrical current generated.
- Ten digital inputs: two encoders, four inputs for a joystick, and four limit switches.
- Two digital outputs: azimuth and elevation motors.

Figure 9. Controller.

The controller calculates the elevation and azimuth angles to be performed at each moment using the state of the system (date, time, position of the sun, position of the tracker, etc.). On the other hand, the error inferred by the FRBS system is added to the calculated angles. The angular movements of elevation or azimuth of the tracker are carried out by means of an algorithm that calculates the activation and braking time of the motor as well as a maximum safety time. The movement made at each angle is verified by means of the encoders.

To measure the real elevation and azimuth error of the tracker in the sun, a Black Photon Tracking Accuracy Sensor measuring instrument (Figure 10) is available. The instrument is able to measure elevation and azimuth errors in the range ± 1.2° with a resolution of 0.0005°. The data obtained from the instrument allow us to check the correct tracker pointing and are not used in tracker control.

Figure 10. Tracking accuracy sensor.

Data generated by the system (sensor measurements, solar radiation, temperature, solar position, tracker position, electrical current generated, etc.) are sent to an Internet IoT cloud platform that stores the data and allows users to monitor the temporal evolution of all variables using a web browser.

The main characteristics of HCPV modules used in the tracker are the following (at 1000 W/m2, 25°C, AM1.5D): short-circuit current 6.35A, open-circuit voltage 18.45 V, DC power 95 W, and needed pointing error < ± 0.6°.

To characterize the HCPV module, a complete exploration was carried out by measuring the short-circuit electrical current generated by varying its position with respect to the sun in an angular sector of ± 3° in elevation and azimuth. Figure 11 shows the obtained surface where it is observed that the maximum electrical current generation is not at the 0° elevation and azimuth point. The maximum current (5.54 A) is at an elevation error of +0.2° and an azimuth error of −0.8°. On the other hand, the surface shows that the current generated falls drastically with a small variation of the elevation and azimuth angles.

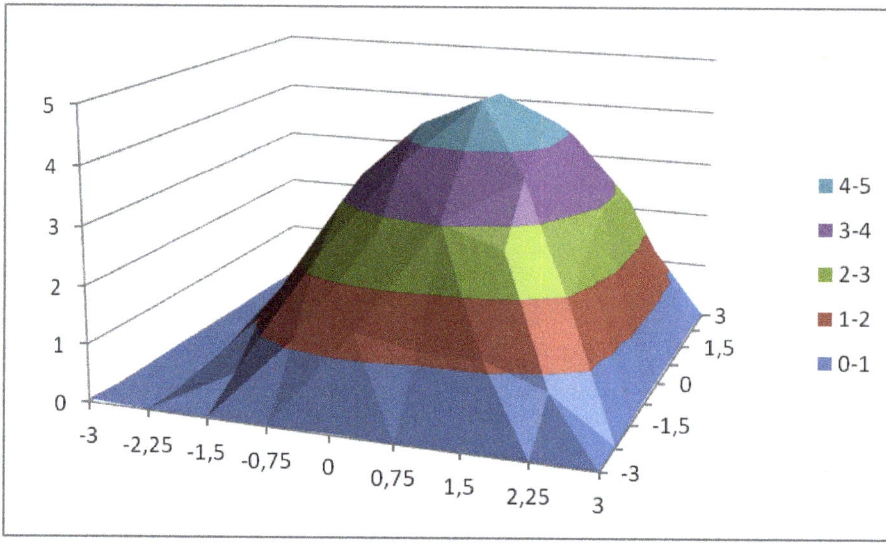

Figure 11. Electrical current generated in an elevation and azimuth sector of ±3°.

4.2. Controller based on an Ephemeris Algorithm

The first part of the experiments carried out has the objective of measuring the effectiveness of an ephemeris algorithm applied to a real tracker in which different imprecision factors can exist according to the reasons stated in Section 2.

Figure 12 shows a simulation in which the position of the sun (angles of azimuth and elevation) is calculated by the ephemeris algorithm as well as the tracker pointing during a day without taking into account the elevation and azimuth errors. In the graph, it can be observed how the tracker would be pointing at the sun practically without error during the period of time in which the sun elevation was greater than 15°. The rest of the day, the tracker would be in a resting position (45° elevation, 180° azimuth).

Figure 13 shows the results obtained when the real tracker is controlled with the ephemeris algorithm. Figure 13a shows the DNI (direct normal irradiance) of March 15, 2019. It is a sunny day with a maximum of 1000 w/m2. Figure 13b shows the Isc (short circuit current) obtained by the HCPV module. Figure 13c shows the evolution of elevation and azimuth errors (measured by the precision instrument) with respect to the sun. Finally, Figure 13d presents the Isc/DNI ratio, which represents a normalization that allows comparison of the currents generated on different days. Figure 13c,d shows the results from approximately 9:00 a.m. at 6:00 p.m. which corresponds to an elevation of the sun greater than 15°.

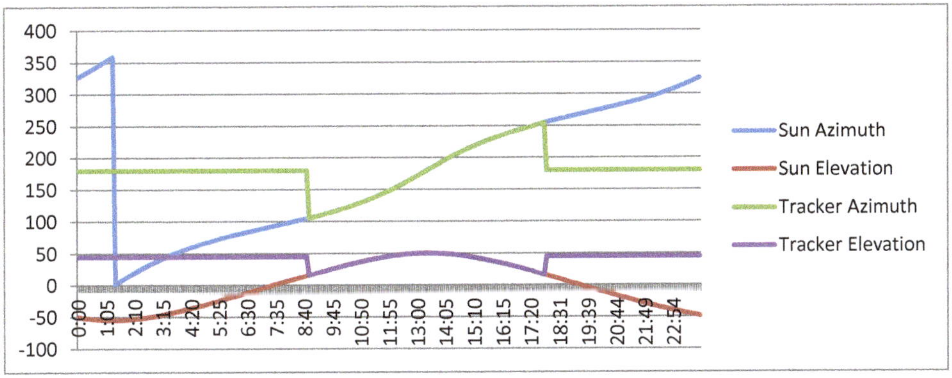

Figure 12. Sun position and tracker pointing (with both elevation and azimuth angles).

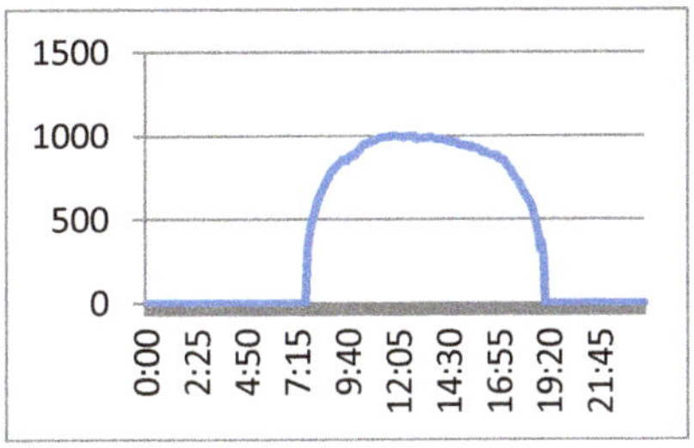

(**a**) DNI(W/m2)

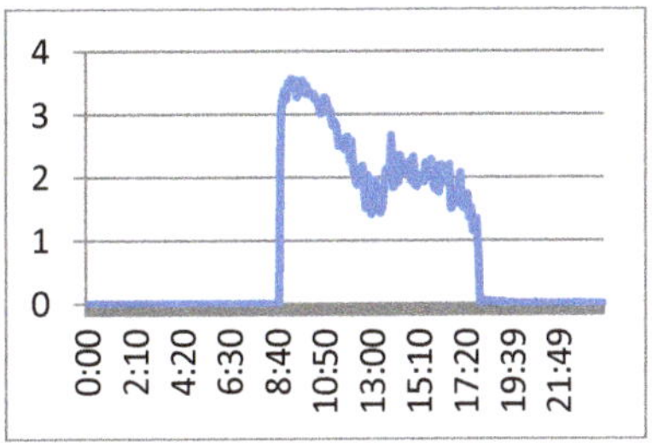

(**b**) Isc(A)

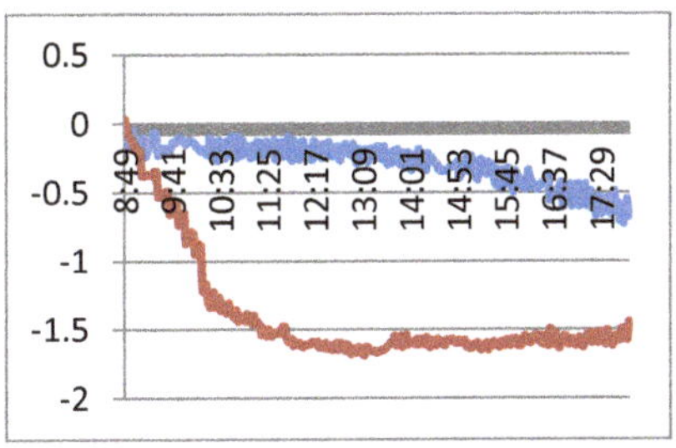

(**c**) Azimuth error (°) (blue)

Elevation error (°) (red)

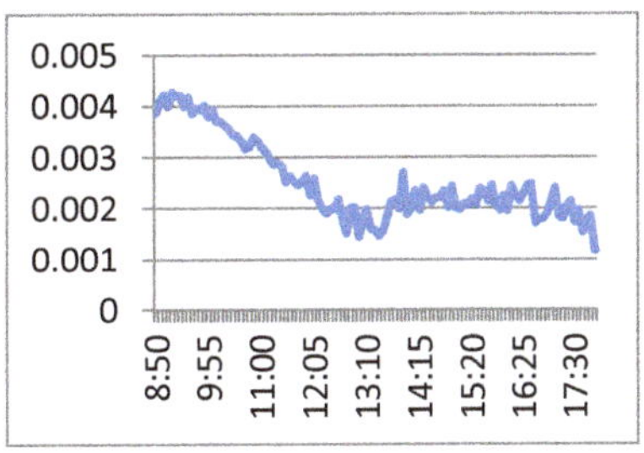

(**d**) I/DNI

Figure 13. Results of the ephemeris controller. March 15, 2019.

Figure 13c shows how at the beginning of the experiment (9 am), it starts with an error of 0° in elevation and azimuth errors and the way in which both errors are increasing and exceeding the values recommended by the manufacturer (elevation error). Due to these errors, the generated current decreases in the first part of the day, although it is stabilized at the end. As a result of the elevation and azimuth pointing errors, the concentration generator does not operate at maximum, and the current generated is much lower than expected.

4.3. Controller Based on an FRBS Sensor with a Pointing Device

To improve the performance of the photovoltaic generator and correct the evolution of elevation and azimuth errors, this work proposes the use of an economic sensor based on an FRBS system and a pointing device that infers the elevation and azimuth errors.

This controller is composed of an optical light/shadow pointing device, a set of four PRs, a signal adaptation stage and an FRBS system. In the case of a pointing error, the pointing device increases the difference in radiation received by some PRs compared to others to detect small pointing errors. The pointing sensor is based on the device shown in Figure 14, and the FRBS system is shown in Figure 5 and Tables 1–4.

Figure 14. Pointing device.

Each PR of the pointing sensor is able to measure the luminosity that they are receiving by means of modifying their electrical resistance. Through an adaptation stage, the microcontroller can measure a proportional voltage (by means of a 12-bit analog input) in such a way that a value of 1,023 is obtained in the maximum luminosity and 0 in the dark.

Figure 15 shows the evolution of the luminosity value obtained in the four PRs. The data in the figure have been obtained with the tracker stopped while pointing to the sun with an approximate error of 0° in elevation and azimuth at a certain time (2:00 p.m). In this way, the evolution of the luminosity measured can be observed during a full sunny day.

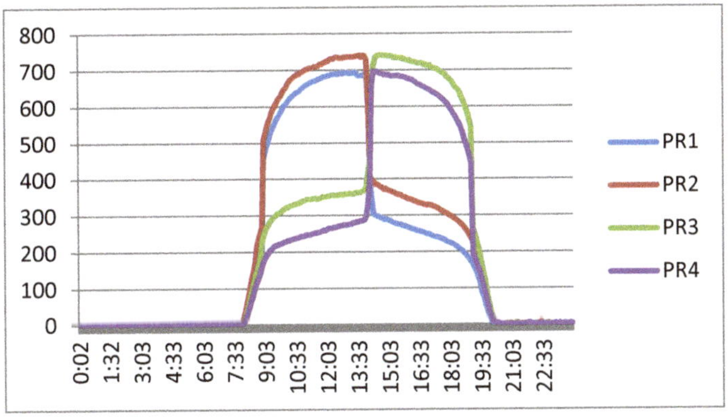

Figure 15. Evolution of the luminosity value obtained in the four PRs.

Figure 16 shows the error inferred by the FRBS system between 13:00 and 15:00. In the initial and final parts, the sensor infers a constant error: a positive degree of error in elevation and azimuth in the initial part and a positive degree in elevation and a negative degree in azimuth. In the central part, the figure shows the inferred error when the device is pointing to the sun, inferring that the sensor has not been correctly oriented (minimum inferred error +0.5° elevation, 0° azimuth).

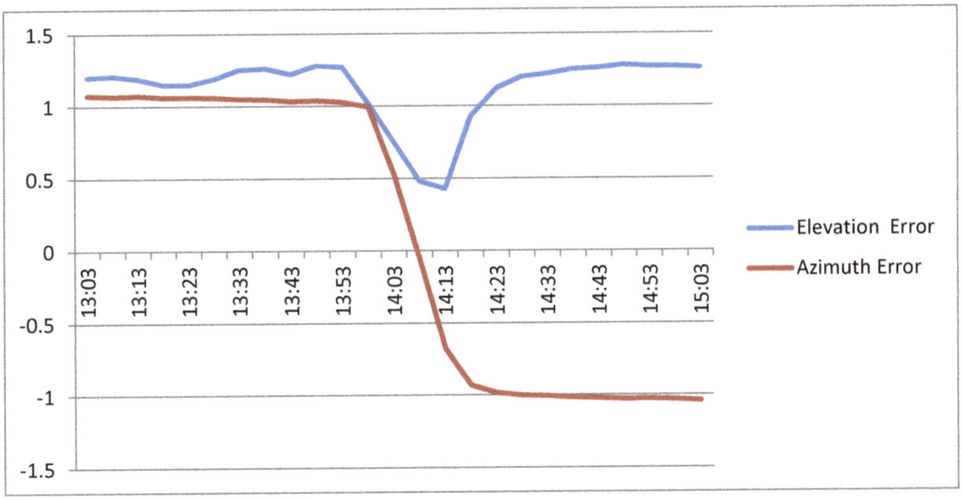

Figure 16. Elevation and azimuth errors inferred by the FRBS.

Figure 17 shows the results obtained when the tracker is controlled by the ephemeris algorithm modified with the FRBS sensor with the pointing device. Figure 17a shows the DNI of March 12, 2019. It is a sunny day with a maximum of 850 w/m2. Figure 17b shows the Isc current obtained by the HCPV module. Figure 17c shows the evolution of elevation and azimuth errors (measured by the precision instrument) with respect to the sun. Finally, Figure 17d presents the Isc / DNI ratio. Figure 17c,d shows the results from approximately 9:00 a.m. at 6:00 p.m. which corresponds to a sun elevation greater than 15°.

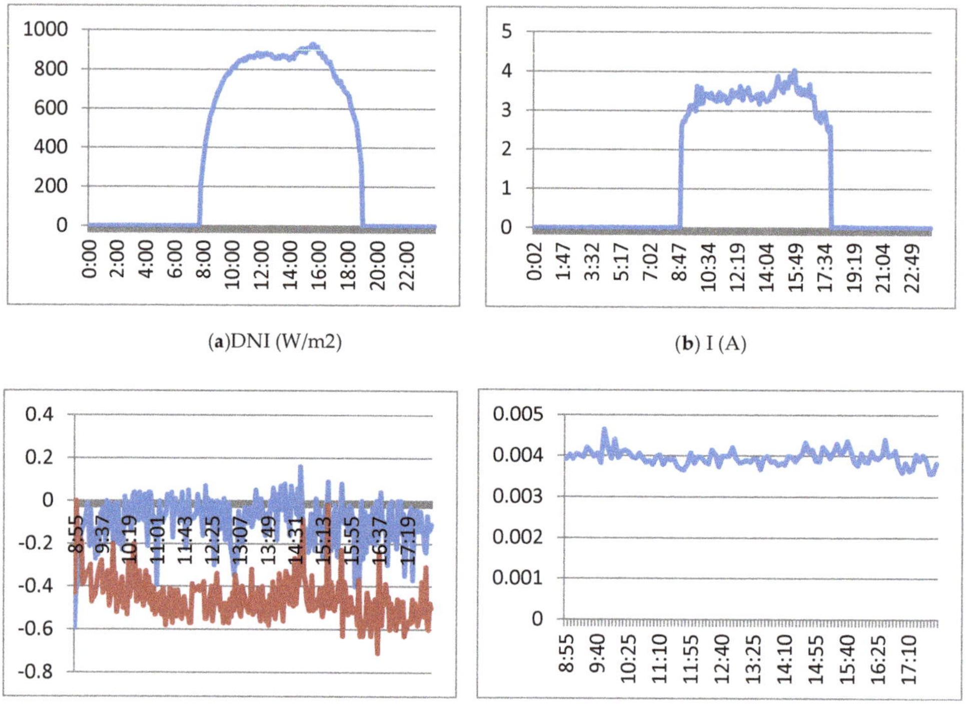

(**a**)DNI (W/m2) (**b**) I (A)

(**c**) Azimuth error (°) (blue)

Elevation error (°) (red)

(**d**) I/DNI

Figure 17. Results of the controller based on an FRBS sensor with the pointing device. March 12, 2019.

Figure 17 shows the following:

(a) Although the maximum electrical current generation (obtained in the characterization of the HCPV module) is not achieved, the generated electrical current is greater when using the controller based on an FRBS sensor with the pointing device.

(b) The elevation and azimuth errors of this controller are quite minor compared to the errors observed in the controller based on an ephemeris algorithm. The average values of the errors measured by the instrument are −0.10° for the azimuth and −0.45° for the elevation, closer to those of the characterization.

(c) The Isc/DNI ratio shows a performance improvement when using this control system.

4.4. Controller based on an FRBS Sensor based on the Electrical Current Generated

Although the inference of the pointing error improves the electrical current generated by the HCPV module, it does not obtain the maximum current generated.

In this section, an FRBS based on the electrical current generated is proposed in order to modify the position of the tracker obtained by the ephemeris algorithm. The controller proposed is based on the calculation of the pointing error by means of the FRBS system described in Section 3.3. In this controller, each time the tracker moves in elevation or azimuth (by means of ephemeris), the pointing error is inferred. The error is corrected in the next movement so that no extraordinary movements are made exclusively to correct the error.

Figure 18a shows the DNI of March 14, 2019. It was a sunny day with a maximum of 1000 w/m2. Figure 18b shows the Isc electrical current obtained by the HCPV module. Figure 18c shows the evolution of elevation and azimuth errors (measured by the precision instrument) with respect to the sun. Finally, Figure 18d presents the Isc / DNI ratio, which represents a normalization that allows comparison of the currents generated on different days. Figure 18c,d shows the results from approximately 9:00 a.m. to 6:00 p.m., which correspond to an elevation of the sun greater than 15°.

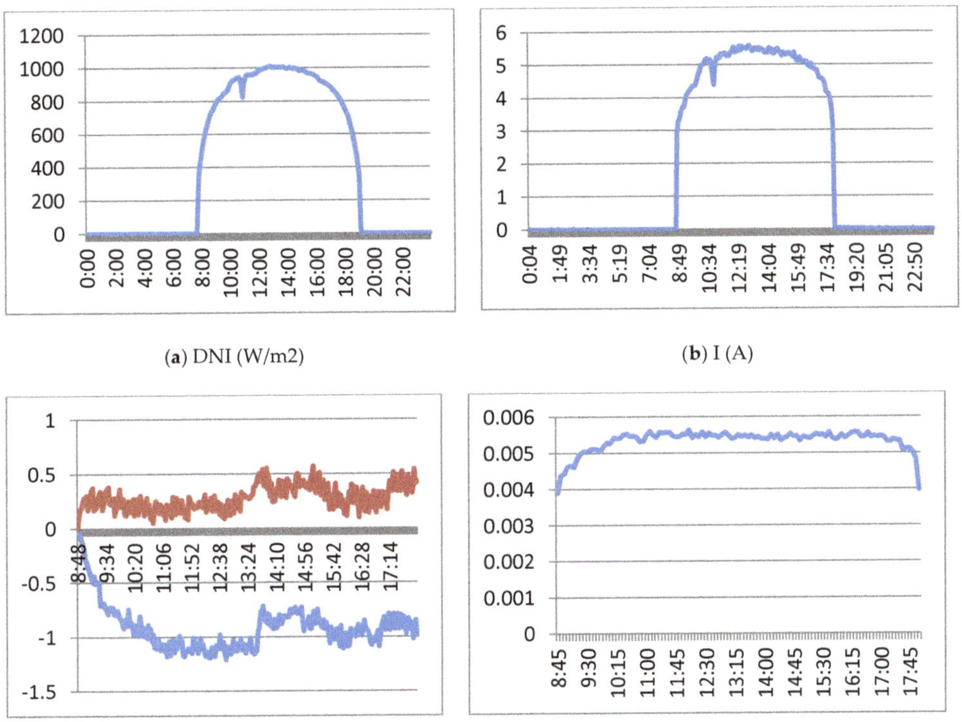

(**a**) DNI (W/m2) (**b**) I (A)

(**c**) Azimuth error (°) (blue)
Elevation error (°) (red)

(**d**) I/DNI

Figure 18. Results of the controller based on the FRBS sensor based on the electrical current generated. 14 March 2019.

Figure 18 shows the following:

(a) The generated current is greater than that obtained with previous controllers, with a value very similar to the value obtained in the characterization of the module. On the other hand, there is less ripple in the current in this case.

(b) Although the electrical current is stabilized and the Isc/DNI ratio is practically flat, it should be noted that the error does not remain constant throughout the day. The control method dynamically modifies the error and tracker position in order to obtain the highest current.

(c) The average values of the errors measured by the instrument (average azimuth −0.91°; average elevation 0.27°) are very similar to those obtained in the characterization of the module (azimuth error −0.8°; elevation error +0.2°).

(d) The Isc/DNI ratio shows a performance improvement over the other control methods.

The main benefits of the proposed controller are as follows:

- The controller showed the best performance, near the maximum current measured in the module characterization;

- A knowledge-based controller with correct adaptation to inaccuracy and uncertainty and easy to understand knowledge;
- The controller can be executed on a resource-constrained cheap device, using IoT technology. A summary of the controller cost is as follows: 32 bits microcontroller—€20, wind sensors—€20, calibrated solar cell—€7, electrical current sensor—€4, miscellaneous material (photoresistors, temperature probe, etc.)—€2;
- It dynamically modifies the tracker position to obtain near the maximum of generated electrical energy with minimal oscillation.

5. Conclusions

IoT technologies were able to execute HCPV controllers and monitor the evolution of variables in a satisfactory way. The constrained resource microcontroller executed the knowledge-based controllers with response times shorter than needed in the application. Although there was some timely loss of data due to the unavailability of communication on the Internet, the cloud computing architecture used in the project was more than sufficient. Data obtained in the system was correctly stored in the platform and monitored by users

Additional factors were presented to those provided in the literature [13], which increase imprecision and uncertainty in HCPV solar tracker installations. The factors presented in the project are the following:

(a) Inaccuracy in the manufacturing process of HCPV modules in relation to the alignment of the different components;
(b) Precision errors in the manufacture of the tracker structure;
(c) Precision errors in the installation of tracker structure (e.g., wrong leveling);
(d) Precision errors in the installation of HCPC modules in the tracker;
(e) Minimum movement of the electrical engines, which may be greater than necessary in some instances;
(f) Inaccurate movement of the electrical engines;
(g) Low resolution of encoders that measure the angle performed by electrical engines;
(h) Other factors such as wind.

The characterization of the HCPV module installed in the tracker verifies that the maximum energy may not be in the zero pointing error of the tracker due to the imprecision and uncertainty factors. In addition, to generate the maximum electrical current, it must be taken into account that a pointing error less than 0.6° is necessary.

The controller based exclusively on the ephemeris algorithm obtains very low performance due to the accumulation of azimuth and elevation errors. In this case, the tracker leveling error was important. When using this kind of controller, it would be necessary to calculate the error made and take it into account in the control algorithm.

The controller based on the FRBS sensor with a pointing device infers the azimuth and elevation error and increases the generated electrical current, improving the performance of the exclusively ephemeris-based controller. This controller requires that the maximum electrical current generation of the HCPV module installed in the tracker and the pointing device be perfectly calibrated (pointing to the same exact angle). In this case, a periodic calibration would be necessary.

The controller based on the FRBS sensor and an electrical current probe showed the best performance, obtaining values similar to those obtained in the module characterization. In this case, calibration is not necessary since the algorithm dynamically locates the maximum current generation.

Regarding future work, we propose the following actions: to use an IoT fog computing architecture in order to avoid punctual data loss; to characterize different HCPV modules in the real tracker; to compare other controllers; and to characterize HCPV systems composed of several modules in order to locate their maximum current generation.

Author Contributions: Conceptualization, J.C.-B. and P.P.-H.; Funding acquisition, J.C.-B. and P.P.-H.; Investigation, J.-A.F.-P. and M.-A.G.-M.; Methodology, J.C.-B., J.-A.F.-P., and M.-A.G.-M.; Project administration, J.C.-B. and P.P.-H.; Resources, J.-A.F.-P. and M.-A.G.-M.; Software, J.C.-B., and M.-A.G.-M.; Supervision, J.C.-B. and P.P.-H.; Validation, J.-A.F.-P. and M.-A.G.-M.; Visualization, J.C.-B., J.-A.F.-P. and M.-A.G.-M.; Writing—original draft preparation, J.C.-B.; Writing—review and editing, J.-A.F.-P., M.-A.G.-M.. All authors have read and agreed to the published version of the manuscript.

Funding: This work forms part of the project "Nuevos conceptos basados en tecnología de concentración fotovoltaica: desarrollo de sistemas de muy alta concentración fotovoltaica" (ENE2013-45242-R) supported by the Spanish Economy Ministry and the European Regional Development Fund/Fondo Europeo de Desarrollo Regional (ERDF/FEDER).

Conflicts of Interest: The authors declare no conflict of interest.

References

1. Jaeger-Waldau, A. *PV Status Report 2019*; EUR 29938 EN; Publications Office of the European Union: Luxembourg, 2019; ISBN 978-92-76-12608-9. [CrossRef]
2. D'Adamo, I. The Profitability of Residential Photovoltaic Systems. A New Scheme of Subsidies Based on the Price of CO_2 in a Developed PV Market. *Soc. Sci.* **2018**, *7*, 148. [CrossRef]
3. Das, U.K.; Tey, K.S.; Seyedmahmoudian, M.; Mekhilef, S.; Yamani, M.; van Deventer, W.; Horan, B.; Stojcevski, A. Forecasting of photovoltaic power generation and model optimization: A review. *Renew. Sustain. Energy Rev.* **2018**, *81*, 912–928. [CrossRef]
4. Luque, A.; Sala, G.; Luque-Heredia, I. Photovoltaic concentration at the onset of its commercial deployment. *Prog. Photovolt. Res. Appl.* **2006**, *14*, 413–428. [CrossRef]
5. Perez-Higueras, P.; Muñoz, G.E.; Almonacid; Vidal, P. High Concentrator PhotoVoltaics efficiencies: Present status and forecast. *Renew. Sustain. Energy Rev.* **2011**, *15*, 1810–1815. [CrossRef]
6. Cristobal, A.; Martí, A.; Luque, A. *Next Generation of Photovoltaics*; Springer: Madrid, Spain, 2012.
7. Luque, A.; Andreev, V. *Concentrator Photovoltaics*; Springer: New York, NY, USA, 2007.
8. Philipps, S.; Bett, A.; Horowitz, K.; Kurtz, S. *Current Status of Concentrator Photovoltaic (CPV) Technology*; National Renewable Energy Laboratory NREL: Golden, CO, USA, 2015. [CrossRef]
9. Talavera, D.L.; Pérez-Higueras, P.; Almonacid, F.; Fernández, E.F. A worldwide assessment of economic feasibility of HCPV power plants: Profitability and competitiveness. *Energy* **2017**, *119*, 408–424. [CrossRef]
10. Lee, C.-Y.; Chou, P.-C.; Chiang, C.-M.; Lin, C.-F. Sun Tracking Systems: A Review. *Sensors* **2009**, *9*, 3875–3890. [CrossRef] [PubMed]
11. Luque-Heredia, I.; Quéméré, G.; Cervantes, R.; Laurent, O.; Chiappori, E.; Chong, J. The sun tracker in concentrator photovoltaics. *Spriger Ser. Opt. Sci.* **2012**, *165*, 61–93.
12. Gómez-Gil, F.J.; Wang, X.; Barnett, A. Energy production of photovoltaic system: Fixed, tracking and concentrating. *Renew. Sustain. Energy Rev.* **2012**, *16*, 306–313. [CrossRef]
13. Kim, Y.S.; Kang, S.; Winton, R. Tracking control of high-concentrator photovoltaic systems for minimizing power losses. *Prog. Photovolt. Res. Appl.* **2014**, *22*, 1001–1009. [CrossRef]
14. Cordón, O.; Herrera, F.; Hoffmann, F.; Magdalena, L. *Genetic Fuzzy Systems: Evolutionary Tunning and Learning of Fuzzy Knowledge Bases;* World Scientific: Singapore, 2001; Volume 141.
15. Zadeh, A. Fuzzy Sets. *Inf. Control* **1965**, *8*, 338–353. [CrossRef]
16. Buyya, R.; Dastjerdi, A. *Internet of Thing. Principles and Paradigms*; Elsevier: Amsterdam, The Netherlands, 2016.
17. Madakam, S.; Ramaswamy, R.; Tripathi, S. Internet of Things (IoT): A Literature Review. *J. Comput. Commun.* **2015**, *3*, 164–173. [CrossRef]
18. Vermesan, O.; Friess, P. *Internet of Things: Converging Technologies for Smart Environments and Integrated Ecosystems;* River Publishers: Aalgorg, Denmark, 2013.
19. Al-Rousan, N.; Isa, N.M.; Desa, M.M. Advances in solar photovoltaic tracking systems: A review. *Renew. Sustain. Energy Rev.* **2018**, *82*, 2548–2569. [CrossRef]
20. Assaf, E. Deslgn and implementation of a two axis solar tracking system using plc tlechniques by an inexpensive method. *Int. J. Acad. Sci. Res.* **2014**, *3*, 54–65.
21. Tudorache, T.; Krelndler, L. Design of a solar tracker system for PV power plants. *Acta Polytech. Hung.* **2010**, *7*, 23–39.

22. Reda, I.; Afshin, A. Solar position algorithm for solar radiation application. *Sol. Energy* **2004**, *76*, 577–589. [CrossRef]
23. Lee, C.; Huang, H.; Yeh, H. The development of sun-tracking system using image processing. *Sensors* **2013**, *13*, 5448–5459. [CrossRef]
24. Alata, M.; Al-Nimr, M.; Qaroush, Y. Developing a multipurpose sun tracking system using fuzzy control. *Energy Convers. Manag.* **2005**, *46*, 1229–1245. [CrossRef]
25. Choi, J.; Kim, D.; Park, K.; Choi, C.; Chung, D. Design of fuzzy controller based on PC for solar tracking system. In Proceedings of the ICSMA 2008—International Conference on Smart Manufacturing Application, Gyeonggi-do, Korea, 9–11 April 2008.
26. Taherbaneh, M.; Fard, H.; Rezaie, A.; Karbasian, S. Combination of fuzzy-based maximum power point tracker and sun tracker for deployable solar panels in photovoltaic systems. In Proceedings of the IEEE International Conference on Fuzzy Systems, London, UK, 23–26 July 2007.
27. Yousef, H. Design and implementation of a fuzzy logic computer-controlled sun tracking system. In Proceedings of the IEEE International Symposium on Industrial Electronics, Bled, Slovenia, 12–16 July 1999.
28. Bendib, B.; Belmili, H.; Krim, F. A survey of the most used MPPT methods: Conventional and advanced algorithms applied for photovoltaic systems. *Renew. Sustain. Energy Rev.* **2015**, *45*, 637–648. [CrossRef]
29. Salas, V.; Olias, E.; Barrado, A.; Lazaro, A. Review of the maximum power point tracking algorithms for stand-alone photovoltaic systems. *Sol. Energy Mater. Sol. Cells* **2006**, *90*, 1555–1578. [CrossRef]
30. Chou, K.-Y.; Yang, S.-T.; Chen, Y.-P. Maximum Power Point Tracking of Photovoltaic System Based on Reinforcement Learning. *Sensors* **2019**, *19*, 5054. [CrossRef]
31. Daliento, S.; Chouder, A.; Guerriero, P.; Pavan, A.M.; Mellit, A.; Moeini, R.; Tricoli, P. Monitoring, diagnosis, and power forecasting for photovoltaic fields: A review. *Int. J. Photoenergy* **2017**, *2017*, 1356851. [CrossRef]
32. Madeti, S.R.; Singh, S.N. Monitoring system for photovoltaic plants: A review. *Renew. Sustain. Energy Rev.* **2017**, *67*, 1180–1207. [CrossRef]
33. Jiménez-Castillo, G.; Muñoz-Rodríguez, F.J.; Rus-Casas, C.; Gómez-Vidal, P. Improvements in Performance Analysis of Photovoltaic Systems: Array Power Monitoring in Pulse Width Modulation Charge Controllers. *Sensors* **2019**, *19*, 2150. [CrossRef] [PubMed]
34. Fernandez, F.; Perez-Higueras, P.; Luoreiro, A.G.; Vidal, P. Outdoor evaluation of concentrator photovoltaic systems modules from different manufacturers: First results and steps. *Prog. Photovolt. Res. Appl.* **2013**, *21*, 693–701. [CrossRef]
35. Kim, Y.; Winston, R. Power conversion in CPV systems. *Prog. Photovolt.* **2013**, *22*, 984–992. [CrossRef]
36. Mamdani, E. Application of fuzzy algorithms for control of simple dynamic plant. *Proc. Inst. Electr. Eng.* **1974**, *121*, 1585–1588. [CrossRef]
37. Mamdani, E.; Assilian, S. An experiment in linguistic synthesis with a fuzzy logic controller. *Int. J. Man Mach. Stud.* **1975**, 1–13. [CrossRef]
38. Takagi, T.; Sugeno, M. Fuzzy Identification of Systems and Its Applications to Modeling and Control. *IEEE Trans. Syst. Man Cybern.* **1985**, *15*, 116–132. [CrossRef]
39. Gadeo-Martos, M.; Yuste-Delgado, A.; Almonacid, F.; Fernández-Prieto, J.; Canada-Bago, J. Modeling a High Concentrator Photovoltaic Module using Fuzzy Rule-Based Systems. *Energies* **2019**, *12*, 567. [CrossRef]
40. Canada-Bago, J.; Fernandez-Prieto, J.; Gadeo-Martos, M.; Velasco, J. A New Collaborative Knowledge-Based Approach for Wireless Sensor Networks. *Sensors* **2010**, *10*, 6044–6062. [CrossRef]
41. Gadeo-Martos, M.; Fernandez-Prieto, J.; Canada-Bago, J.; Velasco, J. An Architecture for Performance Optimization in a Collaborative Knowledge-Based Approach for Wireless Sensor Networks. *Sensors* **2011**, *11*, 9136–9159. [CrossRef] [PubMed]
42. Mariscal-Ramirez, A.; Fernandez-Prieto, J.; Canada-Bago, J.; Gadeo-Martos, M. A new algorithm to monitor noise pollution adapted to resource-constrained devices. *Multimed. Tools Appl.* **2015**, *74*, 9175–9189. [CrossRef]
43. Vermesan, O.; Friess, P.; Furness, A. The Internet of Things 2012 New Horizons. 2012. Available online: http://www.internet-of-things-research.eu/pdf/IERC_Cluster_Book_2012_WEB.pdf (accessed on 9 November 2019).
44. Stallings, W. *Foundations of Modern Networking: SDN, NFV*; QoE, IoT and Cloud, Pearson Education: New York, NY, USA; p. 2016.
45. Hao, Z.; Novak, E.; Yi, S.; Li, Q. Challenges and Software Architecture for Fog Computing. *IEEE Int. Comput.* **2017**, *21*, 44–53. [CrossRef]

46. Freeman, H.; Zhang, T. The Emerging Era of Fog Computing and Networking. *IEEE Commun. Mag.* **2016**, *54*, 4–5.
47. Vermesan, O. *Internet of Things. Strategic Research Roadmap*; European Commission: Brussels, Belgium, 2009; Available online: http://www.internet-of-things-research.eu/pdf/IoT_Cluster_Strategic_Research_Agenda_2009.pdf (accessed on 10 September 2019).
48. Sundmaeker, H.; Guillemin, P.; Friess, P.; Woelffle, S. *Vision and Challenges for Realising the Internet of Things*; Publication office of the European Union: Luxenbourg, 2010.

sensors

MDPI

Article

Machine Learning-Based Ensemble Classifiers for Anomaly Handling in Smart Home Energy Consumption Data

Purna Prakash Kasaraneni [1], Yellapragada Venkata Pavan Kumar [2], Ganesh Lakshmana Kumar Moganti [2,*] and Ramani Kannan [3]

1 School of Computer Science and Engineering, VIT-AP University, Amaravati 522237, Andhra Pradesh, India
2 School of Electronics Engineering, VIT-AP University, Amaravati 522237, Andhra Pradesh, India
3 Department of Electrical and Electronics Engineering, Universiti Teknologi Petronas (UTP), Seri Iskandar 32610, Malaysia
* Correspondence: ganesh.moganti@vitap.ac.in; Tel.: +91-8632370365

Abstract: Addressing data anomalies (e.g., garbage data, outliers, redundant data, and missing data) plays a vital role in performing accurate analytics (billing, forecasting, load profiling, etc.) on smart homes' energy consumption data. From the literature, it has been identified that the data imputation with machine learning (ML)-based single-classifier approaches are used to address data quality issues. However, these approaches are not effective to address the hidden issues of smart home energy consumption data due to the presence of a variety of anomalies. Hence, this paper proposes ML-based ensemble classifiers using random forest (RF), support vector machine (SVM), decision tree (DT), naive Bayes, K-nearest neighbor, and neural networks to handle all the possible anomalies in smart home energy consumption data. The proposed approach initially identifies all anomalies and removes them, and then imputes this removed/missing information. The entire implementation consists of four parts. Part 1 presents anomaly detection and removal, part 2 presents data imputation, part 3 presents single-classifier approaches, and part 4 presents ensemble classifiers approaches. To assess the classifiers' performance, various metrics, namely, accuracy, precision, recall/sensitivity, specificity, and F1 score are computed. From these metrics, it is identified that the ensemble classifier "RF+SVM+DT" has shown superior performance over the conventional single classifiers as well the other ensemble classifiers for anomaly handling.

Keywords: classification; data anomalies; data imputation; energy consumption data; ensemble classifiers; machine learning; smart home data; smart meter data; tracebase dataset

Citation: Kasaraneni, P.P.; Venkata Pavan Kumar, Y.; Moganti, G.L.K.; Kannan, R. Machine Learning-Based Ensemble Classifiers for Anomaly Handling in Smart Home Energy Consumption Data. *Sensors* **2022**, *22*, 9323. https://doi.org/10.3390/s22239323

Academic Editor: Davide Brunelli

Received: 3 November 2022
Accepted: 25 November 2022
Published: 30 November 2022

1. Introduction

Considering the global thrust towards the development of grid-independent and green energy systems for addressing the unrelenting growth of loads as well as environmental pollution, smart home and renewable energy-based microgrid culture has been increasing worldwide. Smart cities are new-era establishments where all the smart homes are jointly operated to consolidate and optimize electricity utilization. As these establishments are realized with a combination of electrical, communication, and information technology, the gathering of quality data is a challenging task. Smart homes connected to the power network continuously generate huge volumes of energy consumption data, which is normally a combination of timestamps and readings. The reading information in this data is a key value that helps in understanding the energy consumption behavior, billing generation, load profiling, forecasting, contingency analysis, device health condition analysis, etc. All these operations rely upon the quality of the data being captured. However, this data often may consist of different anomalies, viz., garbage data, outliers, redundant data, and missing data due to malfunctioning of advanced metering infrastructure, failure of communication channels, unanticipated issues in power networks, etc. If these anomalies are left unhandled in the dataset, there will be an adverse effect on the system operations and further delude the analytics of the energy consumption data. So, handling these anomalies is highly essential to enable analysts to perform accurate energy data analytics. Thus, the multifaceted nature of the smart home data when compared to other datasets gains importance in the data analysis field. So, this becomes an important research focus for data analysts when compared to the datasets of other applications. This is the prime motivation for the proposed work of this paper.

Big data refers to a huge quantity of data. However, the data quality is a more complex and significant aspect than the quantity in the direction of research [1]. Moreover, the issues related to data quality have gained much importance and attention in energy big data analytics [2]. The increased use of several intelligent devices in power system applications has become the major source of big data, which reflects on data storage, data processing, and data quality [3–5]. The failure of these intelligent devices makes data incomplete during the acquisition of energy consumption data. This incomplete data is commonly referred to as missing data [6–8]. Handling rather than ignoring this missing data drives toward better data analytics on energy consumption [9]. Hence, it is essential to analyze and impute the missing data in the smart home energy consumption data. Following this, several state-of-the-art works on missing data imputation and ensemble methods are discussed as follows.

The researchers suggested several ML-based imputation methods as well as thorough benchmarks for the comparison between conventional and modern methods [10]. An imputation algorithm "opt.impute" was introduced in [11] to achieve the finest solutions to the missing data. Further, an extensive review was conducted on the imputation of missing data using ML which helps in understanding the limitations of ML imputation methods [12]. A framework was implemented to improvise the multivariate imputation by chained equations (MICE) in imputing the missing sensor data [13]. A graph-based method was discussed in [14] to impute the missing sensor data. A copy-paste imputation method was introduced in [15] to impute the time-series data of energy. A mixture factor analysis method was discussed to estimate the missing data in the building's energy load [16]. Different imputation methods, viz., MICE, KNN, and RF-based imputation were implemented to impute the missing data in the sensor data of the internet of things [17]. A data splitting-based imputation method named "nullify the missing values before the imputation" was proposed to impute the missing data [18].

A new statistical and ML-based imputation method was implemented in [19] to impute missing data in the applications of power grids. A fuzzy inductive reasoning method was discussed to deal with the missing data during the forecasting process in smart grids [20]. A six-stage particle swarm optimization imputation method was implemented for smart meter data collected from an Indian institution [21]. An imputation method based on a denoising autoencoder was presented in [22]. An imputation model named "bagged averaging of multiple linear regression" was discussed in [23] for imputing missing data in phasor measurement units. A two-stage deep autoencoder-based data imputation method was discussed in [24] for imputing missing data in wind farms. A bagging algorithm was implemented to impute the missing data in time-series data [25]. An autoencoder neural network was presented to impute missing data for classification [26]. The appropriate selection of the best imputation method and classification was discussed in [27]. An extensive study on the packages available in "R" for data imputation was presented in [28]. Electricity theft detection in smart grids using various ML algorithms and deep learning techniques was discussed in [29,30]. An AdaBoost ensemble model was implemented to detect electricity theft [31]. An improvised ensemble model of a general regression neural network and successive geometric transformations model was presented in [32] to recover the partial or fully missed data.

In summary, the abovementioned literature discusses the concepts of big data, sources of big data, and energy data analytics. In addition, the importance of handling anomalies in big data was discussed. To handle the anomalies in energy consumption datasets, a few imputation methods such as data splitting, fuzzy inductive reasoning, denoising autoencoder, and bagging are used. Further, to evaluate their performance, various single classifiers, namely, SVM, neural networks, etc., are used. However, these approaches are found ineffective to address the hidden issues of smart home energy consumption data due to the presence of a variety of anomalies such as garbage data, outlier data, redundant data, missing data, etc.

On the other hand, in recent days, the ensemble classification approach is supporting effective classification in data imputation in different applications, which was not tried for the smart home energy consumption data. With this motivation, this paper proposes ML-based ensemble classifiers to handle all the possible anomalies in smart home energy consumption data. The major contributions of this paper are summarized as follows:

- The proposed approach initially identifies all anomalies and removes them, and then imputes this information. The entire implementation consists of four parts.
  - Part 1 (anomaly detection and removal) considers the original dataset and refines it by removing all the identified anomalies.
  - Part 2 (data imputation) considers this refined dataset and performs the missing data imputation using median, KNN, and bagging imputation methods, thereby producing an anomaly-free dataset.
  - Part 3 (single-classifier approaches) performs the classification of the dataset using the conventional single-classifier approaches such as RF, SVM, DT, NB, KNN, and NNET.
  - Part 4 (ensemble classifiers approaches) performs the classification of the dataset using the proposed ensemble classifier approaches such as RF+SVM+DT, RF+SVM+NB, RF+SVM+KNN, RF+SVM+NNET, RF+DT+NB, RF+DT+KNN, RF+DT+NNET, RF+NB+KNN, RF+NB+NNET, and RF+KNN+NNET.
- To assess the classifiers' performance, various metrics, namely, accuracy, precision, recall/sensitivity, specificity, and F1 score are computed. From these metrics, it is identified that the ensemble classifier "RF+SVM+DT" has shown superior performance over the conventional single classifiers as well the other ensemble classifiers for anomaly handling in smart home energy consumption data.

All these contributions are structured in the paper as follows. Section 2 presents the description of the dataset. Section 3 presents the description and implementation of the proposed approach. Section 4 presents simulation results and their discussion. Finally, Section 5 concludes the outcomes of the paper in a synopsized way.

2. Description of Dataset

To implement the proposed approach, the data of an appliance (refrigerator) from the Tracebase dataset [33] is considered. This dataset consists of 43 different appliances with 158 device IDs that are connected to various smart homes/buildings. Each appliance consists of CSV files that represent the energy consumption data of a day. A detailed description of this dataset can be obtained from [34]. Further, this dataset was considered and used in various literary works. The Tracebase dataset was used in the extensive study of different non-intrusive load monitoring (NILM) power consumption datasets described in [35–37]. The present and the future directions for energy management techniques using NILM datasets are discussed in [38].

The CSV file (dev_98C08A_2011.09.17.csv) data of the refrigerator appliance is prepared with the columns such as CAPTURED_DATE, CAPTURED_HOUR, CAPTURED_MINUTE, CAPTURED_SECOND, and CAPTURED_READING for implementing the proposed ensemble classifier approach.

3. Description and Implementation of the Proposed Approach

The conceptual model of the proposed approach is shown in Figure 1. It consists of four parts, viz., Part 1, Part 2, Part 3, and Part 4. The smart home energy consumption dataset will be given as input to Part 1. In Part 1, an analysis of the missing data will be carried out for understanding the missingness in the original dataset. Further, the identification and removal of different anomalies (viz., garbage data, outliers in the data, and redundant data) will be performed. From this, a dataset with the abovementioned anomalies removed will be produced and given as input to Part 2. In Part 2, the imputation of missing data will be completed. In Part 3, a single-classifier approach will be applied. This will provide a recommendation of the best single-classifier approach as the output. By taking this best single-classifier as the basis, the ensemble classifiers approach will be applied in Part 4. This will provide a recommendation of the best ensemble classifier to perform the imputation.

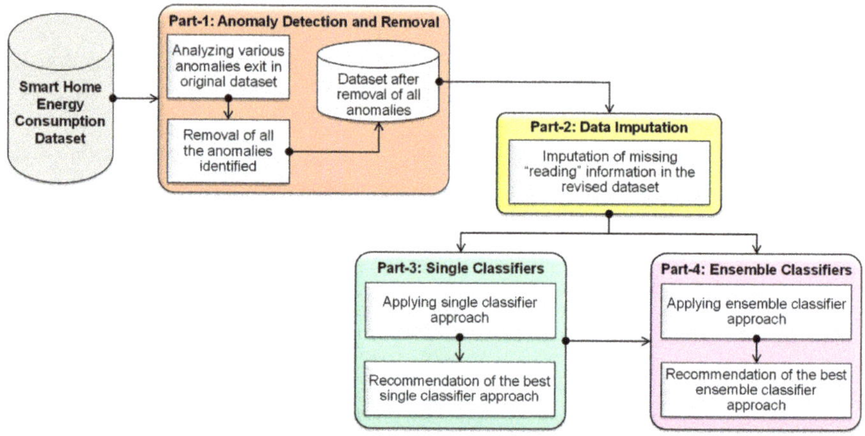

Figure 1. Conceptual model for the proposed approach.

The implementation flow of the proposed ensemble classifiers approach through all the proposed parts is shown in Figure 2. The detailed description and implementation processes are discussed in Sections 3.1–3.4 respectively for Part 1, Part 2, Part 3, and Part 4.

3.1. Implementation of Part 1 (Anomaly Detection and Removal)

The process starts from Part 1 by reading the smart home energy consumption dataset and saving it in an object "shec_dat". Initially, the missing data information in this "shec_dat" is analyzed [6]. Further, the process is continued with the identification of garbage data in the dataset. To identify the garbage data i.e., the data other than the numerical data, a function *grepl("[[:digit:]]* is used on each column of the dataset. If garbage data exists, those records are removed and the remaining data are given as input to identify the outliers data. If there is no garbage data, the existing dataset is used as it is and is given as input to identify the outliers data. The outliers data is the data that does not exist within the expected range. To identify outliers data, a boxplot analysis is applied to the data obtained after removing the garbage data. The boxplot analysis is a standardized approach to showing data distribution in a five-number summary (i.e., minimum, first quartile, median, third quartile, and maximum). The data that lies in between the "minimum" and "maximum" values are considered as the data within the range and useful for the analysis. The data that lies below the "minimum" and above the "maximum" values are considered outliers data and needs to be removed to achieve better analytics. The function boxplot() is applied to the readings column by using *boxplot(shec_dat$CAPTURED_READING, plot=F)$out*. If the outliers exist in the readings column, those records are removed and the remaining data are given as input to identify the redundant data. If there are no outliers, the existing dataset is used as it is to identify the redundant data. In general, redundant data refers to the duplication of the entire record in the dataset. However, in this case, there exist two types of redundant data in the dataset. They are the records with the same timestamp and same reading information, and records with the same timestamp and different reading information. The detailed process of identifying these types of redundant data is discussed in [39]. If the abovementioned types of redundant data exist, those records are removed. If there are no redundant data, the existing data is used as it is to perform the next step. At the end of Part 1, a dataset is obtained after removing all the anomalies (garbage data, outliers data, and redundant data). As several records in the dataset are removed due to the existence of different anomalies, this dataset consists of missing timestamps. Hence, these missing timestamps are filled, and the respective reading information is set to "NA (Not Available)" [8] before proceeding to the implementation of Part 2.

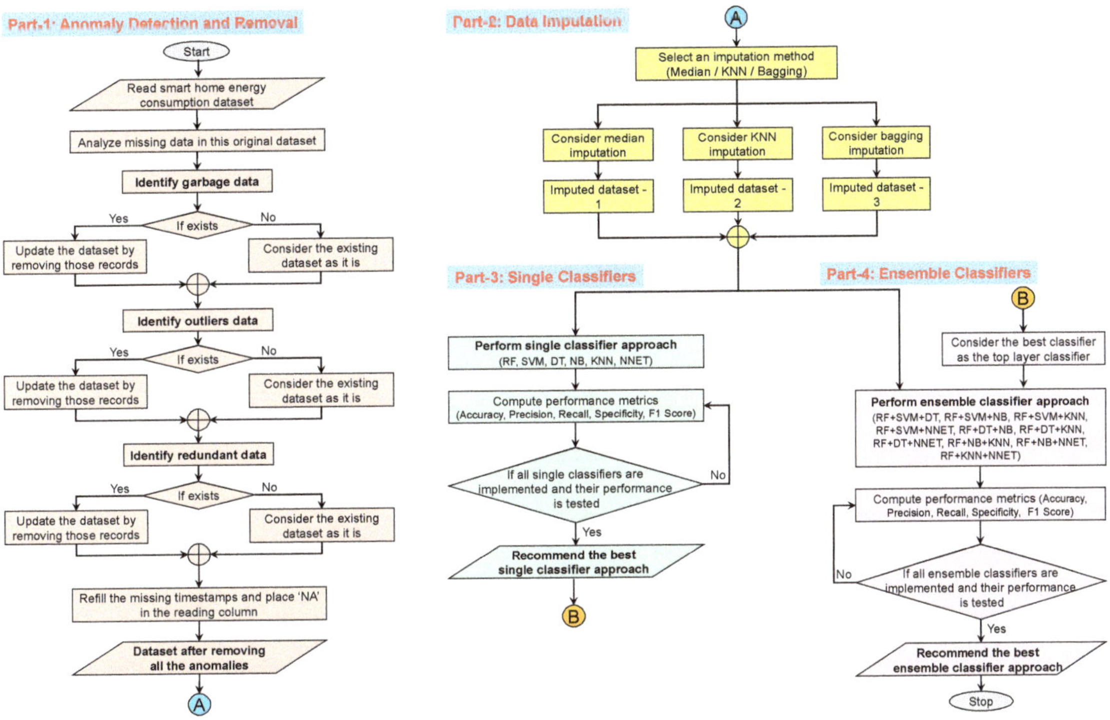

Figure 2. Implementation flow of the proposed approach.

3.2. Implementation of Part 2 (Data Imputation)

Once all the missing records are finalized in the dataset obtained after removing all the anomalies, the imputation methods such as median imputation, KNN imputation, and bagging imputation are applied. The implementation of these imputation methods produces datasets with imputed reading values. Further, the single-classifier approach is applied to these imputed datasets. The implementation of the median, KNN, and bagging imputation methods is discussed in Sections 3.2.1–3.2.3 respectively.

3.2.1. Implementation of the Median Imputation Method

In the median imputation method, the median value of the reading information in the CAPTURED_READING column is calculated, and that value is used for imputing the missing reading information. This imputation method is simple and fast. The process of calculating the median value starts with the ordering of readings information in ascending order. Once the ordering of readings information is done, then the number of values (odd or even) in the CAPTURED_READING is taken into consideration. Here, the number of values plays a major role in calculating the median value of the readings information. The formula for calculating the median value is given in Equation (1).

$$Median(D) = \begin{cases} D\left(\frac{s+1}{2}\right) & if\ s\ is\ odd \\ \frac{D\left(\frac{s}{2}\right)+D\left(\frac{s}{2}+1\right)}{2} & if\ s\ is\ even \end{cases} \quad (1)$$

where D = list of values ordered in the CAPTURED_READING column, and s = number of values in the CAPTURED_READING column.

If the number of values in the CAPTURED_READING column is odd, then the middle value is considered as the median. If the number of values is even in the CAPTURED_READING column, then the average of the middle two values is considered as the median.

3.2.2. Implementation of the KNN Imputation Method

In the KNN imputation method, the distance between the k-nearest neighbor values is calculated by using the Euclidean distance metric. In the CAPTURED_READING column, the distance between the k-closest samples of the readings is calculated and that distance value is used to impute the missing reading information. The formula for calculating Euclidean distance is given in Equation (2).

$$dist(p,q) = \sqrt{\sum_{i=1}^{m} (p_i - q_i)^2} \quad (2)$$

where $dist$ = Euclidean distance, m = number of points, and p_i & q_i are the points.

3.2.3. Implementation of the Bagging Imputation Method

In the bagging imputation method, the term 'bagging' refers to bootstrap aggregation. The bootstrap is a statistical technique of iteratively resampling the data with replacement in the dataset. To perform this, initially, the number of bootstrap samples is to be fixed, and then the sample size. For each sample of bootstrap the following steps are performed: draw the sample with replacement, fit the model, anticipate the performance of the model based on the out-of-bag sample, and calculate the average of the sample of the model. The multiple iterations of sampling improve the prediction performance of the model. The bagging method fits a bagged tree. This method is simple, powerful, and accurate to impute the missing values in the readings information. However, it is computationally high-cost.

3.3. *Implementation of Part 3 (Single-Classifier Approach)*

In this section, the single-classifier approach is performed using various classifiers, viz., RF, SVM, DT, NB, KNN, and NNET, for the classification. All these classifiers are implemented individually on the dataset. To implement these, the dataset is divided into train_set and test_set. These classifiers are trained on the train_set using k-fold cross-validation. Here, the k-value considered is 10. Further, these classifiers are applied to the test_set to predict the classes Yes (Y) or No (N). Here, class 'Y' represents missing data, and class 'N' represents non-missing data. After the implementation, the performance metrics such as accuracy, precision, recall/sensitivity, specificity, and F1 score are computed using a confusion matrix to evaluate each classifier's performance. The confusion matrix is shown in Figure 3 and the formulae for computing the performance metrics are given in Equations (3)–(7).

$$Accuracy = \frac{T.Pos. + T.Neg.}{T.Pos. + T.Neg. + F.Pos. + F.Neg.} \quad (3)$$

$$Precision = \frac{T.Pos.}{T.Pos. + F.Pos.} \quad (4)$$

$$Recall/Sensitivity = \frac{T.Pos.}{T.Pos. + F.Neg.} \tag{5}$$

$$Specificity = \frac{T.Neg.}{T.Neg. + F.Pos.} \tag{6}$$

$$F1Score = \frac{2 * (Precision * Recall)}{(Precision + Recall)} \tag{7}$$

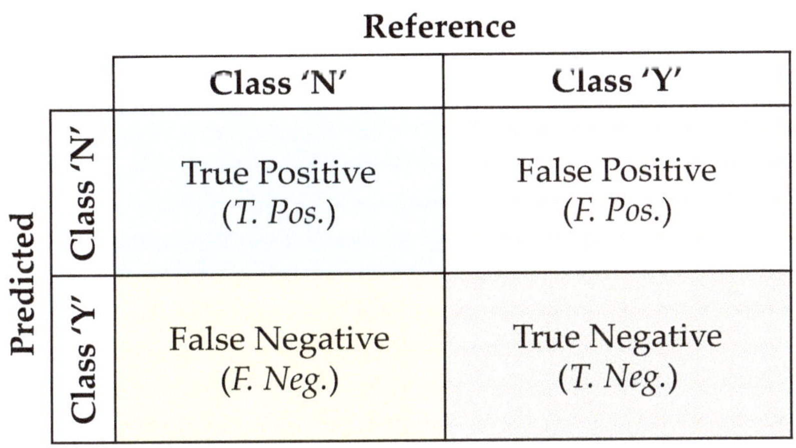

Figure 3. Confusion matrix.

If all the single classifiers are implemented and their performance is verified, then, the best single classifier is recommended. Otherwise, the performance metrics are re-verified.

3.4. Implementation of Part 4 (Ensemble Classifiers Approach)

This section uses the best single classifier recommended in Part3 as the input to develop ensemble classifiers. The ensemble of classifiers is performed using the "stacking" method. In stacking, there are two layers called the top layer and the bottom layer. The top layer consists of a classifier, which is referred to as a base classifier and the bottom layer consists of other classifiers. The output of the bottom layer is given as input to the top layer. The classifier used in the top layer is an ensemble with the output of the bottom layer classifiers, which produces an ensemble classifier. The stacking of classifiers is shown in Figure 4. From this figure, it is seen that the single classifiers used in the bottom layer are an ensemble with the recommended best classifier used in the top layer. For example, the single classifiers SVM and DT are part of the ensemble with the recommended best classifier RF. Similarly, all the other single classifiers form an ensemble with RF and produce ensemble classifiers. To implement these ensemble classifiers, the imputed datasets are given as input. Further, each imputed dataset is divided into train_set and test_set. The ensemble classifiers are trained on the train_set using k-fold cross-validation. Here, the k-value considered is 10. Further, these ensemble classifiers are applied to test_set to predict the classes Y or N. After the implementation, the performance metrics such as accuracy, precision, recall/sensitivity, specificity, and F1 score are computed using a confusion matrix to evaluate each ensemble classifier's performance. If all the ensemble classifiers are implemented and their performance is verified then the best ensemble classifier for the imputation is recommended, otherwise the performance metrics are re-verified.

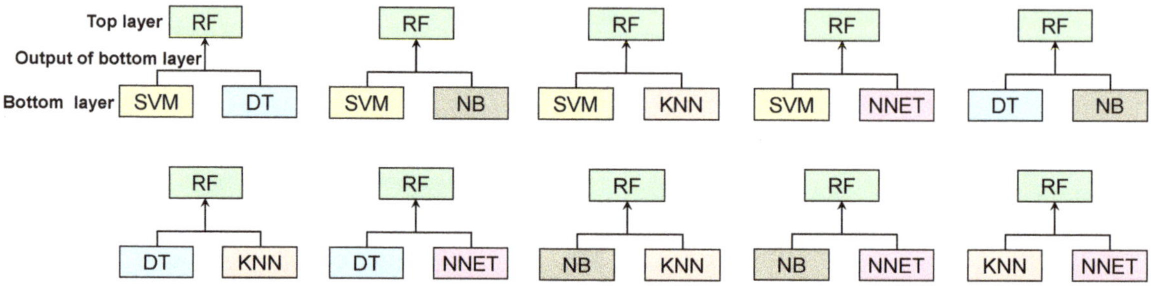

Figure 4. Stacking of classifiers.

4. Simulation Results and Discussion

In keeping with the aims of the paper, the simulation results of the implementation are presented in three subsections. Sections 4.1–4.3 present the results corresponding to anomaly detection and removal, single-classifier approach, and ensemble classifiers approach, respectively.

4.1. Results Corresponding to Anomaly Detection and Removal

This section presents the details of the missing data in the original CSV file (original dataset) and the missing data in the dataset after eliminating the anomalies. The number of records in this original dataset is 155,374. During the analysis of missing data, 700 records are missed in the original dataset [7]. During the identification of garbage data, no garbage data (other than numerical data) are identified in the original CSV file. Hence, no records are removed and the same number of records (155,374) are available. During the identification of outliers data, there are 25 readings identified as outliers and the respective records are removed from the dataset. The removal of records with outliers left the dataset with 155,349 records. During the identification of redundant data, the records with the same timestamp and same reading are identified and those records are removed from the dataset. This removal left the dataset with 98,779 records. Further, the records with the same timestamp and different readings are identified and those records are removed from the dataset. This removal left the dataset with 72,597 records. Once the redundant data are removed, the missing data are filled with the respective timestamps and the respective reading with NA value, as shown in Figure 5 (all the highlighted rows). After this filling, there are 86,400 records in the dataset, out of these, 13,803 records contain missing readings.

The proportions of the available data and missing data in the original dataset and the dataset available after removing anomalies are shown in Figure 6a–c. These figures show the proportion of the missing data and available data in the considered dataset in three different scenarios, namely, (i) consideration of the original dataset, (ii) consideration of the dataset that is obtained after removing the anomalies, and (iii) consideration of the dataset after filling the missing timestamps, ready for the imputation.

From Figure 6a, it is understood that the proportion of available data is 99.55% and missing data is 0.45% in the original dataset. From Figure 6b, it is seen that the proportion of available data is 84% and missing data is 16% in all columns of the dataset obtained after removing anomalies. From Figure 6c, it is evident that there are no missing data in the columns CAPTURED_DATE, CAPTURED_HOUR, CAPTURED_MINUTE, CAPTURED_SECOND and the proportion of data availability is 84%. Further, there are missing readings in the column CAPTURED_READING with a proportion of 16%.

CAPTURED_DATE	CAPTURED_HOUR	CAPTURED_MINUTE	CAPTURED_SECOND	CAPTURED_READING
17-09-2011	0	0	0	NA
17-09-2011	0	0	1	126
17-09-2011	0	0	2	126
17-09-2011	0	0	3	126
17-09-2011	0	0	4	NA
17-09-2011	0	0	5	NA
17-09-2011	0	0	6	126
17-09-2011	0	0	7	126
17-09-2011	0	0	8	128
17-09-2011	0	0	9	NA
17-09-2011	0	0	10	126
17-09-2011	0	0	11	126
17 09 2011	0	0	12	128
17-09-2011	0	0	13	126
17-09-2011	0	0	14	126
17-09-2011	0	0	15	126
17-09-2011	0	0	16	126
17-09-2011	0	0	17	NA
17-09-2011	0	0	18	126

Figure 5. Data after filling missing timestamps and placing NA.

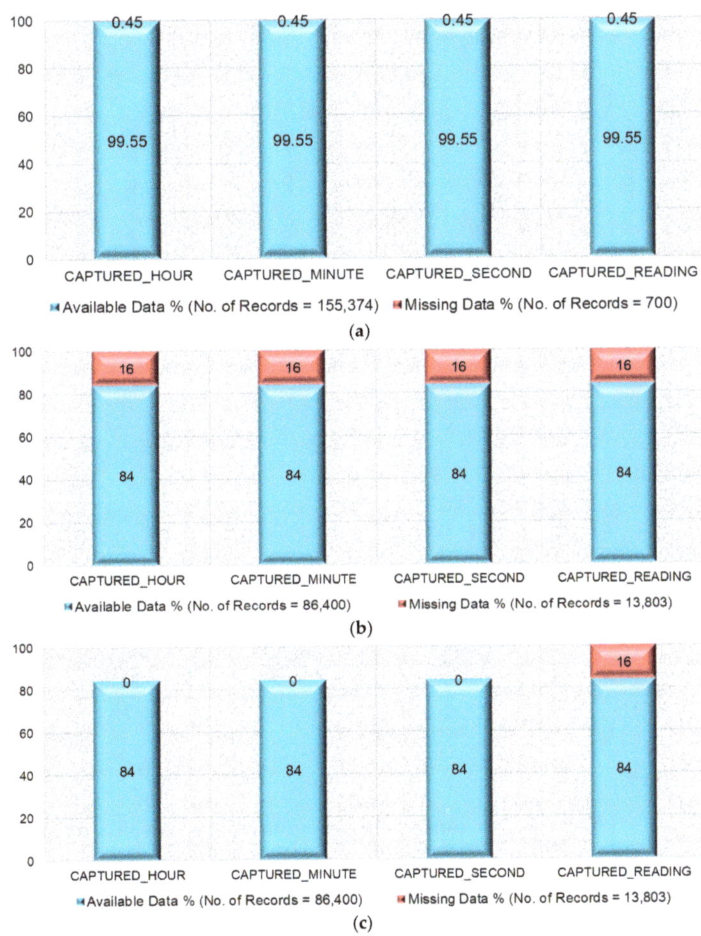

Figure 6. Proportion of missing data in the dataset. (**a**) Original dataset. (**b**) After removing the anomalies. (**c**) After filling missing timestamps.

4.2. Results Corresponding to the Single-Classifier Approach

This section presents the performance of a single-classifier approach on the imputed datasets. The performance of classifiers in the median, KNN, and bagging imputation methods are discussed in Sections 4.2.1–4.2.3, respectively.

4.2.1. Performance of the Single-Classifier Approach in the Median Imputation Method

The performance metrics of each classifier are shown in Figure 7, where the red colored bar(s) indicate the highest value achieved corresponding to that particular metric. From this, the highest accuracy value of 98.1% is observed in RF, while the lowest accuracy value of 76.3% is observed in KNN, as shown in Figure 7a. The highest precision value of 99% is observed in RF, while the lowest precision value of 80.5% is observed in SVM and NB, as shown in Figure 7b. The highest recall value of 100% is observed in SVM and NB, while the lowest recall value of 87.9% is observed in KNN, as shown in Figure 7c. The highest specificity value of 95.9% is observed in RF, while the lowest specificity value of 0% is observed in SVM and NB, as shown in Figure 7d. The highest F1 Score value of 98.8% is observed in RF, while the lowest F1 Score value of 85.7% is observed in KNN, as shown in Figure 7e. From the subplots in Figure 7a–e, it is understood that the classifier RF has outperformed the others. Further, the performance summary of all the single classifiers is given in Table 1.

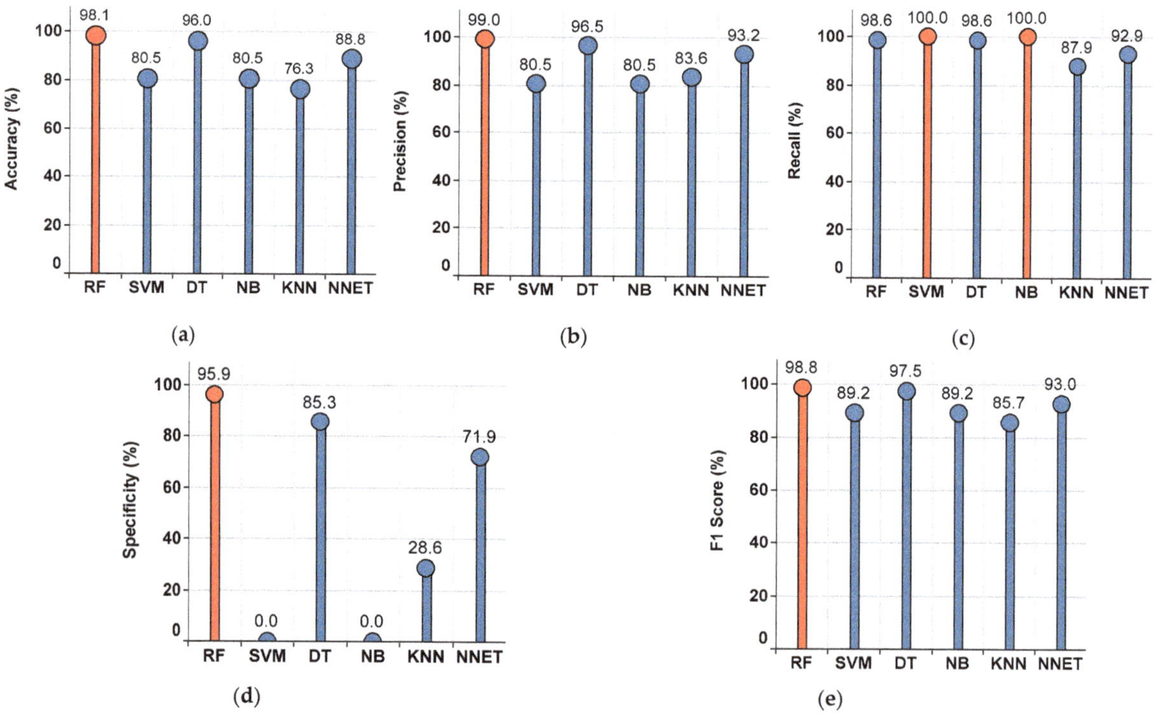

Figure 7. Performance metrics for the single-classifier approach on the median imputed dataset. (**a**) Accuracy (**b**) Precision. (**c**) Recall. (**d**) Specificity. (**e**) F1 Score.

Table 1. Performance comparison of single-classifier approach on the median imputed dataset.

Classifier	Accuracy (%)	Precision (%)	Recall (%)	Specificity (%)	F1 Score (%)
RF	98.1	99	98.6	95.9	98.8
SVM	80.5	80.5	100	0	89.2
DT	96	96.5	98.6	85.3	97.5
NB	80.5	80.5	100	0	89.2
KNN	76.3	83.6	87.9	28.6	85.7
NNET	88.8	93.2	92.9	71.9	93
Superior Classifier	RF	RF	SVM, NB	RF	RF

4.2.2. Performance of the Single-Classifier Approach in the KNN Imputation Method

The performance metrics of each classifier are shown in Figure 8, where the red colored bar(s) indicate the highest value achieved corresponding to that particular metric. From this, the highest accuracy value of 87.7% is observed in RF, while the lowest accuracy value of 68% is observed in NNET, as shown in Figure 8a. The highest precision value of 86.6% is observed in RF, while the lowest precision value of 80.3% is observed in NNET, as shown in Figure 8b. The highest recall value of 100% is observed in RF, SVM, DT, NB, and KNN, while the lowest recall value of 79.9% is observed in NNET, as shown in Figure 8c. The highest specificity value of 36.3% is observed in RF, while the lowest specificity value of 0% is observed in SVM and KNN, as shown in Figure 8d. The highest F1 Score value of 92.8% is observed in RF, while the lowest F1 Score value of 80.1% is observed in NNET, as shown in Figure 8e. From the subplots in Figure 8a–e, it is understood that the classifier RF has outperformed the others. Further, the percentage summary of all classifiers is given in Table 2.

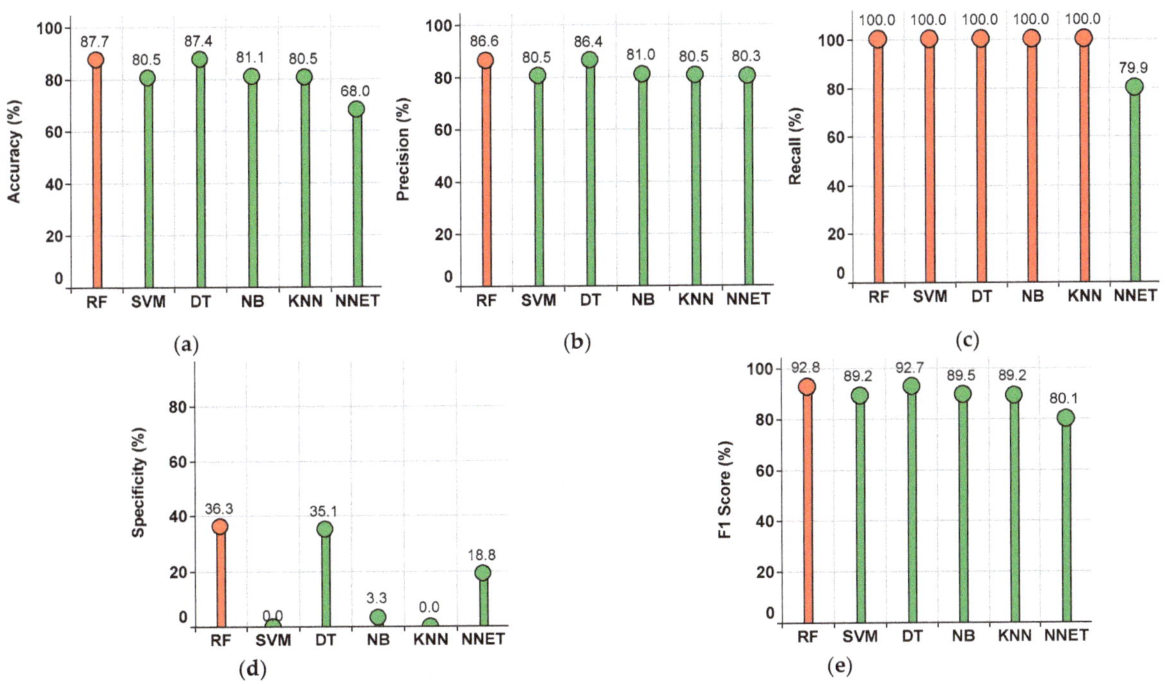

Figure 8. Performance metrics for the single-classifier approach on the KNN imputed dataset. (**a**) Accuracy. (**b**) Precision. (**c**) Recall. (**d**) Specificity. (**e**) F1 Score.

Table 2. Performance comparison of the single-classifier approach on the KNN imputed dataset.

Classifier	Accuracy (%)	Precision (%)	Recall (%)	Specificity (%)	F1 Score (%)
RF	87.7	86.6	100	36.3	92.8
SVM	80.5	80.5	100	0	89.2
DT	87.4	86.4	100	35.1	92.7
NB	81.1	81	100	3.3	89.5
KNN	80.5	80.5	100	0	89.2
NNET	68	80.3	79.9	18.8	80.1
Superior Classifier	RF	RF	All except NNET	RF	RF

4.2.3. Performance of the Single-Classifier Approach in the Bagging Imputation Method

The performance metrics of each classifier are shown in Figure 9, where the red colored bar(s) indicate the highest value achieved corresponding to that particular metric. From this, the highest accuracy value of 95.2% is observed in RF, while the lowest accuracy value of 75.7% is observed in NNET, as shown in Figure 9a. The highest precision value of 100% is observed in RF and DT, while the lowest precision value of 79.5% is observed in NNET, as shown in Figure 9b. The highest recall value of 100% is observed in SVM, and NB, while the lowest recall value of 84.3% is observed in DT, as shown in Figure 9c. The highest specificity value of 100% is observed in RF and DT, while the lowest specificity value of 0% is observed in SVM, NB, and NNET, as shown in Figure 9d. The highest F1 Score value of 96.9% is observed in RF, while the lowest F1 Score value 86.1% is observed in NNET, as shown in Figure 9e. From the subplots Figure 9a–e, it is understood that the classifier RF has outperformed the others. Further, the percentage summary of all classifiers is given in Table 3.

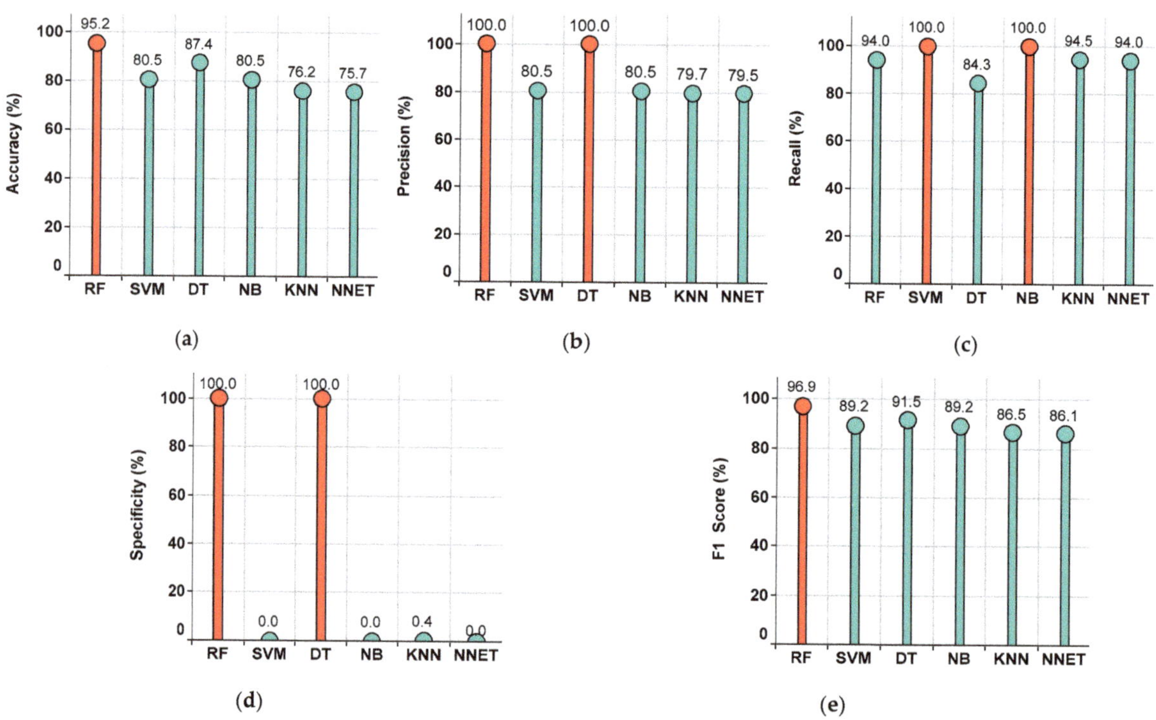

Figure 9. Performance metrics for the single-classifier approach on the bagging imputed dataset. (**a**) Accuracy. (**b**) Precision. (**c**) Recall. (**d**) Specificity. (**e**) F1 Score.

Table 3. Performance comparison of the single-classifier approach on the bagging imputed dataset.

Classifier	Accuracy (%)	Precision (%)	Recall (%)	Specificity (%)	F1 Score (%)
RF	95.2	100	94	100	96.9
SVM	80.5	80.5	100	0	89.2
DT	87.4	100	84.3	100	91.5
NB	80.5	80.5	100	0	89.2
KNN	76.2	79.7	94.5	0.4	86.5
NNET	75.7	79.5	94	0	86.1
Superior Classifier	RF	RF, DT	SVM, NB	RF, DT	RF

4.3. Results Corresponding to the Ensemble Classifiers Approach

This section presents the performance of the ensemble classifiers approaches on the imputed datasets. The performance of ensemble classifiers in the median, KNN, and bagging imputation methods are discussed in Sections 4.3.1–4.3.3 respectively.

4.3.1. Performance of the Ensemble Classifiers Approach in the Median Imputation Method

The performance metrics of each ensemble classifier are shown in Figure 10, where the red colored bar(s) indicate the highest value achieved corresponding to that particular metric. From this, the highest accuracy value of 98.9% is observed in RF+SVM+DT and RF+DT+NNET, while the lowest accuracy value of 72.5% is observed in RF+NB+KNN, as shown in Figure 10a. The highest precision value of 99.4% is observed in RF+SVM+NB, while the lowest precision value of 77.5% is observed in RF+SVM+KNN, as shown in Figure 10b.

The highest recall value of 100% is observed in RF+SVM+DT, RF+DT+NB, RF+DT+NNET, RF+NB+NNET, and RF+KNN+NNET, while the lowest recall value of 81.6% is observed in RF+SVM+NB, as shown in Figure 10c. The highest specificity value of 94.5% is observed in RF+SVM+DT, RF+DT+NB, RF+DT+NNET, and RF+KNN+NNET, while the lowest specificity value of 34.7% is observed in RF+NB+KNN, as shown in Figure 10d.

The highest F1 Score value of 99.3% is observed in RF+SVM+DT, RF+DT+NB, RF+DT+NNET, and RF+KNN+NNET, while the lowest F1 Score value of 82.2% is observed in RF+SVM+KNN and RF+NB+KNN, as shown in Figure 10e. From the subplots in Figure 10a–e, it is understood that the ensemble classifiers RF+SVM+DT and RF+DT+NNET have outperformed the others.

Further, the performance summary of all ensemble classifiers with respect to various parameters is given in Table 4.

Table 4. Performance comparison of the ensemble classifiers approach on the median imputed dataset.

Classifier	Accuracy (%)	Precision (%)	Recall (%)	Specificity (%)	F1 Score (%)
RF+SVM+DT	98.9	98.6	100	94.5	99.3
RF+SVM+NB	81.5	99.4	81.6	76.5	89.7
RF+SVM+KNN	73	77.5	87.5	36.8	82.2
RF+SVM+NNET	91.7	99	91.4	93.6	95
RF+DT+NB	98.8	98.6	100	94.5	99.3
RF+DT+KNN	92.5	98.9	92.4	93.7	95.5
RF+DT+NNET	98.9	98.6	100	94.5	99.3
RF+NB+KNN	72.5	78.9	85.9	34.7	82.2
RF+NB+NNET	88.3	85.5	100	62.6	92.2
RF+KNN+NNET	98.8	98.6	100	94.5	99.3
Superior Classifier	RF+SVM+DT RF+DT+NNET	RF+SVM+NB	RF+SVM+DT RF+DT+NB RF+DT+NNET RF+NB+NNET RF+KNN+NNET	RF+SVM+DT RF+DT+NB RF+DT+NNET RF+KNN+NNET	RF+SVM+DT RF+DT+NB RF+DT+NNET RF+KNN+NNET

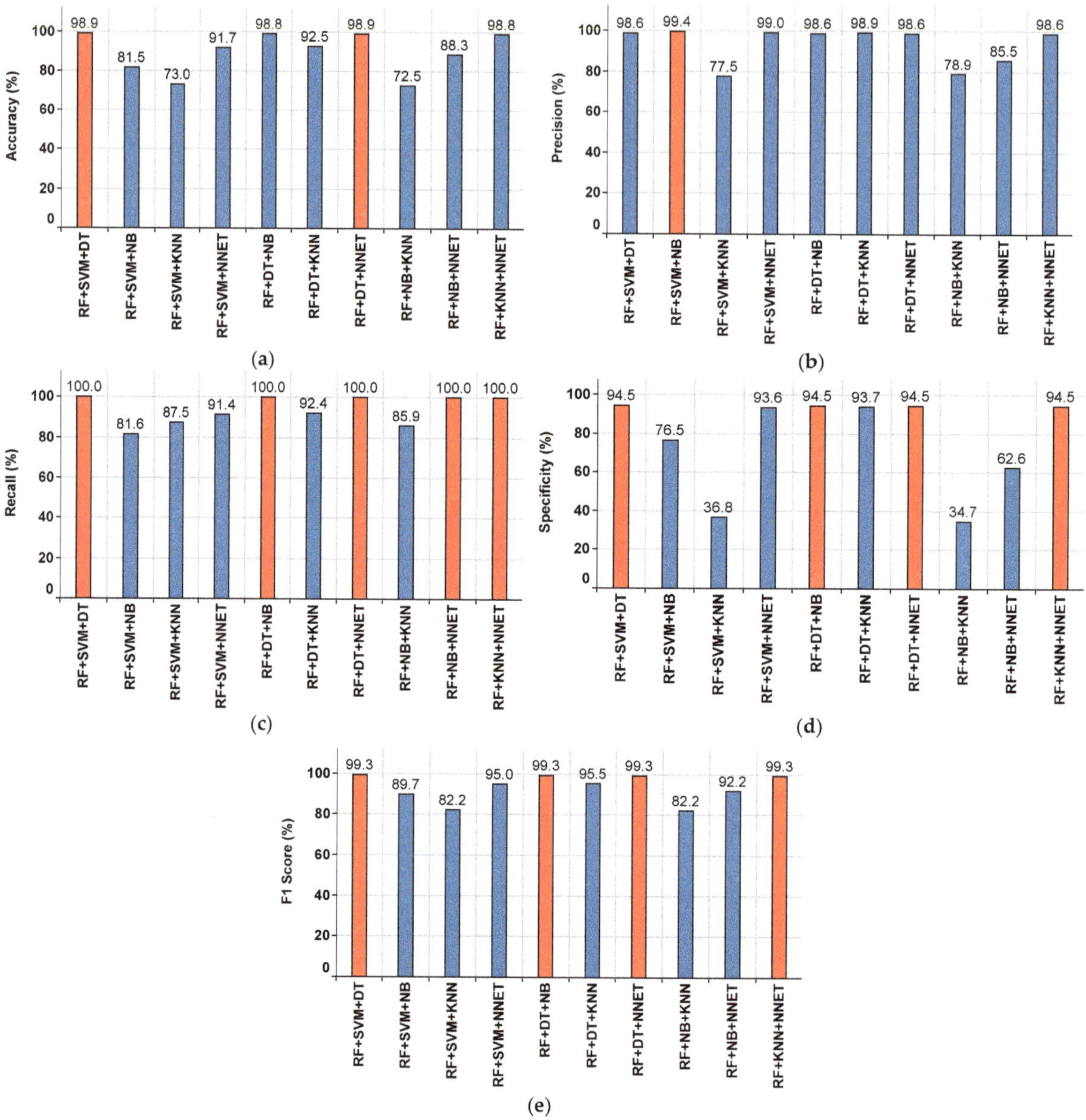

Figure 10. Performance metrics for the ensemble classifiers approach on the median imputed dataset. (**a**) Accuracy. (**b**) Precision. (**c**) Recall. (**d**) Specificity. (**e**) F1 Score.

4.3.2. Performance of the Ensemble Classifiers Approach in the KNN Imputation Method

The performance metrics of each ensemble classifier are shown in Figure 11, where the red colored bar(s) indicate the highest value achieved corresponding to that particular metric. From this, the highest accuracy value of 80.2% is observed in RF+DT+KNN, while the lowest accuracy value of 70.9% is observed in RF+SVM+KNN, as shown in Figure 11a. The highest precision value of 99.3% is observed in RF+DT+KNN, while the lowest precision value of 81.3% is observed in RF+NB+NNET, as shown in Figure 11b.

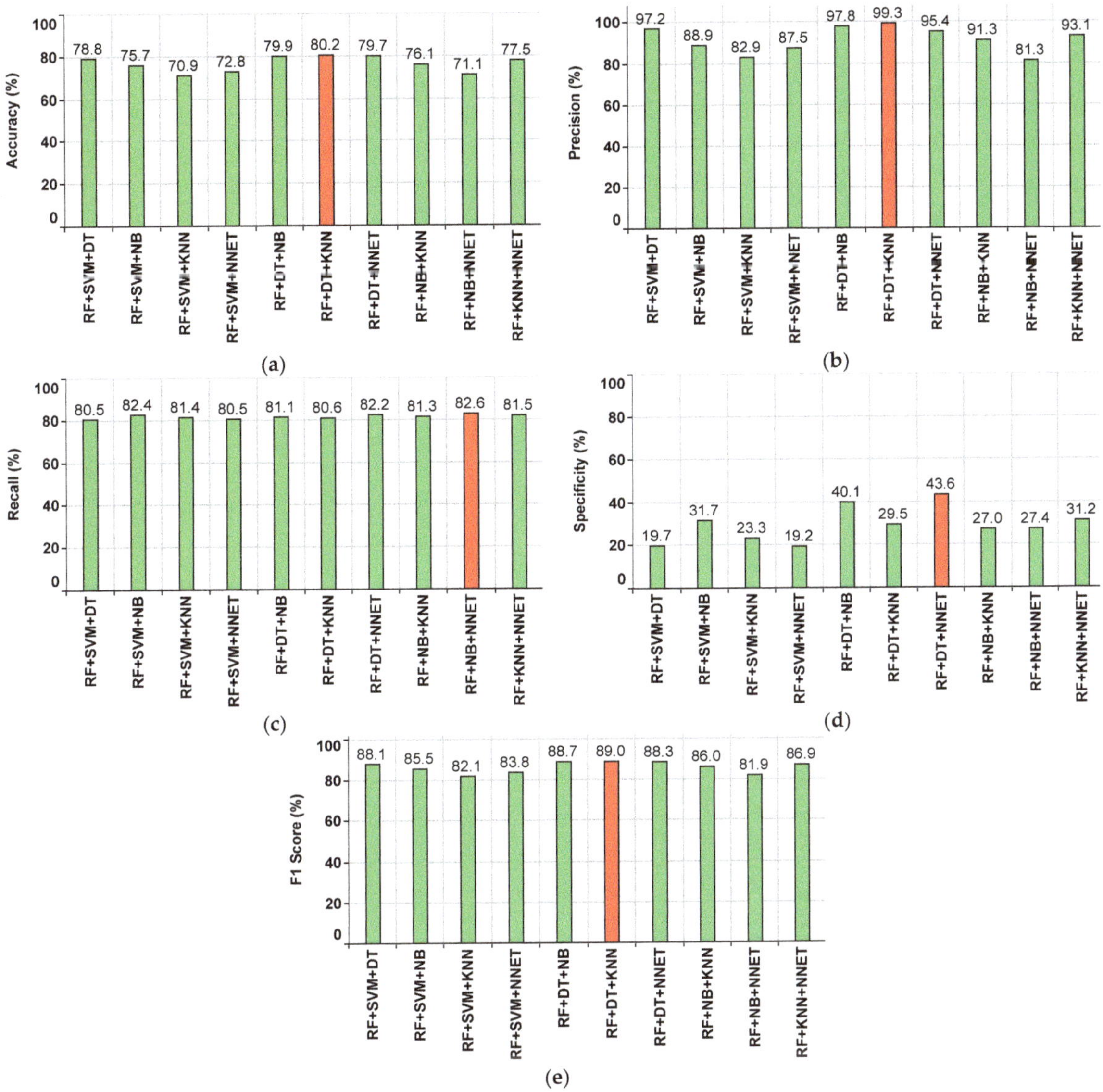

Figure 11. Performance metrics for the ensemble classifiers approach on the KNN imputed dataset. (**a**) Accuracy. (**b**) Precision. (**c**) Recall. (**d**) Specificity. (**e**) F1 Score.

The highest recall value of 82.6% is observed in RF+NB+NNET, while the lowest recall value of 80.5% is observed in RF+SVM+DT, RF+SVM+NNET, as shown in Figure 11c. The highest specificity value of 43.6% is observed in RF+DT+NNET, while the lowest specificity value of 19.2% is observed in RF+SVM+NNET, as shown in Figure 11d. The highest F1 Score value of 89% is observed in RF+DT+KNN, while the lowest F1 Score value of 81.9% is observed in RF+NB+NNET, as shown in Figure 11e. From the subplots in Figure 11a–e, it is understood that the ensemble classifier RF+DT+KNN has outperformed the others. Further, the performance summary of all ensemble classifiers with respect to various parameters is given in Table 5.

Table 5. Performance comparison of the ensemble classifiers approaches on the KNN imputed dataset.

Classifier	Accuracy (%)	Precision (%)	Recall (%)	Specificity (%)	F1 Score (%)
RF+SVM+DT	78.8	97.2	80.5	19.7	88.1
RF+SVM+NB	75.7	88.9	82.4	31.7	85.5
RF+SVM+KNN	70.9	82.9	81.4	23.3	82.1
RF+SVM+NNET	72.8	87.5	80.5	19.2	83.8
RF+DT+NB	79.9	97.8	81.1	40.1	88.7
RF+DT+KNN	80.2	99.3	80.6	29.5	89
RF+DT+NNET	79.7	95.4	82.2	43.6	88.3
RF+NB+KNN	76.1	91.3	81.3	27	86
RF+NB+NNET	71.1	81.3	82.6	27.4	81.9
RF+KNN+NNET	77.5	93.1	81.5	31.2	86.9
Superior Classifier	RF+DT+KNN	RF+DT+KNN	RF+NB+NNET	RF+DT+NNET	RF+DT+KNN

4.3.3. Performance of the Ensemble Classifiers Approach in the Bagging Imputation Method

The performance metrics of each ensemble classifier are shown in Figure 12, where the red colored bar(s) indicate the highest value achieved corresponding to that particular metric. From this, the highest accuracy value of 89.6% is observed in RF+SVM+DT, while the lowest accuracy value of 71.2% is observed in RF+SVM+KNN, as shown in Figure 12a. The highest precision value of 98.8% is observed in RF+SVM+DT, while the lowest precision value of 86.5% is observed in RF+SVM+KNN, as shown in Figure 12b. The highest recall value of 89.4% is observed in RF+SVM+DT, while the lowest recall value of 78.8% is observed in RF+SVM+NNET, as shown in Figure 12c.

The highest specificity value of 91.1% is observed in RF+SVM+DT, while the lowest specificity value of 0.2% is observed in RF+SVM+NNET, as shown in Figure 12d. The highest F1 Score value of 93.9% is observed in RF+SVM+DT, while the lowest F1 Score value 82.8% is observed in RF+SVM+KNN, as shown in Figure 12e. From the subplots in Figure 12a–e, it is understood that the ensemble classifier RF+SVM+DT has outperformed the others.

Further, the performance summary of all ensemble classifiers with respect to various parameters is given in Table 6.

Table 6. Performance comparison of the ensemble classifiers approaches on the bagging imputed dataset.

Classifier	Accuracy (%)	Precision (%)	Recall (%)	Specificity (%)	F1 Score (%)
RF+SVM+DT	89.6	98.8	89.4	91.1	93.9
RF+SVM+NB	75.7	93.3	79.9	8.9	86.1
RF+SVM+KNN	71.2	86.5	79.5	12.5	82.8
RF+SVM+NNET	72.4	89.9	78.8	0.2	84
RF+DT+NB	77.9	96.8	80	0.7	87.6
RF+DT+KNN	75.9	93.2	80.2	14.5	86.2
RF+DT+NNET	75.9	93.2	80.2	14.5	86.2
RF+NB+KNN	77	94.5	80.4	16.9	86.8
RF+NB+NNET	78.9	93.9	82.3	40	87.7
RF+KNN+NNET	75.2	92.6	79.8	9.3	85.7
Superior Classifier	RF+SVM+DT	RF+SVM+DT	RF+SVM+DT	RF+SVM+DT	RF+SVM+DT

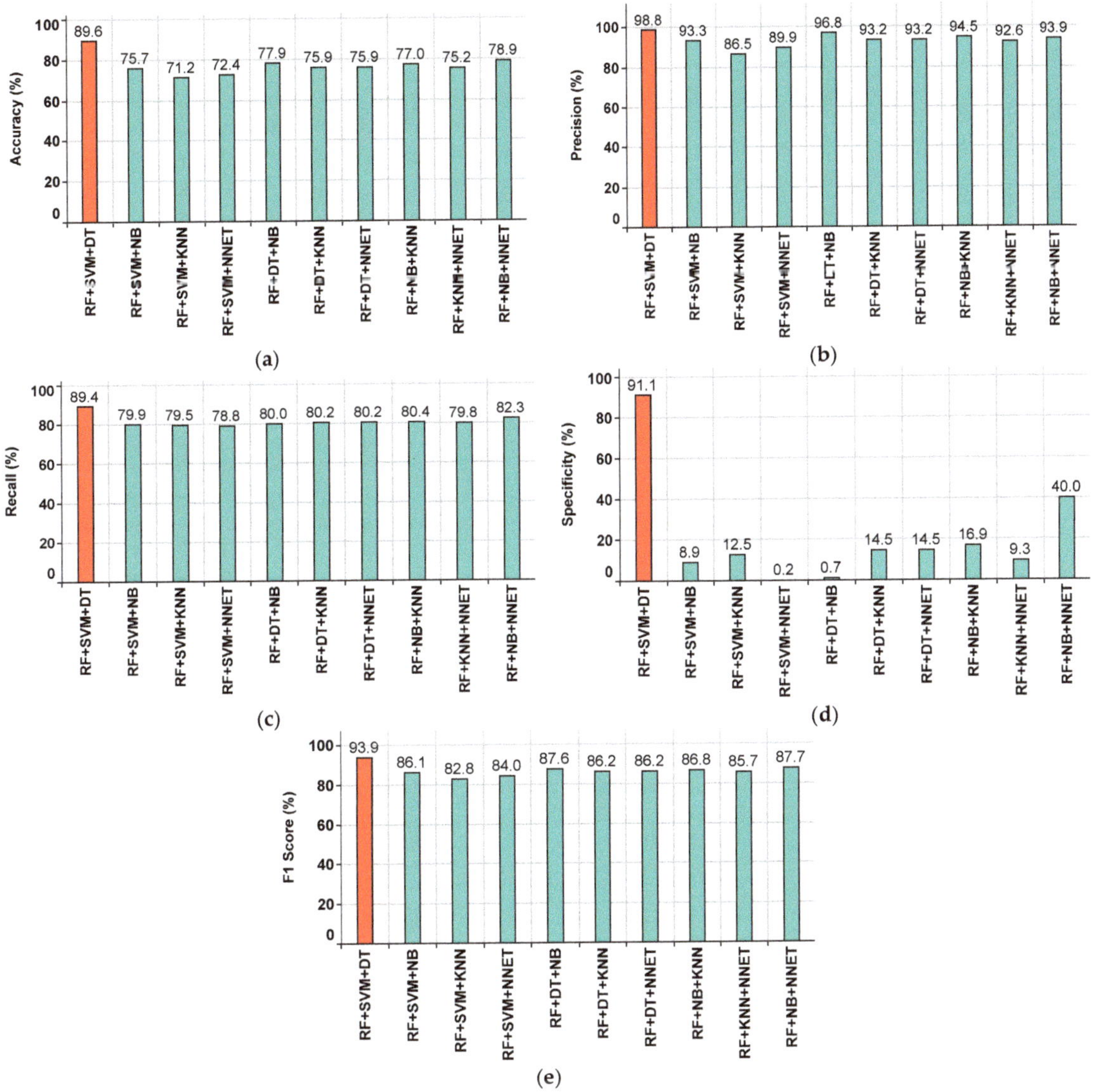

Figure 12. Performance metrics for ensemble classifiers approach on bagging imputed dataset. (**a**) Accuracy. (**b**) Precision. (**c**) Recall. (**d**) Specificity. (**e**) F1 Score.

5. Conclusions

This paper proposes a machine learning-based ensemble classifiers approach to address the anomalies present in smart homes' energy consumption data. This proposed approach has proven to be more effective than the conventional single-classifier approach that is presented in the literature. The salient observations from this work are summarized as follows:

- All the possible anomalies are successfully identified and removed from the dataset. The number of records in the original dataset is 155,374 and the number of records available in the refined dataset after removing anomalies is 86,400, which is the actual expected number of records as per the dataset description.

- Out of 86,400 records, 13,803 records are identified as records with missing data. This missing data has been successfully imputed by using various imputation methods (median, KNN, and bagging).
- To assess the process of imputation, various conventional single-classifier approaches, as well as the proposed ensemble classifiers approaches, are implemented. From the computation of the performance metrics (accuracy, precision, recall/sensitivity, specificity, and F1 score), the RF classifier is identified as the superior single-classifier to all other single classifiers.
- Out of the proposed ensemble classifiers, "RF+SVM+DT" has shown superior performance over the conventionally best single classifier (RF) as well the other ensemble classifiers for imputing the missing reading information.

Thus, the proposed ensemble classifiers approach has successfully handled anomalies that exist in the smart home energy consumption data.

Impacts and Implications of the Work

The proposed work in this paper helps in data preprocessing by the cleansing of data, which is typically essential to carry out precise analytics, and thereby, take superior decisions for energy management in smart buildings. Furthermore, the outcome of this work helps as a ready reference to understand the irregularities of the live data captured in a smart building/home/grid application for better data analytics. This impacts one of the important objectives of "United Nations Sustainable Development Goals (UN SDGs)—SDG 7: Energy" in producing an anomaly-free dataset for providing several customer services.

In addition, the identification of different data anomalies, viz., missing data, outliers data, garbage data, and redundant data in the energy consumption dataset, may be applied to the malfunctioning of metering infrastructure, failure/glitches of communication channels, cyber-attacks, energy thefts, unanticipated situations in power networks, etc.

Author Contributions: Conceptualization, P.P.K.; formal analysis, G.L.K.M.; funding acquisition, G.L.K.M.; investigation, P.P.K.; project administration, R.K.; software, P.P.K. and G.L.K.M.; supervision, Y.V.P.K.; Validation, Y.V.P.K.; writing—original draft, P.P.K.; writing—review & editing, Y.V.P.K. and R.K. All authors have read and agreed to the published version of the manuscript.

Funding: This research received no external funding.

Institutional Review Board Statement: Not applicable.

Informed Consent Statement: Not applicable.

Data Availability Statement: Not applicable.

Acknowledgments: The authors would like to thank VIT-AP University, Amaravati, Andhra Pradesh, India for funding the open-access publication fee for this research work.

Conflicts of Interest: The authors declare no conflict of interest.

Abbreviations

CSV	comma-separated values
DT	decision tree
KNN	K-nearest neighbour
ML	machine learning
MICE	multivariate imputation by chained equations
NA	not available
NB	naive Bayes
NILM	non-intrusive load monitoring
NNET	neural networks
RF	random forest
SVM	support vector machine

References

1. Firmani, D.; Mecella, M.; Scannapieco, M.; Batini, C. On the Meaningfulness of "Big Data Quality". *Data Sci. Eng.* **2016**, *1*, 6–20. [CrossRef]
2. Chen, W.; Zhou, K.; Yang, S.; Wu, C. Data Quality of Electricity Consumption Data in a Smart Grid Environment. *Renew. Sustain. Energy Rev.* **2017**, *75*, 98–105. [CrossRef]
3. Tu, C.; He, X.; Shuai, Z.; Jiang, F. Big Data Issues in Smart Grid—A Review. *Renew. Sustain. Energy Rev.* **2017**, *79*, 1099–1107. [CrossRef]
4. Ghorbanian, M.; Dolatabadi, S.H.; Siano, P. Big Data Issues in Smart Grids: A Survey. *IEEE Syst. J.* **2019**, *13*, 4158–4168. [CrossRef]
5. Bhattarai, B.P.; Paudyal, S.; Luo, Y.; Mohanpurkar, M.; Cheung, K.; Tonkoski, R.; Hovsapian, R.; Myers, K.S.; Zhang, R.; Zhao, P.; et al. Big Data Analytics in Smart Grids: State-of-the-art, Challenges, Opportunities, and Future Directions. *IET Smart Grid* **2019**, *2*, 141–154. [CrossRef]
6. Kasaraneni, P.P.; Yellapragada, V.P.K. Simple and Effective Descriptive Analysis of Missing Data Anomalies in Smart Home Energy Consumption Readings. *J. Energy Syst.* **2021**, *5*, 199–220. [CrossRef]
7. Kasaraneni, P.P.; Yellapragada Venkata, P.K. Analytical Approach to Exploring the Missing Data Behavior in Smart Home Energy Consumption Dataset. *J. Renew. Energy Environ.* **2022**, *9*, 37–48. [CrossRef]
8. Kasaraneni, P.P.; Yellapragada, V.P.K. Systematic Statistical Analysis to Ascertain the Missing Data Patterns in Energy Consumption Data of Smart Homes. *Int. J. Renew. Energy Res.* **2022**, *12*, 1560–1573. [CrossRef]
9. Emmanuel, T.; Maupong, T.; Mpoeleng, D.; Semong, T.; Mphago, B.; Tabona, O. A Survey on Missing Data in Machine Learning. *J. Big Data* **2021**, *8*, 140. [CrossRef]
10. Jäger, S.; Allhorn, A.; Bießmann, F. A Benchmark for Data Imputation Methods. *Front. Big Data* **2021**, *4*, 693674. [CrossRef]
11. Dimitris, B.; Pawlowski, C.; Ying, D.Z. From Predictive Methods to Missing Data Imputation: An Optimization Approach. *J. Mach. Learn. Res.* **2018**, *18*, 1–39. Available online: https://jmlr.csail.mit.edu/papers/volume18/17-073/17-073.pdf (accessed on 2 November 2022).
12. Alabadla, M.; Sidi, F.; Ishak, I.; Ibrahim, H.; Affendey, L.S.; Che Ani, Z.; Jabar, M.A.; Bukar, U.A.; Devaraj, N.K.; Muda, A.S.; et al. Systematic Review of Using Machine Learning in Imputing Missing Values. *IEEE Access* **2022**, *10*, 44483–44502. [CrossRef]
13. Wu, R.; Hamshaw, S.D.; Yang, L.; Kincaid, D.W.; Etheridge, R.; Ghasemkhani, A. Data Imputation for Multivariate Time Series Sensor Data with Large Gaps of Missing Data. *IEEE Sens. J.* **2022**, *22*, 10671–10683. [CrossRef]
14. Jiang, X.; Tian, Z.; Li, K. A Graph-Based Approach for Missing Sensor Data Imputation. *IEEE Sens. J.* **2021**, *21*, 23133–23144. [CrossRef]
15. Weber, M.; Turowski, M.; Cakmak, H.K.; Mikut, R.; Kuhnapfel, U.; Hagenmeyer, V. Data-Driven Copy-Paste Imputation for Energy Time Series. *IEEE Trans. Smart Grid* **2021**, *12*, 5409–5419. [CrossRef]
16. Jeong, D.; Park, C.; Ko, Y.M. Missing Data Imputation Using Mixture Factor Analysis for Building Electric Load Data. *Appl. Energy* **2021**, *304*, 117655. [CrossRef]
17. Okafor, N.U.; Delaney, D.T. Missing Data Imputation on IoT Sensor Networks: Implications for on-Site Sensor Calibration. *IEEE Sens. J.* **2021**, *21*, 22833–22845. [CrossRef]
18. Bhagat, H.V.; Singh, M. NMVI: A Data-Splitting Based Imputation Technique for Distinct Types of Missing Data. *Chemom. Intell. Lab. Syst.* **2022**, *223*, 104518. [CrossRef]
19. Su, T.; Shi, Y.; Yu, J.; Yue, C.; Zhou, F. Nonlinear Compensation Algorithm for Multidimensional Temporal Data: A Missing Value Imputation for the Power Grid Applications. *Knowl.-Based Syst.* **2021**, *215*, 106743. [CrossRef]
20. Jurado, S.; Nebot, À.; Mugica, F.; Mihaylov, M. Fuzzy Inductive Reasoning Forecasting Strategies Able to Cope with Missing Data: A Smart Grid Application. *Appl. Soft Comput.* **2017**, *51*, 225–238. [CrossRef]
21. Hemanth, G.R.; Charles Raja, S. Proposing Suitable Data Imputation Methods by Adopting a Stage Wise Approach for Various Classes of Smart Meters Missing Data—Practical Approach. *Expert Syst. Appl.* **2022**, *187*, 115911. [CrossRef]
22. Ryu, S.; Kim, M.; Kim, H. Denoising Autoencoder-Based Missing Value Imputation for Smart Meters. *IEEE Access* **2020**, *8*, 40656–40666. [CrossRef]
23. Le, N.T.; Benjapolakul, W. A Data Imputation Model in Phasor Measurement Units Based on Bagged Averaging of Multiple Linear Regression. *IEEE Access* **2018**, *6*, 39324–39333. [CrossRef]
24. Liu, X.; Zhang, Z. A Two-Stage Deep Autoencoder-Based Missing Data Imputation Method for Wind Farm SCADA Data. *IEEE Sens. J.* **2021**, *21*, 10933–10945. [CrossRef]
25. Andiojaya, A.; Demirhan, H. A Bagging Algorithm for the Imputation of Missing Values in Time Series. *Expert Syst. Appl.* **2019**, *129*, 10–26. [CrossRef]
26. Choudhury, S.J.; Pal, N.R. Imputation of Missing Data with Neural Networks for Classification. *Knowl. Based Syst.* **2019**, *182*, 104838. [CrossRef]
27. Sim, J.; Lee, J.S.; Kwon, O. Missing Values and Optimal Selection of an Imputation Method and Classification Algorithm to Improve the Accuracy of Ubiquitous Computing Applications. *Math. Probl. Eng.* **2015**, *2015*, 538613. [CrossRef]
28. Yadav, M.L.; Roychoudhury, B. Handling Missing Values: A Study of Popular Imputation Packages in R. *Knowl. Based Syst.* **2018**, *160*, 104–118. [CrossRef]
29. Banga, A.; Ahuja, R.; Sharma, S.C. Accurate Detection of Electricity Theft Using Classification Algorithms and Internet of Things in Smart Grid. *Arab. J. Sci. Eng.* **2022**, *47*, 9583–9599. [CrossRef]

30. Khan, I.U.; Javeid, N.; Taylor, C.J.; Gamage, K.A.A.; Ma, X. A Stacked Machine and Deep Learning-Based Approach for Analysing Electricity Theft in Smart Grids. *IEEE Trans. Smart Grid* **2022**, *13*, 1633–1644. [CrossRef]
31. Qu, Z.; Liu, H.; Wang, Z.; Xu, J.; Zhang, P.; Zeng, H. A Combined Genetic Optimization with AdaBoost Ensemble Model for Anomaly Detection in Buildings Electricity Consumption. *Energy Build.* **2021**, *248*, 111193. [CrossRef]
32. Izonin, I.; Tkachenko, R.; Verhun, V.; Zub, K. An Approach towards Missing Data Management Using Improved GRNN-SGTM Ensemble Method. *Eng. Sci. Technol. Int. J.* **2021**, *24*, 749–759. [CrossRef]
33. The Tracebase Data Set. Available online: http://www.tracebase.org (accessed on 30 September 2022).
34. Purna Prakash, K.; Pavan Kumar, Y.V.; Reddy, C.P.; Pradeep, D.J.; Flah, A.; Alzaed, A.N.; Al Ahamdi, A.A.; Ghoneim, S.S.M. A Comprehensive Analytical Exploration and Customer Behaviour Analysis of Smart Home Energy Consumption Data with a Practical Case Study. *Energy Rep.* **2022**, *8*, 9081–9093. [CrossRef]
35. Himeur, Y.; Alsalemi, A.; Bensaali, F.; Amira, A. Building Power Consumption Datasets: Survey, Taxonomy and Future Directions. *Energy Build.* **2020**, *227*, 110404. [CrossRef]
36. Iqbal, H.K.; Malik, F.H.; Muhammad, A.; Qureshi, M.A.; Abbasi, M.N.; Chishti, A.R. A Critical Review of State-of-the-Art Non-Intrusive Load Monitoring Datasets. *Electr. Power Syst. Res.* **2021**, *192*, 106921. [CrossRef]
37. Pipattanasomporn, M.; Chitalia, G.; Songsiri, J.; Aswakul, C.; Pora, W.; Suwankawin, S.; Audomvongseree, K.; Hoonchareon, N. CU-BEMS, Smart Building Electricity Consumption and Indoor Environmental Sensor Datasets. *Sci. Data* **2020**, *7*, 241. [CrossRef]
38. Gopinath, R.; Kumar, M.; Prakash Chandra Joshua, C.; Srinivas, K. Energy Management Using Non-Intrusive Load Monitoring Techniques–State-of-the-Art and Future Research Directions. *Sustain. Cities Soc.* **2020**, *62*, 102411. [CrossRef]
39. Kasaraneni, P.P.; Yellapragada, V.P.K.; Moganti, G.L.K.; Flah, A. Analytical Enumeration of Redundant Data Anomalies in Energy Consumption Readings of Smart Buildings with a Case Study of Darmstadt Smart City in Germany. *Sustainability* **2022**, *14*, 10842. [CrossRef]

sensors

MDPI

Article

Smart Control and Energy Efficiency in Irrigation Systems Using LoRaWAN

Francisco Sánchez-Sutil and Antonio Cano-Ortega *

Department of Electrical Engineering, University of Jaen, Campus Lagunillas s/n, Edificio A3, 23071 Jaén, Spain; fssutil@ujaen.es

* Correspondence: acano@ujaen.es, Tel.: +34-953-212343

Abstract: Irrigation installations in cities or agricultural operations use large amounts of water and electrical energy in their activity. Therefore, optimising these resources is essential nowadays. Wireless networks offer ideal support for such applications. The long-range wide-area network (LoRaWAN) used in this research offers a large coverage of up to 5 km, has low power consumption and does not need additional hardware such as repeaters or signal amplifiers. This research develops a control and monitoring system for irrigation systems. For this purpose, an irrigation algorithm is designed that uses rainfall probability data to regulate the irrigation of the installation. The algorithm is complemented by checking the sending and receiving of information in the LoRa network to reduce the loss of information packets. In addition, two temperature and humidity measurement devices for LoRaWAN (THMDLs) and an electrovalve control device for LoRaWAN (ECDLs) were developed. The hardware and software were also designed, and prototypes were built with the development of the electronic board. The wide coverage of the LoRaWAN allows the covering of small to large irrigation areas.

Keywords: LoRaWAN; smart irrigation systems; smart energy

Citation: Sánchez-Sutil, F.; Cano-Ortega, A. Smart Control and Energy Efficiency in Irrigation Systems Using LoRaWAN. *Sensors* **2021**, *21*, 7041. https://doi.org/10.3390/s21217041

Academic Editor: Ivan Andonovic

Received: 20 September 2021
Accepted: 21 October 2021
Published: 24 October 2021

Publisher's Note: MDPI stays neutral with regard to jurisdictional claims in published maps and institutional affiliations.

1. Introduction

In modern irrigated agricultural facilities, the competitiveness of the sector, combined with rising global temperatures, has necessitated the development of new and more sustainable agricultural techniques and crops to help reduce water consumption in these facilities, coupled with optimal water and energy management strategies. An efficient farming system is defined by the right amount of water at the right time, resulting in improved crop yields through efficient energy consumption. The use of innovative irrigation technologies is necessary to ensure an optimal amount of irrigation water. Optimisation of the irrigation system involves improving crop development conditions by planning the installation: optimal water and energy quantity and management. This requires variable monitoring and decision-making systems that allow us to optimise current irrigation installations.

The need for optimisation in agriculture began in the last century. In the beginning, design solutions with wired electronics were used but had numerous problems. Since then, the development and optimisation of irrigation systems have been linked to the rise and evolution of ICT (Information and Communication Technology). It is important to design sustainable models capable of supplying energy through renewable sources based on solar photovoltaic (PV) energy. Another fundamental part is given by the communication network, which is currently realised through wireless networks with low energy consumption, such as Low-Power Wide-Area Networks (LPWANs).

This article describes the design of an intelligent system to implement the irrigation control of a facility located on the university campus of the University of Jaén through wireless communication and low energy consumption powered by solar PV panels. This system consists of a wireless network with sensors and actuators that send the collected data, which are subsequently analysed in the cloud. This research focuses on optimising an irrigation system and reducing its energy consumption with an LPWAN supplied by a PV system.

2. Related Work

In the last decade, there has been a tendency to implement intelligent irrigation management systems based on wireless sensor networks, which have also been used in other areas such as industry, cities, housing, etc. The advantages of these wireless networks in the agricultural sector have been analysed by several authors, such as Goumopoulos et al. [1], who described the design of an intelligent system based on a wireless sensor and actuator network used for irrigation control in greenhouses. Doko Bandur et al. [2] analysed the energy consumption of the different components of the wireless sensor network, indicating the main energy consumers as well as how energy efficiency should be improved. The study of greenhouse crops with wireless technology used for sensor communication as well as the transmission rate was presented by Kochhar et al. [3]. Hamami et al. [4] reviewed the wireless sensor networks used in irrigation systems. This type of technology is ideal for system management and reducing water consumption.

Nowadays, the integration of devices with long-range wide-area networks (LoRaWAN) stores data in the cloud, where they are processed, analysed through Big Data and interact with other networks. These technologies enable the design of the Internet of Things (IoT) and cloud computing systems applicable to agriculture. Froiz-Míguez et al. [5] detailed an IoT system that develops a smart irrigation system covering large areas through a network (LPWAN) with soil temperature and humidity and air temperature sensors. Valente et al. [6] presented the development of a low-cost system and analysed energy consumption with a maximum of 400μA using a LoRaWAN network. Ameloot et al. [7] developed and analysed a wireless network with six wireless nodes to characterise the temperature and relative humidity of suburban areas using a long-range network (LoRa) at various locations in the city of Ghent (Belgium). It has also been used by Cano-Ortega et al. [8], who developed an optimal LoRa network using ABC algorithms to reduce Package Loss Rate (PLR) and dispatch time to determine the load profiles of a dwelling. Smart street lighting systems using an LPWAN control was realised by Sánchez-Sutil et al. [9,10]. Finally, Cruz et al. [11] monitored the filling level of urban waste containers in Lisbon (Portugal) using LPWAN technology. Ritesh-Kumar et al. [12] applied LoRaWANs to implement a greenhouse control system that enables energy and water savings through continuous monitoring of the installation.

The advances of ICTs in irrigation systems have been quite important, as can be seen. Nam et al. [13] discussed the use of ICTs in water management in agriculture and irrigation facilities. Goap et al. [14] presented an intelligent system that obtains soil moisture data through sensors together with current meteorological information to optimise the irrigation of an agricultural facility. To minimise water losses, Canales-Ide et al. [15] analysed a set of techniques and criteria aimed at optimal irrigation management that determines the water needs of plants and the optimal efficiency of the irrigation systems. By studying each plant from existing databases, Munir et al. [16] proposed an optimal irrigation system based on daily needs, considering the time of day, soil moisture and humidity. Migliaccio et al. [17] developed a smartphone application for scheduling urban lawn irrigation using evapotranspiration data from weather stations.

Among the most important advantages of using IoT systems in agriculture is automated irrigation, as measurements can be taken by sensors (humidity, temperature, irradiation, etc.) and actions (solenoid valves, pumps) through the different devices that make up this system. Methodologies have also been developed for the analysis and development of scientific networks based on the evaluation of the needs of different crops, soil attributes, climate, etc. Some authors have developed different IoT devices, as Fernández-Ahumada et al. [18] presented a low-cost device for automatic irrigation based on an ESP32-LoRa microcontroller and Internet connection through the Sigfox network. Fraga-Lamas et al. [19] proposed an IoT smart irrigation system specifically designed for remote urban areas. Chazarra-Zapata et al. [20] presented an IoT device that optimises battery power consumption, GPRS (General Packet Radio Service) with an Nb-IoT (Narrow-band Internet of Things) system for communication and sending information every two hours to reduce energy consumption. López-Morales et al. [21] proposed an IoT system that enables decision making on pumping efficiency in an irrigation community by easily integrating heterogeneous data sources, which improves the energy efficiency of pumping with higher economic, environmental and social returns in a sustainable way. Additionally, Glória et al. [22] developed a sustainable irrigation system that allows for improving natural resources, both water and energy, and reducing the economic cost through an IoT system with a network with batteries that has communication times of two hours. The monitoring of climatic parameters, soil moisture, vegetation health, plant diseases and crop yields while using IoT systems with wireless networks was developed by Khan et al. [23]. Additionally, Mohammed et al. [24] developed an IoT system for the control of date palms in arid regions using an underground irrigation system that remotely controlled climatic parameters and water volume in the soil. Tiglao et al. [25] presented a low-cost system that has a soil moisture sensor, a temperature sensor, a humidity sensor and a valve actuator within a mesh configuration that regulates drip irrigation. Finally, Sánchez Sutil et al. [26] performed the design of an intelligent system for measuring electrical variables to obtain load profiles in households.

Different control systems applied to irrigation have been developed. Al-Ali et al. [27] presented a microcontroller based on fuzzy logic algorithms for drip irrigation control, and Sudharshan et al. [28] studied a solenoid valve control system using fuzzy logic data from temperature, humidity and soil moisture sensors. Nawandar et al. [29] proposed a greenhouse, garden and farm control system and an automatic irrigation system capable of tracking the water needs of the crop, providing real-time and historical data of the farm. Liao et al. [30] performed the design of an automatic irrigation system with real-time soil moisture data to estimate the depth of water absorption. Finally, Eltohamy et al. [31] analysed how phosphorus released from the soil surface in paddy fields is influenced under different irrigation scenarios for different soil moistures.

The literature review found the following technological aspects.

- Wireless technologies are used in the works Wi-Fi [1,17,27–30], NRF [25], and RFID [13].
- Different works use LoRa technology [5–7,12,12,18,20,21].
- The algorithms used are optical algorithm [5], multi-objective function [18], and fuzzy logic [27,28].
- The following works use open-source platforms [5,13,29].

Based on the weaknesses and opportunities identified, the main contributions of this research are:

- Irrigation algorithm that connects to the Internet to obtain the probability of precipitation and does not irrigate if the probability of precipitation is greater than specified.
- Change of parameters in real time that allows the system to be much more dynamic and can be adjusted to the needs of the installation at any given moment.
- Routine checking of sending and receiving messages to minimise the number of packets of information lost in the LoRaWAN network.
- Development of low cost and open-source prototypes, which allow the system to be adapted to the particular needs of each installation.

The rest of the document is organised as follows: Section 3 provides an overview of the developed system, including the modular architecture of the system, the developed LPWAN network and the supported sensors and their interconnection with the platform. Section 4 details the technological results obtained. An evaluation of the developed system in terms of a prototype, the analysis of the performance of the LPWAN wireless network, the agronomic impact of the system and an evaluation of the energy consumption of the system are provided. Finally, Section 5 presents the conclusions drawn from this work.

3. Methodology and Design

3.1. Network Scheme

The proposed network scheme has two distinct parts. The first corresponds to the LoRaWAN and the second to the Wide-Area Network (WAN), which can be either wired using Ethernet protocol or wireless using Wi-Fi (Wireless Fidelity) protocol. The PG1301 concentrator manufactured by Dragino Technology Co., LTD., Shenzhen, China, is used as the link between the two networks. PG1301 is mounted on a Raspberry 3 or higher, which provides support for the WAN network. In addition, it can host up to 1000 LoRaWAN devices, which is sufficient for most applications. If a larger number is required, it is sufficient to install more concentrators to cover the required needs. Figure 1 shows the network scheme.

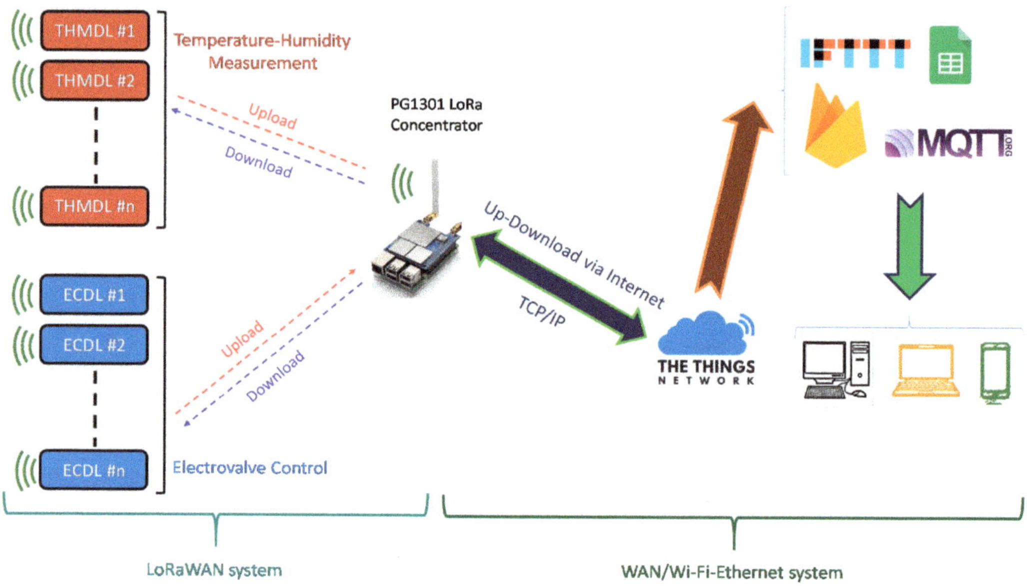

Figure 1. Proposed network scheme.

Within the LoRaWAN, communication is bidirectional between the Temperature and Humidity Measurement Device for LoRaWAN (THMDLs) and Electrovalve Control Device for LoRaWAN (ECDLs) with PG1301 since both data messages (upstream) and command messages (downstream) are needed. The information is concentrated on The Things Network (TTN) server [32]. This service is specially designed to work with LoRaWANs and supports upstream and downstream messages to LoRaWAN devices, such as the ones developed in this research (THMDL and ECDL). Currently, TTN has just implemented the new v3 version, which is much more powerful than the previous one. From TTN, it is possible to send and receive information to different IoT services through the available integrations. These include (i) AWS IoT [33], (ii) Akenza core [34], (iii) Datacake [35], (iv) deZem [36], (v) InfluxDB Cloud 2. 0 [37], (vi) Microsoft Azure [38], (vii) Qubitro [39], TagoIO [40], (ix) thethings.iO [41], (x) ThingsBoard [42], (xi) ThingSpeak [43], (xii) Ubidots [44] and (xiii) UIB [45].

In addition to the above integrations, it is possible to use other options such as (i) Message Queue Telemetry Transport (MQTT) [46], (ii) LoRa cloud [47], (iii) Node-RED [48] and If This, Then That (IFTTT) [49]. Finally, TTN has available the HTTP Webhooks integration that allows sending data to any server using POST and GET. From these, IoT services such as Google Sheets [50], Google Firebase [51], etc., can be accessed. As can be seen, there is a wide range of possibilities that allows the developer or user to find the service that best suits each situation at any given time.

3.2. *Hardware Design*

3.2.1. Design Challenges and Objectives

In order to obtain fully functional devices that perfectly fulfil their assigned tasks, it is necessary to perfectly define the performance objectives to be met by the devices. These objectives will have a decisive influence on the choice of components and technologies to be implemented in THMDL and ECDL. They will also have an important bearing on the software that will run inside these devices. The hardware design objectives are listed below:

- Low power consumption: The devices are placed in the field (THMDL) or where the electrovalves are located (ECDL), and a mains power supply is not always available. It is necessary to use batteries and Solar Panels (SPs) to ensure the power supply of the equipment. In this sense, low power consumption is essential for batteries and SPs to be as small as possible.
- Small size: The devices must be installed in the smallest possible space. In the case of the THMDL, the goal is to be as imperceptible as possible in landscaped areas. For the ECDL, the goal is to be close to the electrovalve, but it is not always possible to have large spaces. This design objective is indivisibly linked to the previous one.
- Component integration, modular design and fault response: If one of the components has a problem and develops a malfunction, the device must be able to maintain the other features that have not been affected by the malfunction. The modular design is of vital importance in these scenarios since it allows components to be changed without the system ceasing to function. This results in highly fault-tolerant devices that provide a high degree of security against device malfunctions.
- Operational safety: The devices are designed to operate autonomously and continuously 24/7. It is, therefore, necessary for the design to be as robust as possible in order to minimise operating problems. This, together with the previous objective, gives the designed devices a high tolerance to failures.
- Low price: In addition to meeting all of the above objectives, the devices must have a final cost that is as low as possible. Thus, achieving designs that can be mass-produced and that are accessible to the majority of users is essential.

The aforementioned objectives entail overcoming a series of challenges and difficulties, the resolution of which will result in the development of fully functional devices. The following is the list of elements to be considered:

- Component selection: In achieving the design objectives, the selection of the components to be implemented in the devices is of particular importance. They have a decisive influence on the proper functioning of the devices and on achieving a final system that is fully functional and safe in its operation.
- Modular design: Combined with the design objectives of component integration, modular design, fault response and operational safety, the objective must provide robustness to the devices. Thus, devices cannot be taken out of service in the event of a malfunction. Rather, all features unaffected by the problem must continue to function correctly.
- Evaluation of alternatives leading to an optimal design: It is essential to evaluate the different implementation possibilities for each of the devices and the final system. The choice of the most correct, optimal and appropriate solution will lead to the fulfilment of the objectives set for the design.
- Printed Circuit Board (PCB) design: The design must be optimised to achieve a minimum size that allows the integration of all the selected components in each of the devices. In this case, two PCBs will be created, one for the THMDL and one for the ECDL.

3.2.2. Components

An appropriate selection of the components to be implemented in the devices will have a decisive influence on the optimisation of the devices. The design objectives of low power consumption, small size, modularity and operational safety must be addressed in the approach to component selection.

Microcontroller

In order for a device to function properly and perform all assigned tasks, it is necessary to have a core. This core is the microcontroller that must drive the interaction between components within the device. Thus, the microcontroller must have different elements such as the microprocessor, memory, ports for communications with other components, digital and analogue inputs and outputs, etc.

Once the design objectives have been studied, it can be concluded that the Arduino family of microcontrollers is an ideal platform for use in the construction of the devices created in this research. The Arduino platform is endorsed by its use in a multitude of projects with industrial and domestic applications.

Table 1 shows some of the most essential features of the various microcontrollers in the Arduino family. This table is the basis for choosing the microprocessor applied to the devices developed in this research.

The Arduinos shown have sufficient memory capacity for program code and data. Therefore, the decision to use one or the other microcontroller depends on other design objectives, which are mainly power consumption and size. In this sense, the microcontroller that best meets the requirements defined in the design is the Arduino Nano (AN), whose specifications can be found in [52].

Table 1. Comparison of the Arduino family.

Component	Surface (mm^2)	Microcontroller	Current Consumption (mA)	Flash Memory (kB)	Clock Speed (MHz)	Unit Price (€)
Arduino Uno [53]	3663.24	ATmega328P	46	32	16	20.00
Arduino Mega [54]	5421.17	ATmega2560	93	256	16	35.00
Arduino Nano [52]	810.00	ATmega328	15	32	16	20.00
Arduino Micro [55]	864.00	ATmega32U4	15	32	16	18.00

LoRa Wireless System

Regarding the LoRa communication system, it should be noted that two components are required: (i) the end device (to be installed in the THMDL and ECDL) and (ii) the gateway responsible for communication with the cloud.

LoRa communication chips are diverse, including (i) Semtech (Semtech Corporation, Camarillo, CA, USA) SX1308 [56], SX1301 [57], SX1276 [58], SX1278 [58] and SX1257 [59]; (ii) HOPERF chip RFM95/96/97/98 [60] (HOPERF, Shenzhen, China); and (iii) Murata CMWX1ZZABZ (Murata Manufacturing, Nagaokakyo, Japan) [61].

Commercially available LoRaWAN-compatible models are built from these chips. Five models were analysed. From this analysis, the model to be implemented in the THMDL and ECDL was chosen. The models analysed were the following: (i) Arduino MKRWAN 1310 (Arduino AG, Ivrea, Italy) [62]; (ii) Monteino (LowPowerLab, Canton-Michigan, USA) [63]; (iii) Libelium (Libelium, Zaragoza, Spain) [64]; (iv) Lopy4 (Pycom, Bucharest, Romania) [65]; and (v) Dragino LoRa Bee (DLB) (Dragino Technology Co., LTD., Shenzhen, China) [66]. These models use the following chips: (i) Murata CMWX1ZZABZ for the Arduino MKR WAN 1310; (ii) HOPERF chip RFM95/96/97/98 for the Monteino; (iii) SX1276 and SX1278 for the Lopy4 and DLB, respectively; and (iv) Semtech SX1272 for the Libelium. The characteristics of the models analysed are similar. Therefore, the DLB was chosen in this research as the component to be installed in the THMDL and ECDL due to its reduced price. Table 2 illustrates the characteristics of the components analysed.

Table 2. Comparison of LoRa end-devices.

Component	Surface (mm^2)	Current Consumption (A)	RSSI Range (dBm)	Sensitivity (dBm)	Blocking Immunity	Unit Price (€)
Lopy4 [65]	1100.00	Rx 12 mA–0.2 µA register retention	−126	−148	High	33.06
Monteino [63]	240.05	RX 10.3 mA–200 nA register retention	−127	−148	Excellent	22.95
Libelium [63]	775.00	RX 10.3 mA–200 nA register retention	−127	−148	Excellent	32.35
MKR WAN 1310 [62]	1693.75	Rx 23.5 mA	−117.5	−133.5	High	33.00
Dragino LoRa Bee [66]	775.00	RX 10.3 mA–200 nA register retention	−127	−148	Excellent	14.50

Once the LoRa component was selected, it was necessary to choose the gateway. This component is responsible for handling the upstream and downstream messages sent back and forth between the THMDL and ECDL to TTN. Although there are different options on the market, we chose to use the Dragino family to ensure better compatibility with the chosen LoRa component. Table 3 shows the four gateways tested, from which the LoRa PG1301 [67] concentrator was chosen. PG1301 is capable of handling up to 1000 devices with 10 channels of communication, which is more than enough for most systems. If one needs to control more than 1000 LoRa devices, additional gateways can be added. PG1301 was mounted on a Raspberry Pi computer that provides support for Internet access, either via Ethernet cable or Wi-Fi.

Table 3. Comparison of LoRa gateways and concentrators.

Component	Number of Channels	Communication Paths	Number of LoRa Devices	Unit Price (€)
Dragino OLG01 [68]	1	Ethernet—Wi-Fi—3G/4G	300	85.79
Dragino OLG02 [69]	2	Ethernet—Wi-Fi—3G/4G	300	95.89
LoRa concentrator [67]	10	Ethernet—Wi-Fi provided by Raspberry	1000	100.19
LoRa GPS Hat [70]	1	Ethernet—Wi-Fi provided by Raspberry	300	35.90

Electrical Variables Meter

For the measurement of DC variables, there are fewer options with sufficient quality. There are three possibilities: (i) FZ0430 [71]; (ii) ACS712 [72]; and (iii) INA219 [73]. FZ0430 is capable of measuring only voltage up to 25 V in direct current. ACS712 measures currents in ranges of 5, 20 or 30 A, depending on the version. To obtain the power consumption, it is necessary to include a unit of these sensors and then perform the necessary calculations. A complementary option is to use the INA219 m, which is capable of measuring voltage and voltage in the same component. It also provides a direct reading of the power consumed. In this research, INA219 has been chosen as it involves making all voltage, current and power measurements on the same component. The characteristics of the components analysed are shown in Table 4.

Table 4. Comparison of electrical sensors.

Component	Measured Variable	Surface (mm^2)	Price (€)
FZ0430 [71]	Voltage	378	1.73
ACS712 [72]	Current	420	1.28
IN219 [73]	Voltage, current, PF and power	2211	1.70

SHT30 Temperature and Humidity Sensor

After searching for temperature and humidity sensors that could work with the Arduino platform, five families of sensors were found. These families are the following:

(i) SHT1x [74]; (ii) SHT2x [75]; (iii) SHT3x [76]; (iv) DHT11 [77]; and (v) DHT22 [78]. Measurement ranges, accuracy, power consumption, supply voltage and communications paths are diverse. Table 5 shows the comparison of the analysed sensor models.

Table 5. Comparison of temperature and humidity sensors.

Sensor	Humidity Accuracy (%)	Temperature Accuracy (°C)	Supply Voltage (V)	Energy Consumption (µW)	Humidity Range (%)	Temperature Range (°C)	Interface
SHT10 [74]	±4.5	±0.5					
SHT11 [74]	±3	±0.4	2.4–5.5	80	0–100	−40/125	SBus
SHT15 [74]	±2	±0.3					
SHT20 [75]	±2	±0.3					
SHT21 [75]	±2	±0.3	2.1–3.6	3.2	0–100	−40/125	I2C PWM SDM
SHT25 [75]	±1.8	±0.2					
SHT30 [76]	±2	±0.2					
SHT31 [76]	±2	±0.2	2.15–5.5	4.8	0–100	−40/125	I2C
SHT35 [76]	±1.5	±0.1					
DHT11 [77]	±5	±5	3.3–5	100	20–80	0–50	Digital pin
DHT22 [78]	±5	±5	3.3–5	100	0–100	−40/125	Digital pin

It can be seen that the family offering the best performance is the SHT3x. Accuracy, power consumption and measurement ranges are outstanding. Moreover, the supply voltage and the I2C (Inter-Integrated Circuit) bus are ideal for use in conjunction with AN. Finally, within the SHT3x family, the SHT30 sensor was chosen for implementation in the THMDL.

Charge Regulator

SeeedStudio controllers offer a wide range of use in the charge control of batteries with SPs. Of the three models analysed, the Lipo Rider Pro (LiPo) [79] model was chosen for implementation in the devices. This model offers ideal characteristics for the 3.7 V battery and the 4.8 V SP used. Moreover, it is also perfectly suited to the supply voltage of the AN board. Table 6 shows the characteristics of the models tested.

Table 6. Comparison of charge regulators.

Sensor	V_{in} Solar (V)	I_{charge} (mA)	I_{load} (mA)	V_{batt} (V)	V_{source} (%)	$V_{destination}$ (°C)
Lipo Rider Pro [79]	5	500	1000	4.2	5	5
Lipo Rider Plus [80]	5	250	250	100	3.3	3.3
Lipo Rider v1.3 [81]	5	800	600	4.2	5	5

Solar Panel

Solar energy is clean, renewable and simple to use. In this sense, it is of great interest to be used as a source of energy for equipment working outdoors, such as those used in this research. The chosen SP has a high transformation efficiency of around 17%. It is made of monocrystalline material and coated with a thin layer of resin on the surface that protects it from atmospheric agents and makes it ideal for outdoor use. The dimensions of the SP are 138 × 160 mm. The nominal output voltage is 5.5 V, with an output of 540 mA, depending on the luminous intensity received. The open-circuit voltage is 8.4 V, and the maximum load voltage is 6.4. The main characteristics can be found in [82].

Battery

For use in THMDL and ECDL, a 3.7 V, 7800 mAh, 28.86 Wh lithium-ion battery was chosen. This is more than enough to power the designed devices. It should be noted that the THMDL has an average consumption of 166.5 Ah, which ensures 31 h of operation with a fully charged battery. In the case of the ECDL, the average consumption is 31 Ah, with a battery life of 174.5 h if the battery is fully charged.

The battery pack has dimensions of 68 × 55 × 19 mm and consists of three individual batteries, with an operating temperature of between −20 °C and +60 °C. The characteristics of the battery are available for consultation in [83].

3.2.3. Hardware Implementation for the THMDL

In the design of the THMDL, two AN microcontroller was used. This is because the communications paths system chosen (LoRaWAN) and the chip that gives access to the DLB are not compatible with the I2C communications paths bus used to read the INA219 and SHT30 sensors. To perform the measurements, AN1 sends the reading request via the serial port, and AN2 performs the measurement of electrical variables, temperature and humidity and returns them via the serial port. Therefore, AN1 takes care of the communication with the LoRaWAN network and controls the measurement request, and AN2 takes care of the necessary measurements.

Due to the fact that the devices developed in this research work autonomously with no possible wired connection to the electrical and Ethernet networks, it is necessary to implement power supply systems and access to communications paths that do not depend on wired networks. For the electrical network, two solutions have been considered: (i) a battery and (ii) a battery and SP controlled by a regulator. This results in two different versions of the THMDL. The aim is to monitor the power consumption of the joint system and the contribution of the battery and the SP in order to control the system and replace or charge the battery to keep the system running.

For the implementation of the wireless system, there are several applicable technologies. These include the following: (i) Bluetooth; (ii) SigFox; (iii) ZigBee; (iv) Nb-IoT; and (v) Wi-Fi. The coverage offered by each of these is variable, in many cases not exceeding tens of metres, as well as requiring repeaters to extend their coverage. SigFox is owned by a company, which is why all services must be contracted with it. NB-IoT requires a data contract and a SIM card to send and receive information.

This research uses LoRaWAN because communications paths can be achieved up to 10 km, with an average of 5 km, which are sufficient distances for application in most landscaped areas in cities. If larger extensions are required, it is only necessary to install more gateways to ensure the necessary system coverage. Figure 2 shows the block diagrams of the two versions of the THMDL. These diagrams express the relationships that are established between the different components of the devices and how they share information and electrical characteristics.

To complement the block diagrams, the wired connections made between the different components are included. This allows any researcher to be able to clone the devices presented in this research. The power supply, serial port communication, I2C bus communication and LoRaWAN network connections can be seen in the block diagrams. Figure 3 shows the wiring diagram for the THMDL in both versions.

The THMDLs PCB board was also designed in its two versions: battery power supply and battery power supply plus SP. The board integrates all the components used, avoids wiring as much as possible and gives solidity to the whole. The dimensions are 95 × 58 mm for the battery-only version and 191 × 71 mm for the version with battery and SP. Figures 4 and 5 show the design of the PCB boards. Figure 6 shows the electronic schematic of the PCB board for THMDL.

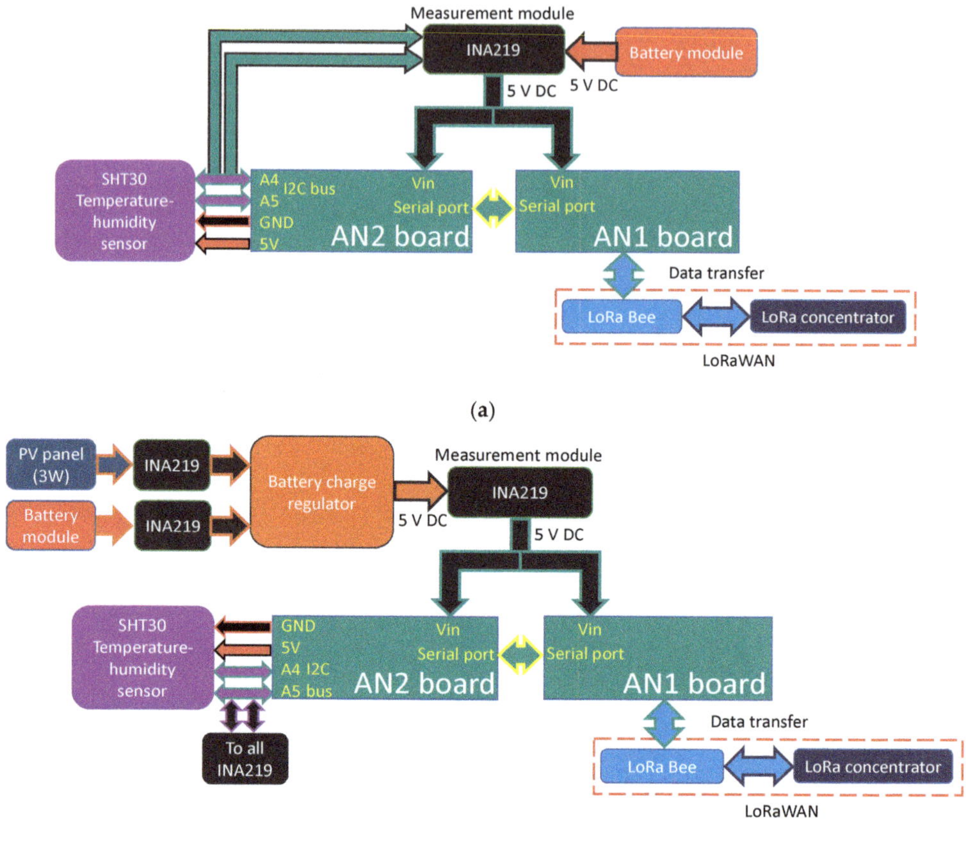

Figure 2. Block diagram of the THMDL: (**a**) battery power supply and (**b**) battery and solar panel power supply.

It is important to perform the economic valuation of the THMDL in its two versions to check whether the reduced-price target is met. In this regard, Tables 7 and 8 show the price of each product and the price of the final set for the two versions of the THMDL. It should be noted that the price of the products is obtained from the official shops of the manufacturers. On the other hand, the fact that the components are licence free means that there are compatible components on the market that can further reduce the price of the set.

Table 7. Cost of components for the THMDL with a battery power supply.

Description	Number	Unit Price (€)	Total (€)
Microcontroller Arduino Nano	2	20.00	40.00
Dragino LoRa Bee	1	14.50	14.50
INA219	1	1.70	1.70
Battery	1	32.96	32.96
PCB board	1	0.40	0.40
SHT30 sensor	1	6.80	6.80
Box container	1	2.54	2.54
Auxiliary material and wiring	-	1.05	1.05
		Total cost	99.95

Table 8. Cost of components for the THMDL with a battery and solar panel power supply.

Description	Number	Unit Price (€)	Total (€)
Microcontroller Arduino Nano	2	20.00	40.00
Dragino LoRa Bee	1	14.50	14.50
INA219	3	1.70	5.10
Battery	1	32.96	32.96
Solar panel	1	12.28	12.28
Li-Po Rider Pro	1	15.33	15.33
PCB board	1	0.40	0.40
SHT30 sensor	1	6.80	6.80
Box container	1	3.02	3.02
Auxiliary material and wiring	-	1.27	1.27
		Total cost	131.66

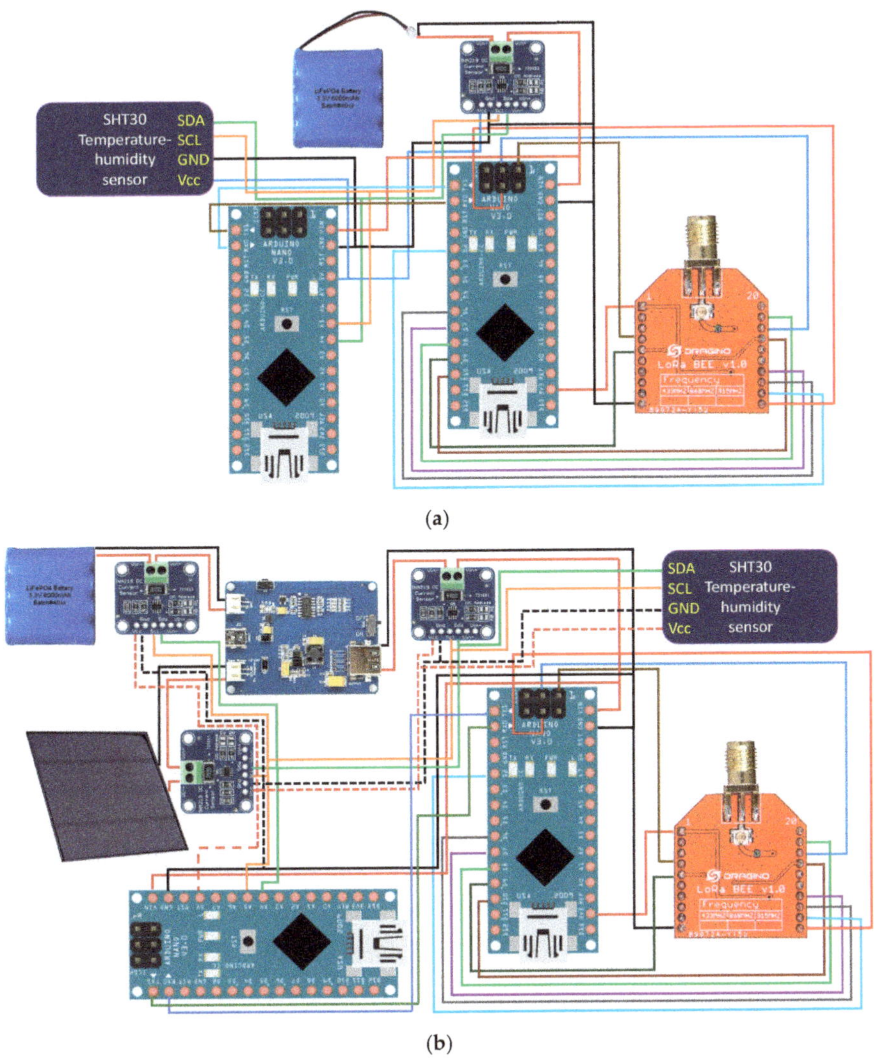

Figure 3. Wiring diagram of the THMDL: (**a**) battery power supply and (**b**) battery and solar panel power supply.

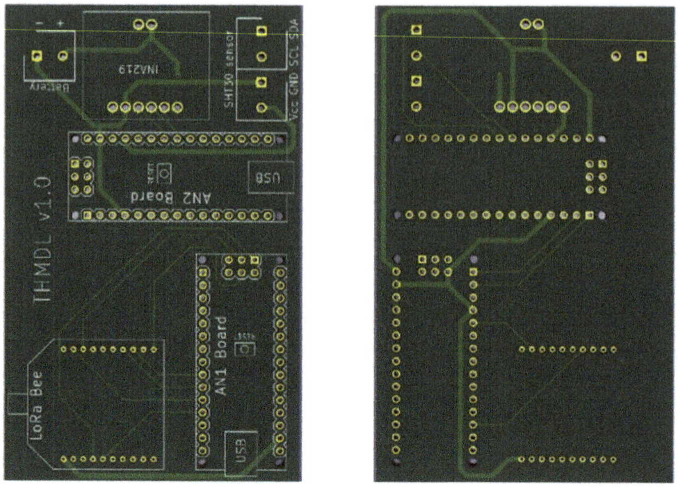

(a) (b)

Figure 4. PCB of the THMDL with battery power supply: (**a**) front side and (**b**) back side.

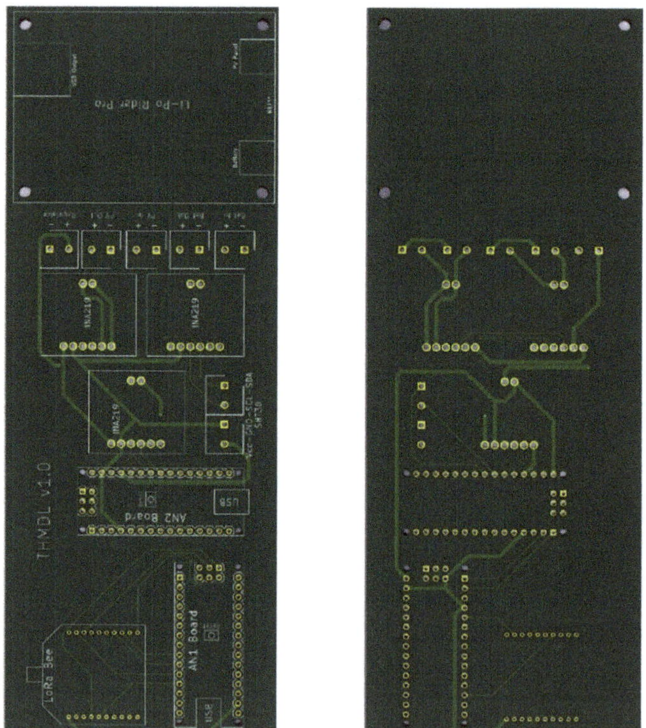

(a) (b)

Figure 5. PCB of the THMDL with battery and solar panel power supply: (**a**) front side and (**b**) back side.

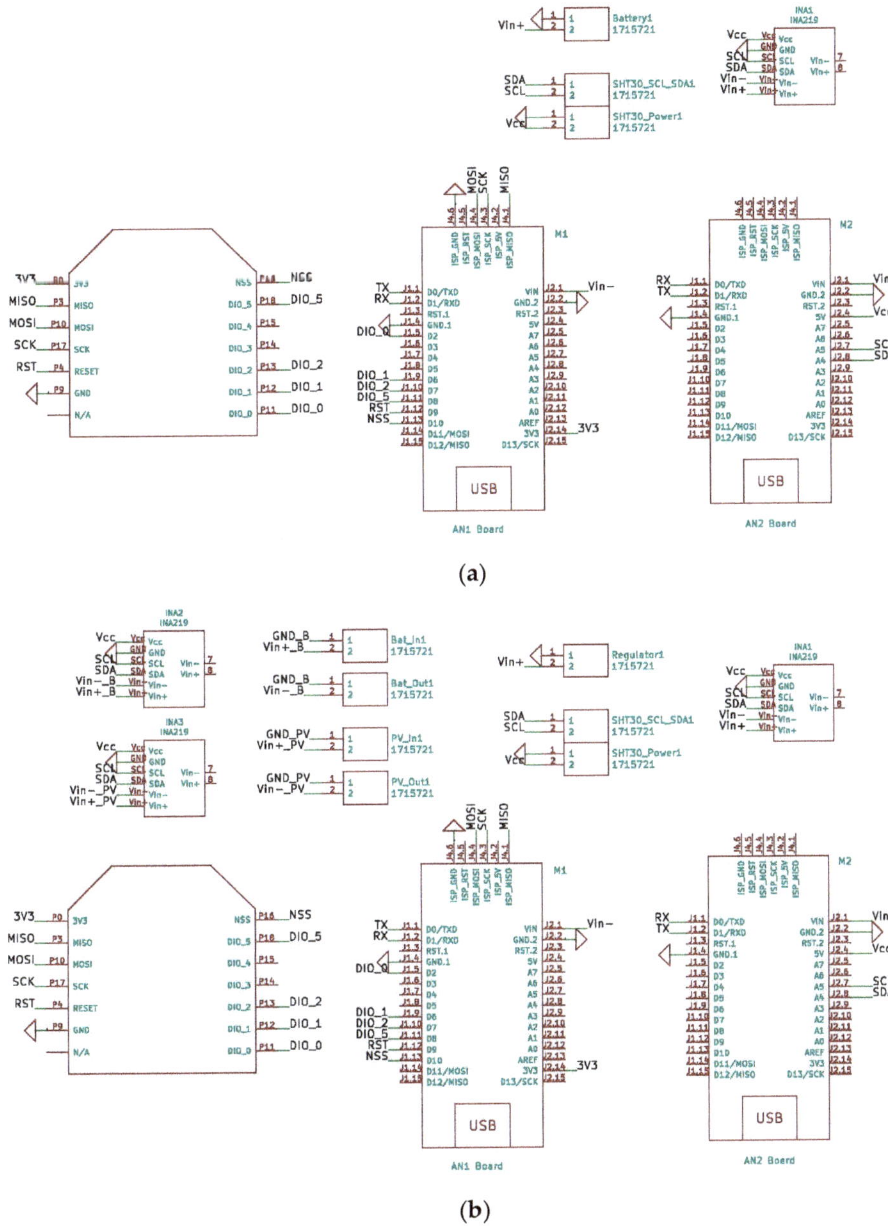

Figure 6. Schematic of the THMDL: (**a**) battery power supply and (**b**) battery and solar panel power supply.

3.2.4. Hardware Implementation for the ECDL

Similar to the THMDL, the ECDL has two ANs due to the incompatibility of the I2C bus with the LoRaWAN system. In this case, a relay is available to operate the electrovalve. This relay is connected to digital output 3 of AN1. The supply voltage of the relay is therefore 5 V, and it can withstand currents of up to 10 A.

AN2 takes care of the electrical measurements made by the INA219. In the SP versions of the THMDL and ECDL, 3 INA219 m are required. It is, therefore, necessary to assign an address on the I2C bus in order to be able to read the contents of each one. The INA219 output meter of the charge controller is assigned the default address 0 × 40. The battery measurement INA219 was assigned the address 0 × 41. For this reason, it is necessary to make a bridge between the A0 contacts of INA219. Finally, the INA219 m of the SP was assigned the address 0 × 44, bridging, in this case, the two A1 contacts. This allows access to the individual measurements without any interference between the measuring device addresses.

The system for access to the LoRaWAN network via the DLB is the same as explained for the THMDL. The difference is the messages exchanged with the network because the ECDL performs different functions. To understand the functional relationships between the components used in the two versions of the ECDL, see Figure 7.

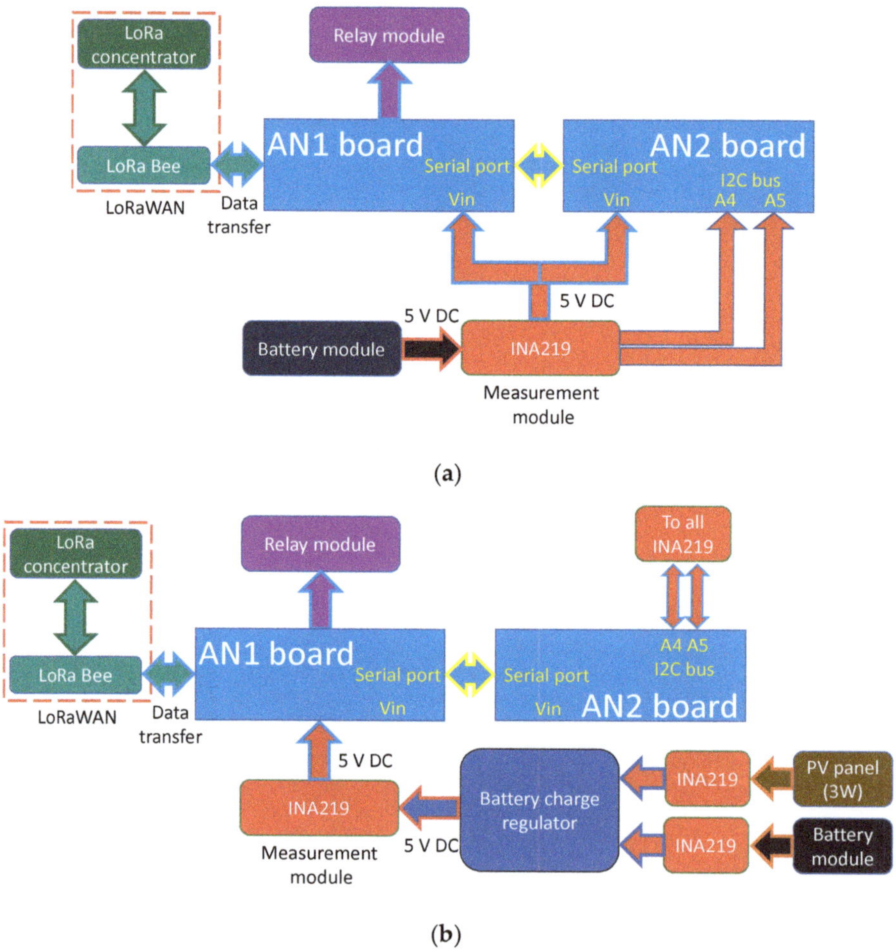

Figure 7. Block diagram of the ECDL: (**a**) battery power supply and (**b**) battery and solar panel power supply.

For the ECDL, it is also necessary to include the wiring diagrams that show the electrical connections to be made in order to build this device. The wiring diagrams are complemented with the block diagrams for a complete definition of the ECDL. With all the information provided, the ECDL is reproducible for any interested researcher. Figure 8 shows the wiring diagrams for the ECDL in its two versions.

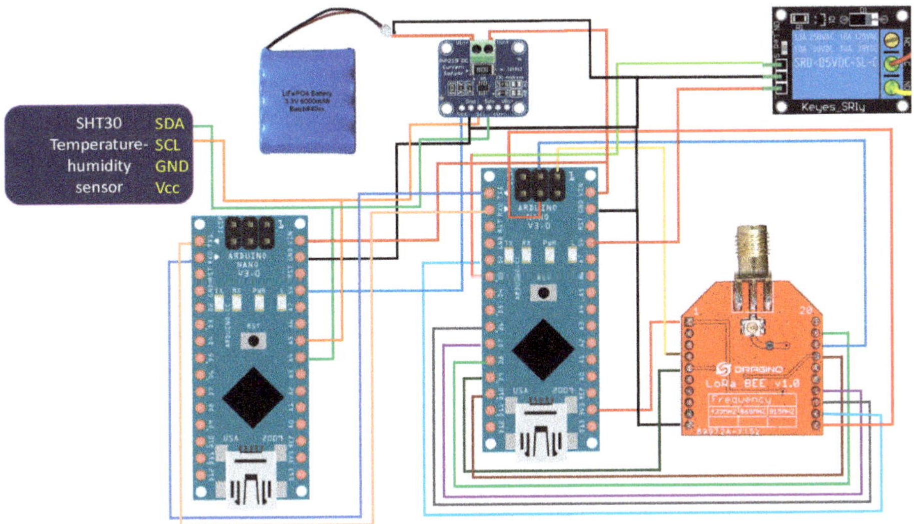

(**a**)

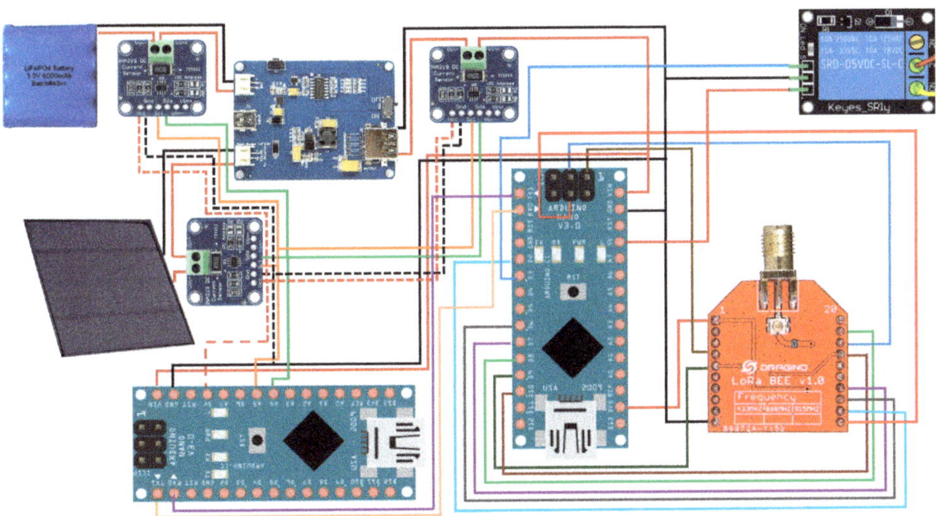

(**b**)

Figure 8. Wiring diagram of the ECDL: (**a**) battery power supply and (**b**) battery and solar panel power supply.

Finally, Figures 9 and 10 show the electronic board developed for the ECDL in its two versions. In this case, the dimensions of the boards are 125 × 59 and 220 × 70 for the ECDL versions with and without an SP. The electronic schematic of the PCB board for THMDL is shown in Figure 11.

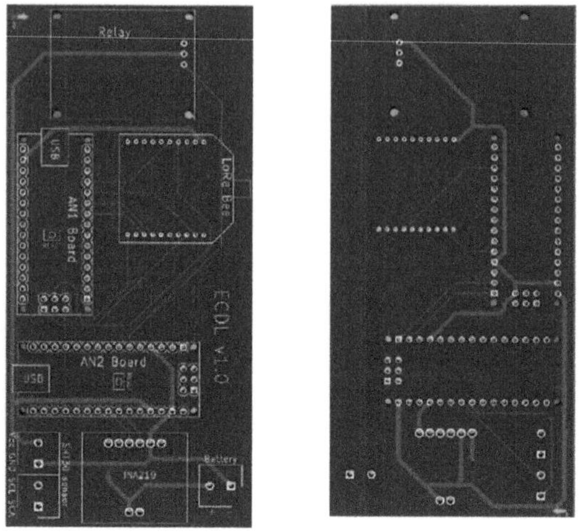

(a) (b)

Figure 9. PCB of the ECDL with battery power supply: (**a**) front side and (**b**) back side.

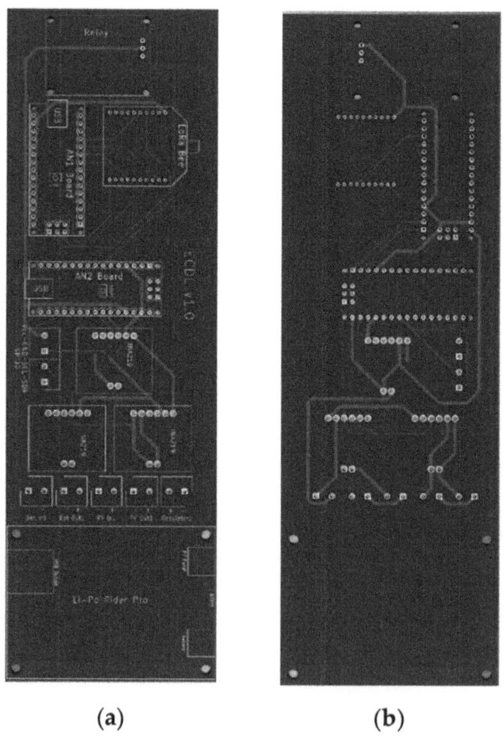

(a) (b)

Figure 10. PCB of the ECDL with battery power supply: (**a**) front side and (**b**) back side.

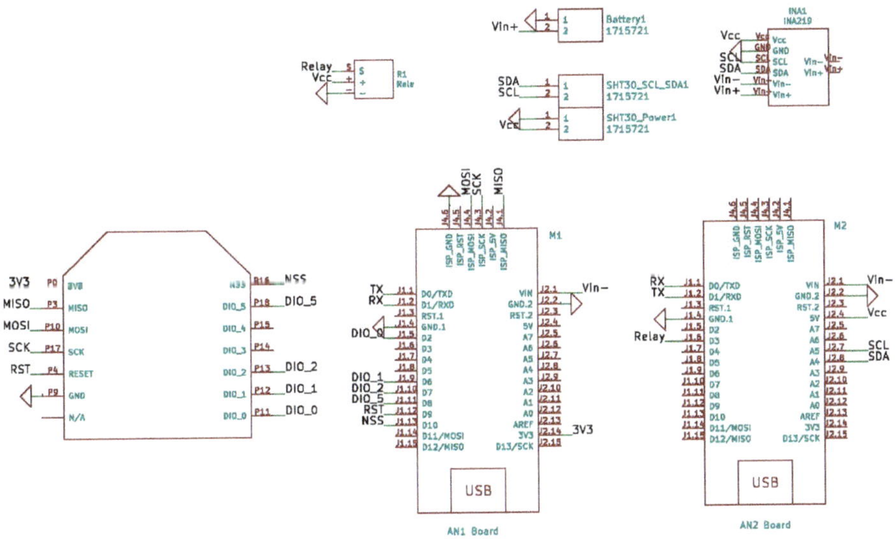

(a)

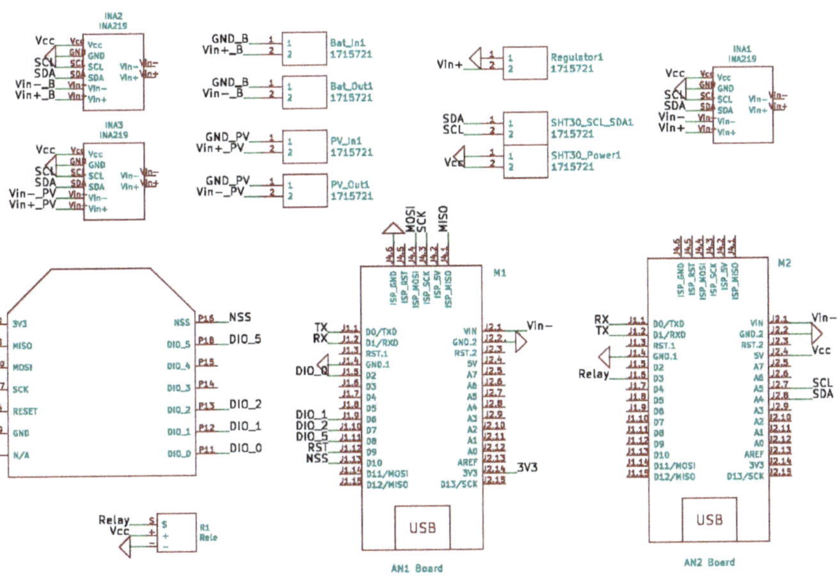

(b)

Figure 11. Schematic of the ECDL: (**a**) battery power supply and (**b**) battery and solar panel power supply.

The economic valuation for the ECDL versions has also been done to verify that the reduced-price target has been met. Tables 9 and 10 give the approximate cost of the two ECDL versions.

Table 9. Cost of components for the ECDL with a battery power supply.

Description	Number	Unit Price (€)	Total (€)
Microcontroller Arduino Nano	2	20.00	40.00
Dragino LoRa Bee	1	14.50	14.50
INA219	1	1.70	1.70
Battery	1	32.96	32.96
PCB board	1	0.40	0.40
Relay	1	0.27	0.27
Box container	1	2.54	2.54
Auxiliary material and wiring	-	1.05	1.05
		Total cost	93.62

Table 10. Cost of components for the ECDL with a battery and solar panel power supply.

Description	Number	Unit Price (€)	Total (€)
Microcontroller Arduino Nano	2	20.00	40.00
Dragino LoRa Bee	1	14.50	14.50
INA219	3	1.70	5.10
Battery	1	11.50	11.50
Solar panel	1	12.28	12.28
Li-Po Rider Pro	1	15.33	15.33
PCB board	1	0.55	0.55
Relay	1	0.27	0.27
Box container	1	3.02	3.02
Auxiliary material and wiring	-	1.27	1.27
		Total cost	122.28

3.3. *Software Design*

The system designed in this research is intended to operate continuously in a 24/7 mode. This allows the automated system to be permanently under control. In addition, the devices have to perform all their functionalities again when there is any problem, e.g., battery change.

Several functionalities have been implemented in the system: (i) battery charge level control; (ii) watering routine based on the weather forecast and humidity level; (iii) complete electrical (v, I, p) and environmental (temperature, humidity) measurements; and (iv) parameter change.

3.3.1. THMDL Software

The THMDL program is structured in two main sections: (i) initialisation and (ii) command control. The initialisation tasks must prepare the components used and the communication ports to start the continuous process.

The command control routine must continuously scan the network for messages sent from the system. These messages are of two types: (i) measurement message and (ii) irrigation message. When the measurement message is received, the measurement routine that measures the electrical and environmental parameters is executed. Subsequently, the battery check routine is executed to check the state of charge of the battery. If the message sent is for irrigation, the need to irrigate is checked, and the necessary order is sent to the system.

Once the task required by the system has been performed, a task completed confirmation message is sent so that the system is notified of the completion of the task. Once the confirmation message is sent, the system returns to the initial step of scanning the LoRaWAN network for new messages. The above process for the command control routine is performed continuously as long as the THMDL is connected. Figure 12 shows the flowchart for the main THMDL program.

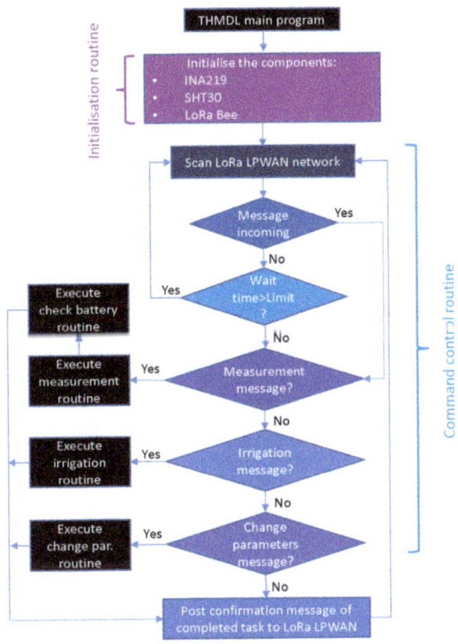

Figure 12. Flow chart of the main program for the THMDL.

Figure 13 shows the flowcharts for the measurement routines in the two versions of the THMDL. The routine is divided into two parts: (i) measurement of electrical variables and (ii) measurement of environmental variables. First, the measurement phase of the electrical part is performed. In this phase, the corresponding INA219 m is called, which returns the variables v, i and p. In the case of the battery-only version, only one measurement is taken. The version with an SP performs three measurements in this order: (i) regulator, (ii) battery and (iii) SP.

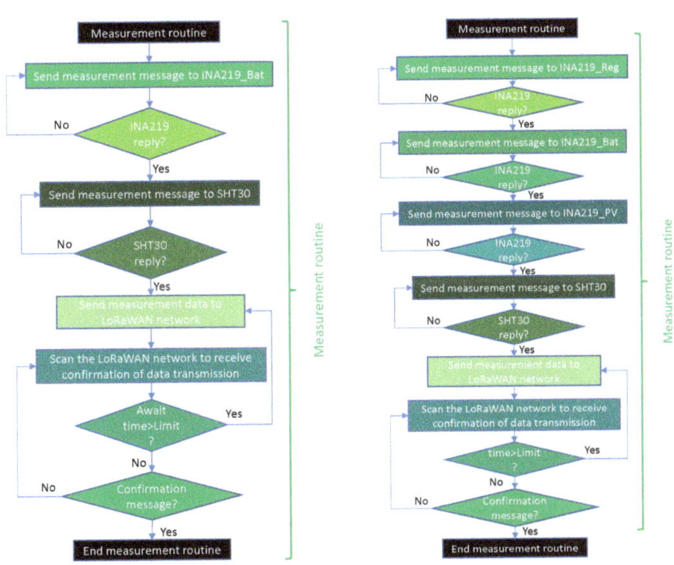

(**a**) (**b**)

Figure 13. Flow chart of the measurement routines for the THMDL: (**a**) battery power supply and (**b**) battery and solar panel power supply.

Once the measurement of electrical variables has been completed, the measurement of environmental variables is performed. To do this, SHT30 is called and returns the temperature and humidity data recorded at that moment. Finally, all the measured data are sent to the LoRaWAN network. In this sense, a message security system has been implemented in order to minimise the loss of information in the system.

Figure 14 shows the flow chart of the irrigation routine. It is important to note that this routine has been implemented based on weather forecasts so that water expenditure is minimised. When the routine is started, a request for rainfall forecast data is sent. The forecast can be extended over the necessary time horizon to be estimated in each application. Once the probability of precipitation is received, it is compared with the minimum probability assigned. If the probability received is higher, the message that it is not necessary to irrigate is sent. On the other hand, if the probability of precipitation is lower than the minimum, it is passed to the part of comparison with the measured humidity level.

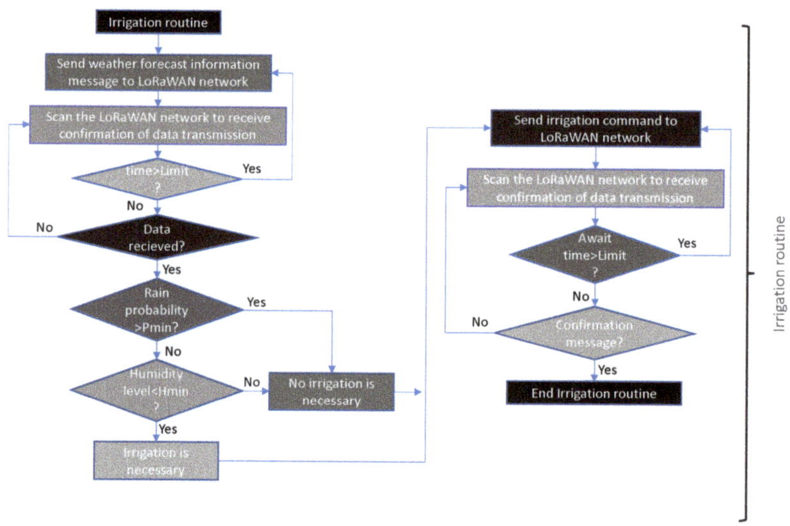

Figure 14. Flow chart of the irrigation routine for THMDL.

If the humidity level is lower than the minimum level set for watering, the watering message is sent. Otherwise, the message that watering is not necessary is sent. The levels of precipitation probability and minimum humidity can be changed by the user at any time, making the system much more efficient and dynamic.

The battery check routine is shown in Figure 15a. Here, it can be seen that the check is performed in relation to the battery voltage. If the above voltage falls below the defined minimum, a low battery message is sent to the system for action by the maintenance staff.

To make the system more dynamic, it is necessary to be able to change the action limits at any time. This allows it to adapt to new situations or approaches in system policies. It also allows sectorisation of the system, making it possible to have different parameters in each zone reflecting the particular characteristics. For this purpose, Figure 15b shows the THMDL parameter change routine. Parameters can be changed together, either individually or in groups, as the routine is prepared for this.

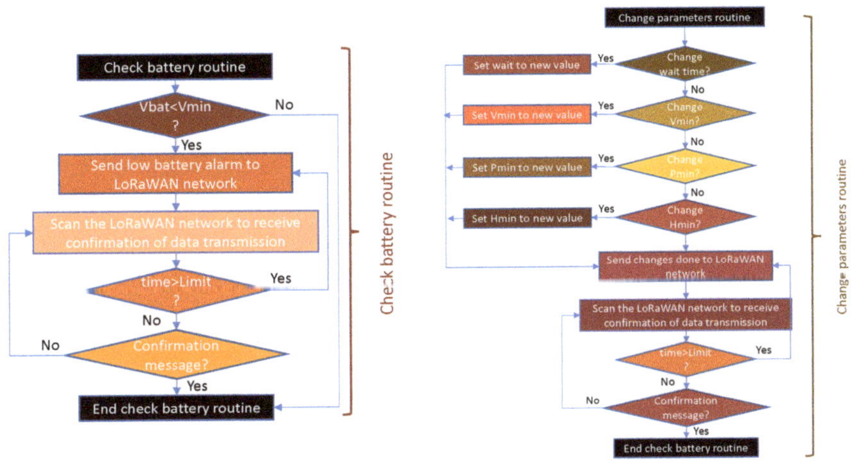

(**a**) (**b**)

Figure 15. Flow chart of the (**a**) check battery routine and (**b**) change parameters routine for the THMDL.

3.3.2. ECDL Software

The design of the main ECDL programme follows a similar philosophy to that of the THMDL. It starts with the initialisation phase of all components. Subsequently, in continuous mode, it executes the following tasks: (i) scanning the LoRaWAN network for new messages; (ii) if a measurement message arrives, it performs the measurement, checks the battery state of charge and sends data; (iii) if a parameter change message arrives, it executes the parameter change routine and sends confirmation; (iv) if the message is an electrovalve on-or-off message, it activates or deactivates the relay. Figure 16 shows the flowchart for the main ECDL program.

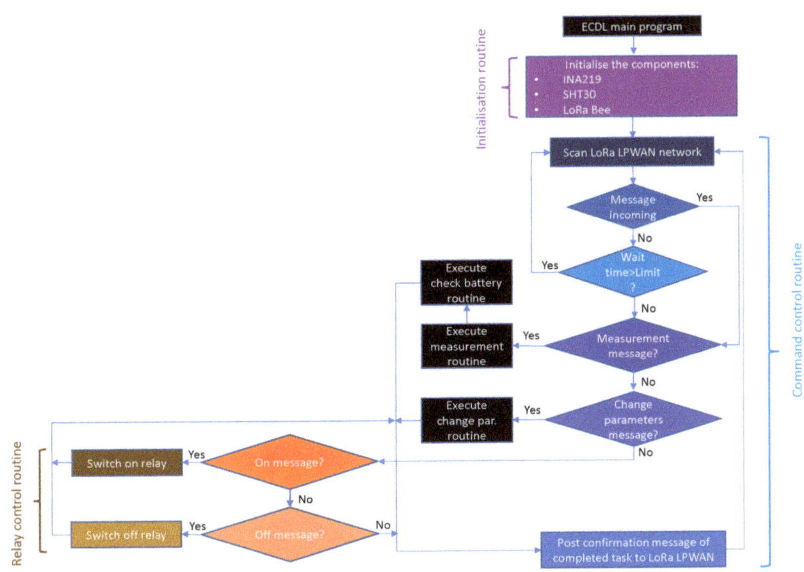

Figure 16. Flow chart of the main program for the ECDL.

Figure 17 shows the flowcharts of the measurement routines for the two versions of the ECDL. The routines work in the same way as those shown above for the THMDL. In this case, the temperature and humidity measurement part is omitted, as this device does not need to take these measurements. It is also enabled to send messages until receiving confirmation of the arrival of the data.

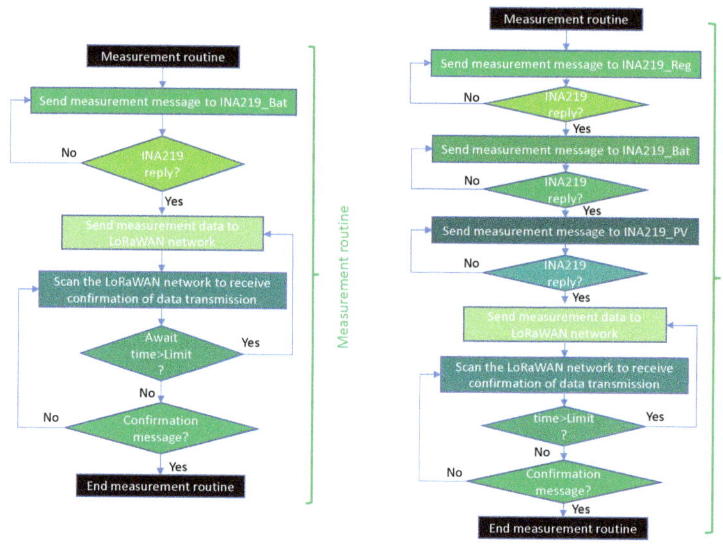

(a) **(b)**

Figure 17. Flow chart of the measurement routines for the ECDL: (**a**) battery power supply and (**b**) battery and solar panel power supply.

As with other routines, the data modification routine is based on the one described for the THMDL. In this case, only two variables are needed: (i) timeout for receiving and sending messages and (ii) minimum voltage for sending low battery warnings. Figure 18 shows the parameter change routine for the ECDL.

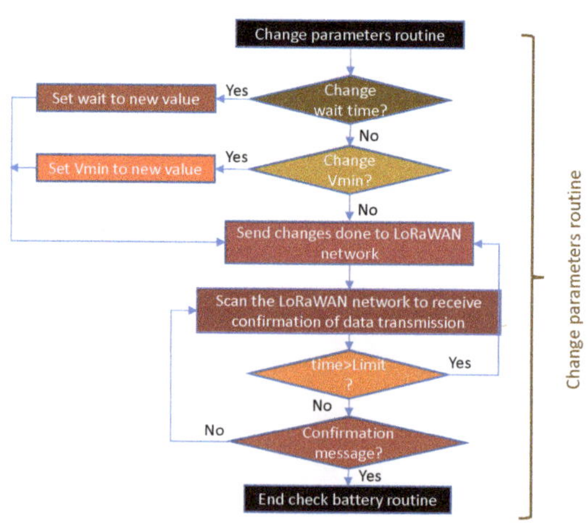

Figure 18. Flow chart of the change parameters routine for the ECDL.

4. Results and Discussion

This section shows the tests conducted to check the devices created in this research. Data were collected in Jaén, Andalusia, Spain, during different times of the year in order to validate the design and implementation.

4.1. Case Study

Figure 19 shows the distribution of irrigation zones on the campus. The number of zones is 22, in which a THMDL device for temperature and humidity measurement has been installed in each zone. An ECDL actuator has also been installed in each zone to control the irrigation electrovalve. This allows the areas to be separated and only irrigate those that really need water. This avoids unnecessary water wastage in areas with sufficient humidity.

Figure 19. Distribution of irrigation zones on the campus.

For applications in cities, the necessary zones will be distributed according to the characteristics of the area to be irrigated automatically. In each zone, a THMDL device must be installed to monitor the zone. The THMDL will communicate with the LoRaWAN system, sending all the recorded data and irrigation orders required. As for the ECDL devices, these will be installed in each of the electrovalves that irrigate the automated zones. It may happen that several zones monitored with the THMDL are irrigated with the same electrovalve. Therefore, the number of THDML devices does not always coincide with the number of ECDL devices. Aerial photographs, photographs taken with drones, maps of the area, etc., can be used to perform the study. These tools allow a detailed study of the area to be done and the best possible system to be implemented.

4.2. LoRaWAN Configuration

In this research, we chose to send messages every minute, which is sufficient for an installation of this type. The initial configuration chosen in this case is BW125 (Band Width 125 kHz), SF7 (Spreading Factor) and CR4/5 (Code Rate). The length of the payload is 16 bytes plus 13 bytes for the header. The 16 bytes of the payload are distributed in 2 bytes for each variable, 2 bytes for temperature, 2 bytes for humidity, 2 bytes for battery voltage, 2 bytes for battery current, 2 bytes for PV module voltage, 2 bytes for PV module current,

2 bytes for bus voltage and 2 bytes for bus current. The header layout is 1 byte for MAC header (MHDR), 4 bytes for LoRaWAN device address, 1 byte for FCtrl (control bit), 2 bytes for Fcont (count bits), 4 bytes for the message integrity code (MIC) and 1 byte for Fport (port number).

Due to the 60 s send time and the chosen payload length, it is only possible to use SFs between 7 and 10, with a 1% duty cycle duration of 41.2 s for SF10 and less for the other SFs. SF7 is chosen as it has the shortest duty cycle duration. The location of the installation is in Europe, so only BW125 and BW250 are possible. The speed of BW250 is higher, but the transmission distance is shorter. To ensure less PLR with the existing distances, a BW125 has been chosen.

Table 11 shows the calculation of the on-air times for the header and payload used. Here, you can see the minimum time for sending messages according to the 1% duty cycle. With the chosen configuration, it is 6.7 s. Table 11 shows all possible combinations for the EU868 zone in Europe. In other regions of the world with different frequency plans, other results can be easily obtained.

Table 11. Airtime parameters for the LoRaWAN in EU868 zone.

Data Rate	Parameters	Airtime	Duty Cycle (1% max)		Fair Access Policy		
			Time (s)	Msg/Hour	Avg/s	Avg/Hour	Msg/24h
DR5	SF7-BW125	66.8	6.7	538	192.4	18.7	448
DR4	SF8-BW125	123.4	12.3	291	355.4	10.1	243
DR3	SF9-BW125	226.3	22.6	159	651.8	5.5	132
DR2	SF10-BW125	441.6	41.2	87	1185.5	3.0	72
DR1	SF11-BW125	905.2	90.5	39	2607.0	1.4	33
DR0	SF12-BW125	1646.2	164.7	21	4742.2	0.8	18
DR6	SF6-BW250	33.4	3.3	1077	96.2	37.4	897

4.3. Measurement of Soil Temperature and Humidity

This section shows the temperature and humidity measurements for all days of the four seasons taken in one of the zones in the year 2020. It can be seen that temperature increases and humidity decreases in the seasons of the year. The season with the highest average temperature corresponds to summer with an average of 29.31 °C, the lowest average temperature of 12.14 °C and the annual average of 19.03 °C.

The highest average humidity occurs in winter with a value of 67.81%, the lowest average in summer with 35.40% and the annual average is 56.09%. Figure 20 shows the data obtained during the meteorological stations of the year 2020.

The location of the campus is defined by its UTM coordinates referenced to zone 30: X = 431,582 and Y = 4,182,595. The geographical location of Jaen has a continental Mediterranean climate. As it is located near the Guadalquivir river valley, this has a decisive influence on the climatic conditions. The temperature variation that can occur is around 20 °C.

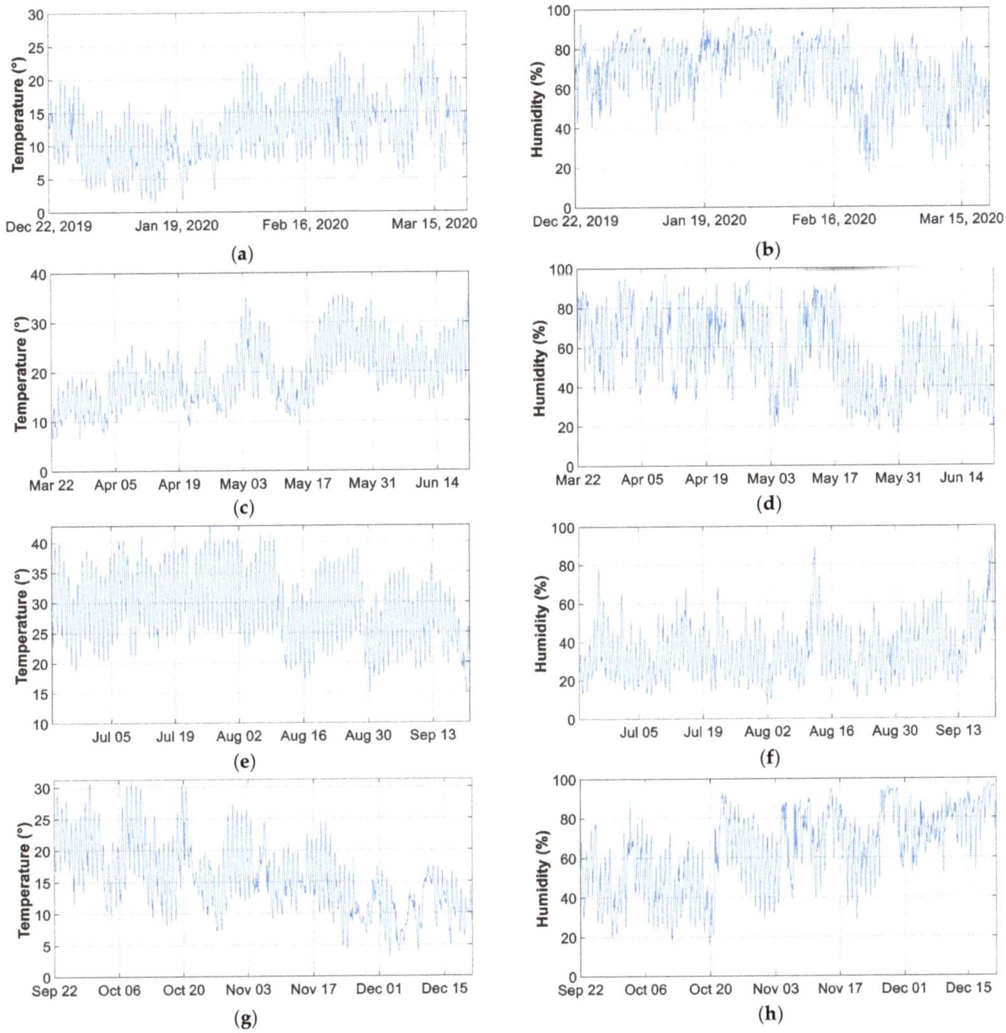

Figure 20. Temperature and humidity graphs for the year 2020: (**a**) temperature in winter; (**b**) humidity in winter; (**c**) temperature in spring; (**d**) humidity in spring; (**c**) temperature in summer; (**f**) humidity in summer; (**g**) temperature in autumn; and (**h**) humidity in autumn.

4.4. Battery Charge

It is possible to charge the battery in two different ways: (i) via the USB port and (ii) via the SP. The first option is possible via the mini USB port of the LiPo.

This port can be connected to various devices, such as a computer USB port, a mobile phone charger, or any other type of charger that provides 5 V and 500 mA DC. Figure 21a,b shows the voltage and current of a complete charging process through the USB port, in this case, of a laptop. From the voltage curve, it can be seen that the voltage increases as the accumulated battery charge increases. The voltage evolves from 2.2 V to 2.6 V at the end of the charging process. The charging current averages 350 mA until the seventh hour of charging. Thereafter, the current decreases to 100 mA at the end of the charging process, which drops abruptly to zero when the battery is fully charged. The complete charging process takes 10 h and 50 min.

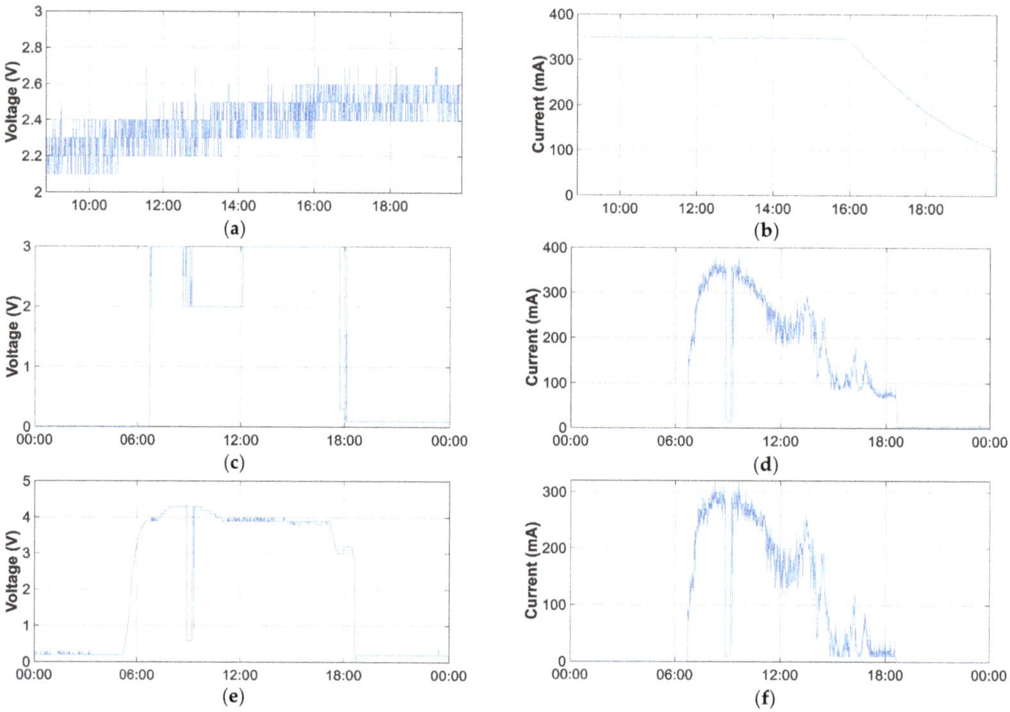

Figure 21. Electrical variables measurement in battery charge: (**a**) battery voltage in USB charge; (**b**) battery current in USB charge; (c) battery voltage in PV charge; (**d**) battery current in PV charge (**e**) PV voltage in PV charge; and (**f**) PV current in PV charge.

Figure 21c–f shows the battery charging process using the SP. It can be seen that although the SP is live, LiPo only switches on the battery charging when the SP voltage is close to 4V. The battery charge remains constant at 3 V, with some periods of 2 V. The charging current depends on the radiation that the SP is receiving. A voltage drop is also observable around 10 am due to passing clouds.

As can be seen, charging the battery with a USB is much faster and more recommendable when the battery is discharged. With a full day of charging with the SP, the battery was not fully charged. This affirms that the function assigned to the SP is to extend the duration of the battery charge by providing charging during sunshine hours. Thus, the SP replenishes the energy consumed during the night and makes the equipment autonomous for long periods of time without the need to charge the battery.

4.5. Battery Discharge

The full discharge test has been performed on a fully charged battery. Figure 22 shows the results for the voltages and currents of the battery and at the output of the LiPo controller board. The time required for full discharge was 166.5 h, which ensures a week of operation with only one battery charge without using an SP.

The battery voltage remains constant at 3 V until the point of full discharge. On the other hand, the regulator card maintains an output voltage between 4 and 5 V. The average current consumption is around 33 mA. In this case, it can be seen that the regulator card maintains an output current of between 204 and 210 mA until the end of the charge.

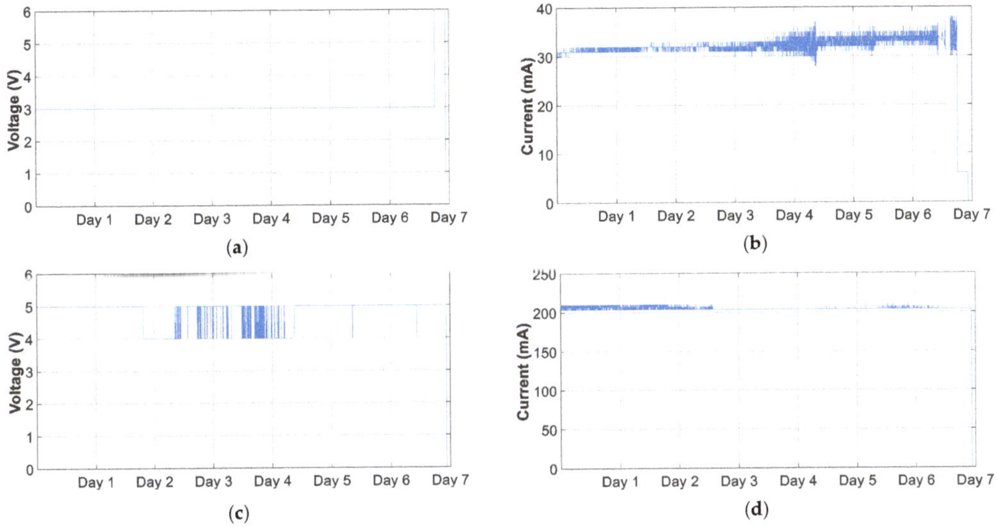

Figure 22. Discharge measurement: (**a**) battery voltage; (**b**) battery current; (**c**) LiPo out voltage; and (**d**) LiPo out current.

4.6. Energy Consumption Comparative

In addition to the wide coverage provided by the LoRaWAN equipment, there is the advantage of the reduced power consumption of these devices. At this point, the consumption of the developed equipment was tested in relation to other wireless technologies.

The comparison was conducted during the month of January 2020 with data taken from THMDL and a Wi-Fi device. The Wi-Fi device tested was an Arduino Wemos D1 mini [84] without a connection to any of the electrical sensors, temperature and humidity sensors or the drive relay. The use of these components would increase the power consumption of the device. It should be noted that the Wemos D1 mini board is one of the boards with the lowest power consumption among those with Internet access via Wi-Fi and the ESP8266 chip. Other boards with this chip have higher power consumption, such as NodeMCU [85], Wemos D1 mini pro [86], Wemos D1 R1 [87], etc.

Figure 23 shows the result of the voltage and current measurements on the battery at the output of the LiPo board. It can be seen that the power consumption of the Wemos D1 mini is approximately three times higher than that of the THMDL, which would be increased by adding the sensors. In addition to this high consumption, the necessary Wi-Fi repeaters or routers would have to be added, as the Wi-Fi coverage is much lower than that offered by the LoRaWAN network, which would increase the final consumption of the whole.

The average THMDL consumption is 33.02 mA, with a standard deviation of 1.76. Wemos D1 mini has a mean of 98.01 mA and a standard deviation of 2.28. The energy consumed by THMDL in January 2020 is 73.6329 Wh and 218.61 Wh for Wemos D1 mini. The total energy consumed in 2020 by THMDL was 869.0219 Wh, corresponding to 2.3762 Wh/day.

As for the output of the LiPo card, it can be seen that, as mentioned above, it maintains an output of between 204 and 210 mA, regardless of the battery consumption. In view of the consumption results, together with the wireless coverage, this supports the use of LoRaWAN technology in systems such as the one developed in this research.

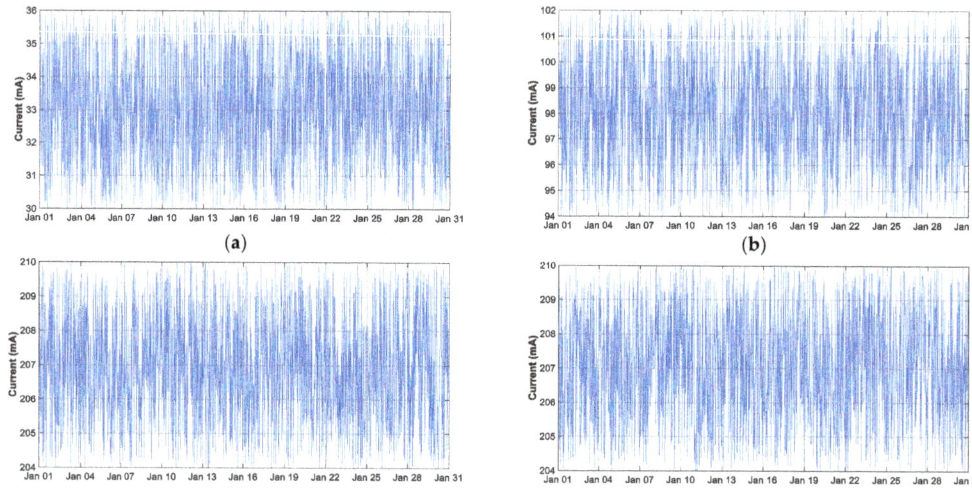

Figure 23. Consumption comparative in January 2020: (**a**) battery current with THMDL connected; (**b**) battery current with Arduino Wemos D1 mini connected; (**c**) LiPo current out with THMDL connected; and (**d**) LiPo current out with Arduino Wemos D1 mini connected.

Statistics for the annual consumption of THMDL have been done. Since the average consumption is 33.02 mA and almost constant, the mean is almost equal in all months at around 2.3750, and a standard deviation of 0.0026 is almost zero. The skewness is practically close to zero, with some positive and negative values but in the region of zero, indicating a symmetrical distribution curve. On the other hand, the kurtosis reflects a mesokurtic distribution with a curve with uniformly distributed values in the symmetrical distribution. Table 12 shows the results obtained.

Table 12. Descriptive statistics of the power THMDL consumption in year 2020.

Month	Energy Generated (Wh)	Sample Daily Mean (Wh)	Sample Variance (Wh^2)	Sample Skewness (Wh^3)	Sample Kurtosis (Wh^4)
January	73.6329	2.3768	0.0026	0.0143	1.7544
February	68.7263	2.3715	0.0025	0.0418	1.8301
March	73.5223	2.3764	0.0025	−0.0118	1.8161
April	71.2977	2.3782	0.0026	−0.0391	1.7955
May	73.6111	2.3761	0.0025	0.0044	1.8280
June	71.2468	2.3765	0.0025	0.0084	1.8135
July	73.4652	2.3714	0.0025	0.0613	1.8834
August	73.5958	2.3756	0.0026	0.0163	1.7741
September	71.3757	2.3808	0.0025	−0.0511	1.8063
October	73.5985	2.3757	0.0025	−0.0037	1.8309
November	71.1906	2.3746	0.0025	0.0261	1.8079
December	73.7580	2.3808	0.0025	−0.0592	1.8529
Year	869.0216	2.3762	0.0025	0.0008	1.8140

4.7. Solar Energy Generated

This section shows the power and energy generated by the 3 W SP used in THMDL and ECDL. The data were collected during 2020 from one of the THMDLs with SP. Figure 24 shows the energy and power for the month of January 2020, which is the lowest generation month together with December and for the whole year.

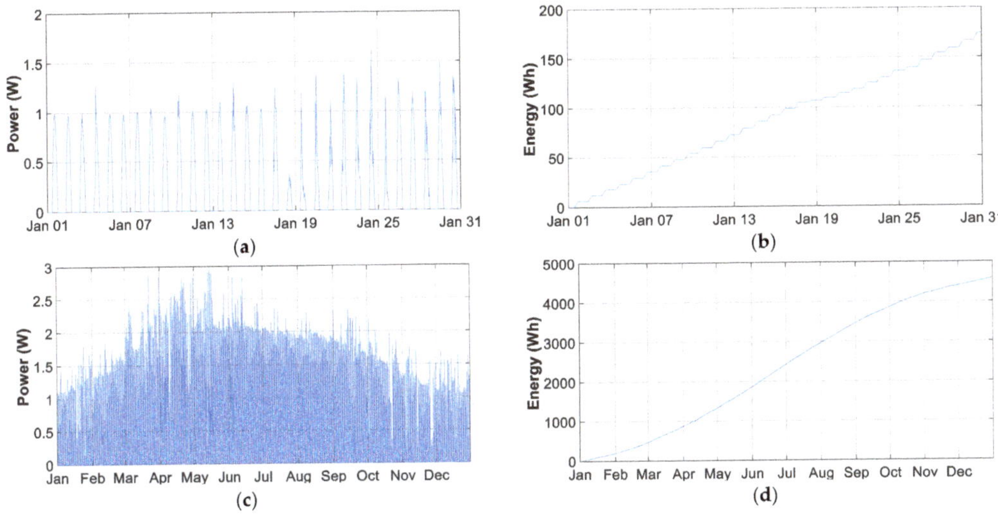

Figure 24. Consumption comparative: (**a**) power obtained in January; (**b**) energy obtained in January; (**c**) power obtained in the year 2020; and (**d**) energy obtained in the year 2020.

The energy generated in January is 202.19 Wh, with a daily average of 6.5266 Wh. The annual photovoltaic generation amounts to 4594.73 Wh and a daily average of 12.5639 Wh.

Table 13 summarises the statistical results of the annual empirical distributions for SP power generation. To compare the results, the monthly and annual average was taken. Thus, the month of maximum generation is June with a mean pf 18.9210 Wh/day. On the other hand, December has the lowest generation with a mean of 6.2873 Wh/day. The generation in December is 33.22% over June and 50.04% over the annual average.

Table 13. Descriptive statistics of the power SP generation in year 2020.

Month	Energy Generated (Wh)	Sample Daily Mean (Wh)	Sample Variance (Wh^2)	Sample Skewness (Wh^3)	Sample Kurtosis (Wh^4)
January	202.19	6.5266	0.2142	1.3553	3.4122
February	276.04	9.5255	0.2751	0.9939	2.3714
March	394.35	12.7469	0.3600	1.0546	2.6887
April	452.60	15.0974	0.4086	1.0099	2.5663
May	538.46	17.3816	0.4194	0.7060	1.9756
June	567.23	18.9210	0.4213	0.4987	1.6505
July	542.87	17.5239	0.3927	0.4846	1.5712
August	496.76	16.0353	0.3786	0.5693	1.6509
September	399.08	13.3120	0.3475	0.8171	2.0916
October	326.09	10.5262	0.2974	0.9640	2.3564
November	204.23	6.8126	0.2220	1.3580	3.4074
December	194.77	6.2873	0.2103	1.3909	3.5401
Year	4594.73	12.5639	0.3502	1.0595	2.7639

Positive skewness values indicate that the tail of the distribution is longer on the right for values above the mean, and the values are concentrated more to the left of the mean, with only January, November and December located to the right of the mean.

The months of January, November and December present a leptokurtic kurtosis because their coefficient is greater than 3, indicating that the values are concentrated around the mean. The rest of the months are mesokurtic because they have a coefficient lower than 3 and their values are further away from the mean.

4.8. Analysis of Consumption, Photovoltaic Generation and Battery Life

Using the data in Tables 12 and 13, a comparison can be made between the energy consumed by THMDL and SP. Starting from the month of lowest generation, which is December with 194.77 Wh and 6.2873 Wh/day and comparing them with the THMDL consumption of 73.7580 Wh and 2.3808 Wh/day, it can be concluded that the energy generated in the most unfavourable month covers the total daily consumption required. The ratio of generation to consumption is 2.64 times higher.

Considering the results obtained for the total discharge of the battery with an average of 33 mA, 166.5 h were needed, which is much longer than the time needed to recover the sunlight the next day and generate the energy consumed during the hours of no photovoltaic generation.

In this sense, fully discharging the battery would take 6.9375 days, corresponding to 1 day 14.41% of the total charge of the battery in the absence of sunlight. As in the most unfavourable month, there are 6.2873 Wh/day of generation and 2.3808 Wh/day are consumed, the PV generation largely covers the maximum of 14.41% of the battery charge consumed. In this way, the lifetime of the battery is extended to a large extent, as complete charge and discharge cycles are not necessary.

4.9. LoRaWAN Measurements

The LoRaWAN was implemented with the network optimisation algorithm developed by Cano-Ortega et al. [8]. The algorithm allows adapting the network parameters in real time in order to obtain the smallest possible ratio of lost packets so as to minimise the loss of information. The algorithm was implemented in the Raspberry that supports the LoRa concentrator.

Figure 25 shows a part of the measurements made on the network. Two hours of measurements are shown. In the graphs, one can see the changes made by the algorithm marked by the points in the graph, where one can see the change of parameters and the reduction in the rate of loss of information. It should be noted that the location of the devices with respect to the concentrator has a decisive influence on the rate of packets lost. The THMDL device is closer to the concentrator than the ECDL device shown. As can be seen, this has a clear influence on the rate of lost data.

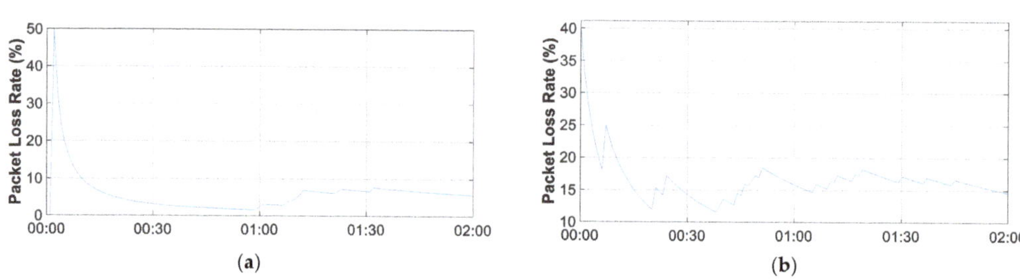

Figure 25. LoRaWAN PLR: (**a**) THMDL measurement and (**b**) ECDL measurement.

4.10. ThingSpeak Integration

As discussed in Section 3.1, data are sent to TTN, and from there, they can be derived to multiple services operating in the cloud. A large number of possible integrations supported by TTN are available. Among them, the integration with the MathWorks ThingSpeak service was chosen to be shown in this research. ThingSpeak allows sending information in its free version of up to four channels of eight fields with a data latency of 15 s. If the needs of the system are greater, it is possible to switch to the paid version, which allows latency times to be reduced to 1 s.

The configuration chosen for the LoRaWAN has the following parameters: BW125, SF7, CR4/5, message header 13 bytes and 16 bytes payload. As can be seen in Table 11, the minimum information sending time is 6.7 s, calculated from the European Telecommunications Standards Institute (ETSI) standard [88], complying with the 1% maximum duty cycle rule. As a data sending time of 60 s was chosen, it complies with the current regulations. Figure 26 shows three examples of integration: (a) temperature and humidity data collection; (b) electrical variables of battery charging in USB mode; and (c) data collection of the electrical variables of the THMDL in the battery-only version.

Figure 26. ThingSpeak Integration: (**a**) temperature and humidity data; (**b**) battery USB charge; and (**c**) THMDL battery power supply operation.

4.11. Future Work

The high power of LoRaWAN is limited by the 1% duty cycle time for sending data, which for the payload used in this research is 6.7 s, together with the limitation of sending data to ThingSpeak every 15 s in the free version may limit the message sending. If the payload is increased with more sensors added to THMDL or ECDL, the minimum time for sending data would increase. In this sense, a future line of research would be to develop a LoRa concentrator that works outside the LoRaWAN specification using technology that avoids data delivery limitations for systems with data latency of less than 1% of the duty cycle. Complementing the previous line, higher data latency systems should be used in the cloud, such as Google's Firebase, which allows up to 0.2 s of data upload time.

It would also be interesting to create a web page and an application for mobile devices where the monitored and controlled installations are collected in real time. Finally, further studies should provide algorithms with machine learning intelligence to improve its performance through the experiences collected during the operation of the installations.

5. Conclusions

This research develops a complete irrigation system based on wireless communication over a LoRaWAN. It meets the objectives of low power consumption, small size, integration, modular design, fault response, operational safety and low price. These objectives have been achieved by overcoming a number of technical challenges, including component selection, modular design, evaluation of alternatives and PCB design to integrate the components used in the THMDL and ECDL.

LoRaWAN has a wide coverage of up to 10 km, with coverage in urban environments reaching up to 5 km. In addition, up to 1000 devices can be integrated with a single gateway or hub, reducing the infrastructure to be installed. If Wi-Fi, Bluetooth, etc., devices were used, a multitude of repeater devices would be required, which would greatly complicate the complexity of the installation. On the other hand, the power consumption of LoRaWAN devices is extremely low compared to Wi-Fi and similar devices. This reduced power consumption increases battery life and extends system uptime.

The system incorporates battery management for low battery warnings with an adjustable warning level. The irrigation routine allows the minimum moisture level for watering to be set. This routine also incorporates a rainfall forecast query that offers the possibility of not watering if the probability of rainfall is higher than the set value and no watering message is sent. The system is equipped with redundant message sending, which minimises the loss of information in the system. On the other hand, all minimum battery voltage levels, minimum rain probability, minimum humidity and waiting time for receiving and sending messages can also be set. This makes for a dynamic, robust and fault-tolerant system that can be installed in a multitude of locations.

A comparison of the prototypes used with other wireless technologies was performed. In this sense, the average consumption of THDML is 2.3762 Wh/day, and the average consumption of the Wi-Fi device studied is 7.0519 Wh/day without sensors and other components, which is 2.96 times higher. The generation of the SP used is 6.2873 Wh/day in the month of lowest generation, which is well within the THMDL average consumption of 2.3808 Wh/day, which is 14.41% of the battery capacity. This contributes to decreasing the charge and discharge cycles of the battery and extending the battery life.

The use of TTN opens up a wide range of possibilities for the development of system functionalities and adaptation to the needs of each implementation. TTN integrates a large set of cloud services, such as the one presented at the end of the Results section (ThingSpeak). In each implementation of the system, the needs of each problem can be studied, and the most suitable service can be used to offer the best solution. On the other hand, the system can also be reprogrammed by adding new functionalities to improve its performance characteristics. By using Arduino as the basis for the devices, the system benefits from the advantages of the open-source platform of this family.

Author Contributions: Conceptualization, F.S.-S. and A.C.-O.; methodology, F.S.-S. and A.C.-O.; software, F.S.-S. and A.C.-O.; validation, F.S.-S. and A.C.-O.; formal analysis, F.S.-S. and A.C.-O.; investigation, F.S.-S. and A.C.-O.; resources, F.S.-S. and A.C.-O.; data curation, F.S.-S. and A.C.-O.; writing—original draft preparation, F.S.-S. and A.C.-O.; writing—review and editing, F.S.-S. and A.C.-O.; visualization, F.S.-S. and A.C.-O.; supervision, F.S.-S. and A.C.-O.; project administration, F.S.-S. and A.C.-O.; funding acquisition, F.S.-S. and A.C.-O. All authors have read and agreed to the published version of the manuscript.

Funding: This research received no external funding.

Acknowledgments: The authors would like to thank the Department of Electrical Engineering of the University of Jaen for allowing the use of their laboratories and material in the development of this research.

Conflicts of Interest: The authors declare no conflict of interest.

Abbreviations

The following abbreviations are used in this manuscript.

AN	Arduino Nano
BW	Bandwidth
CR	Code rate
DLB	Dragino LoRa Bee
ECDL	Electrovalve Control Device for LoRaWAN
GPRS	General Packet Radio Service
I2C	Inter-Integrated Circuit
IFTTT	If This, Then That
IoT	Internet of Things
LiPo	Lipo Rider Pro
LoRa	Long range
LoRaWAN	Long-range wide-area network
LPWAN	Low-power wide-area network
MQTT	Message Queue Telemetry Transport
NB-IoT	Narrow-band Internet of Things
PCB	Printed Circuit Board
SF	Spread factor
SP	Solar panel
THMDL	Temperature and Humidity Measurement Device for LoRaWAN
TTN	The Things Network
Wi-Fi	Wireless Fidelity

References

1. Goumopoulos, C.; O'Flynn, B.; Kameas, A. Automated zone-specific irrigation with wireless sensor/actuator network and adaptable decision support. *Comput. Electron. Agric.* **2014**, *105*, 20–33. [CrossRef]
2. Banđur, Đ.; Jakšić, B.; Banđur, M.; Jović, S. An analysis of energy efficiency in Wireless Sensor Networks (WSNs) applied in smart agriculture. *Comput. Electron. Agric.* **2019**, *156*, 500–507. [CrossRef]
3. Kochhar, A.; Kumar, N. Wireless sensor networks for greenhouses: An end-to-end review. *Comput. Electron. Agric.* **2019**, *163*, 104877. [CrossRef]
4. Hamami, L.; Nassereddine, B. Application of wireless sensor networks in the field of irrigation: A review. *Comput. Electron. Agric.* **2020**, *179*, 105782. [CrossRef]
5. Froiz-Míguez, I.; Lopez-Iturri, P.; Fraga-Lamas, P.; Celaya-Echarri, M.; Blanco-Novoa, Ó.; Azpilicueta, L.; Falcone, F.; Fernández-Caramés, T.M. Design, implementation, and empirical validation of an IoT smart irrigation system for fog computing applications based on Lora and Lorawan sensor nodes. *Sensors* **2020**, *20*, 6865. [CrossRef]
6. Valente, A.; Silva, S.; Duarte, D.; Pinto, F.C.; Soares, S. Low-cost lorawan node for agro-intelligence iot. *Electronics* **2020**, *9*, 987. [CrossRef]
7. Ameloot, T.; van Torre, P.; Rogier, H. Variable link performance due to weather effects in a long-range, low-power lora sensor network. *Sensors* **2021**, *21*, 3128. [CrossRef]
8. Cano-Ortega, A.; Sánchez-Sutil, F. Performance optimization Lora network by artificial bee colony algorithm to determination of the load profiles in dwellings. *Energies* **2020**, *13*, 517. [CrossRef]
9. Sánchez-Sutil, F.; Cano-Ortega, A. Design and testing of a power analyzer monitor and programming device in industries with a lora lpwan network. *Electronics* **2021**, *10*, 453. [CrossRef]
10. Sanchez-Sutil, F.; Cano-Ortega, A. Smart regulation and efficiency energy system for street lighting with LoRa LPWAN. *Sustain. Cities Soc.* **2021**, *70*, 102912. [CrossRef]
11. Cruz, N.; Cota, N.; Tremoceiro, J. Lorawan and urban waste management—A trial. *Sensors* **2021**, *21*, 2142. [CrossRef]
12. Singh, R.K.; Aernouts, M.; De Meyer, M.; Weyn, M.; Berkvens, R. Leveraging LoRaWAN Technology for Precision Agriculture in Greenhouses. *Sensors* **2020**, *20*, 1827. [CrossRef]
13. Nam, W.-H.; Kim, T.; Hong, E.-M.; Choi, J.-Y.; Kim, J.-T. A Wireless Sensor Network (WSN) application for irrigation facilities management based on Information and Communication Technologies (ICTs). *Comput. Electron. Agric.* **2017**, *143*, 185–192. [CrossRef]

14. Goap, A.; Sharma, D.; Shukla, A.K.; Krishna, C.R. An IoT based smart irrigation management system using Machine learning and open source technologies. *Comput. Electron. Agric.* **2018**, *155*, 41–49. [CrossRef]
15. Canales-Ide, F.; Zubelzu, S.; Rodríguez-Sinobas, L. Irrigation systems in smart cities coping with water scarcity: The case of valdebebas, Madrid (Spain). *J. Environ. Manag.* **2019**, *247*, 187–195. [CrossRef]
16. Munir, M.S.; Bajwa, I.S.; Naeem, M.A.; Ramzan, B. Design and implementation of an IoT system for smart energy consumption and smart irrigation in tunnel farming. *Energies* **2018**, *11*, 3427. [CrossRef]
17. Migliaccio, K.W.; Morgan, K.T.; Fraisse, C.; Vellidis, G.; Andreis, J.H. Performance evaluation of urban turf irrigation smartphone app. *Comput. Electron. Agric.* **2015**, *118*, 136–142. [CrossRef]
18. Fernández-Ahumada, L.M.; Ramírez-Faz, J.; Torres-Romero, M.; López-Luque, R. Proposal for the design of monitoring and operating irrigation networks based on IoT, cloud computing and free hardware technologies. *Sensors* **2019**, *19*, 2318. [CrossRef]
19. Fraga-Lamas, P.; Celaya-Echarri, M.; Azpilicueta, L.; Lopez-Iturri, P.; Falcone, F.; Fernández-Caramés, T.M. Design and Empirical Validation of a LoRaWAN IoT Smart Irrigation System. *Proceedings* **2020**, *42*, 62. [CrossRef]
20. Chazarra-Zapata, J.; Parras-Burgos, D.; Arteaga, C.; Ruiz-Canales, A.; Molina-Martínez, J.M. Adaptation of a traditional irrigation system of micro-plots to smart agri development: A case study in murcia (Spain). *Agronomy* **2020**, *10*, 1365. [CrossRef]
21. López-Morales, J.A.; Martínez, J.A.; Skarmeta, A.F. Improving energy efficiency of irrigation wells by using an iot-based platform. *Electronics* **2021**, *10*, 250. [CrossRef]
22. Glória, A.; Dionísio, C.; Simões, G.; Cardoso, J.; Sebastião, P. Water management for sustainable irrigation systems using internet-of-things. *Sensors* **2020**, *20*, 1402. [CrossRef]
23. Khan, N.; Ray, R.L.; Sargani, G.R.; Ihtisham, M.; Khayyam, M.; Ismail, S. Current progress and future prospects of agriculture technology: Gateway to sustainable agriculture. *Sustainability* **2021**, *13*, 4883. [CrossRef]
24. Mohammed, M.; Riad, K.; Alqahtani, N. Efficient iot-based control for a smart subsurface irrigation system to enhance irrigation management of date palm. *Sensors* **2021**, *21*, 3942. [CrossRef] [PubMed]
25. Tiglao, N.M.; Alipio, M.; Balanay, J.V.; Saldivar, E.; Tiston, J.L. Agrinex: A low-cost wireless mesh-based smart irrigation system. *Measurement* **2020**, *161*, 107874. [CrossRef]
26. Sanchez-Sutil, F.; Cano-Ortega, A. Smart public lighting control and measurement system using lora network. *Electronics* **2020**, *9*, 124. [CrossRef]
27. Al-Ali, A.R.; Nabulsi, A.A.; Mukhopadhyay, S.; Awal, M.S.; Fernandes, S.; Ailabouni, K. IoT-solar energy powered smart farm irrigation system. *J. Electron. Sci. Technol.* **2020**, *30*, 100017. [CrossRef]
28. Sudharshan, N.; Karthik, A.V.S.K.; Kiran, J.S.S.; Geetha, S. Renewable Energy Based Smart Irrigation System. *Procedia Comput. Sci.* **2019**, *165*, 615–623. [CrossRef]
29. Nawandar, N.K.; Satpute, V.R. IoT based low cost and intelligent module for smart irrigation system. *Comput. Electron. Agric.* **2019**, *162*, 979–990. [CrossRef]
30. Liao, R.; Zhang, S.; Zhang, X.; Wang, M.; Wu, H.; Zhangzhong, L. Development of smart irrigation systems based on real-time soil moisture data in a greenhouse: Proof of concept. *Agric. Water Manag.* **2021**, *245*, 106632. [CrossRef]
31. Eltohamy, K.M.; Liu, C.; Khan, S.; Niyungeko, C.; Jin, Y.; Hosseini, S.H.; Li, F.; Liang, X. An internet-based smart irrigation approach for limiting phosphorus release from organic fertilizer-amended paddy soil. *J. Clean. Prod.* **2021**, *293*, 126254. [CrossRef]
32. The Things Network. Available online: https://www.thethingsnetwork.org (accessed on 10 September 2021).
33. AWS IoT. Available online: https://aws.amazon.com/es/iot/ (accessed on 10 September 2021).
34. Akenza.io. Available online: https://akenza.io/features (accessed on 10 September 2021).
35. DaraCake. Available online: https://datacake.co (accessed on 10 September 2021).
36. deZem. Available online: https://www.dezem.de/en/data-acquisition/lorawan/ (accessed on 10 September 2021).
37. Influx, DB. Available online: https://docs.influxdata.com/influxdb/v2.0/get-started/ (accessed on 10 September 2021).
38. Microsoft Azure. Available online: https://azure.microsoft.com/es-es/ (accessed on 10 September 2021).
39. Qubitro. Available online: https://www.qubitro.com (accessed on 10 September 2021).
40. Tago, IO. Available online: https://tago.io (accessed on 10 September 2021).
41. The Things.io. Available online: https://thethings.io (accessed on 10 September 2021).
42. Things Board. Available online: https://thingsboard.io (accessed on 10 September 2021).
43. ThingSpeak. Available online: https://thingspeak.com (accessed on 10 September 2021).
44. Ubidots. Available online: https://ubidots.com (accessed on 10 September 2021).
45. UIB. Available online: https://uib.ai (accessed on 10 September 2021).
46. MQTT. Available online: https://mqtt.org (accessed on 10 September 2021).
47. LoRa Cloud. Available online: https://www.loracloud.com (accessed on 10 September 2021).
48. Node-RED. Available online: https://nodered.org (accessed on 10 September 2021).
49. IFTTT. Available online: https://ifttt.com (accessed on 10 September 2021).
50. Google Sheets. Available online: https://www.google.es/intl/es/sheets/about/ (accessed on 10 September 2021).
51. Firebase. Available online: https://firebase.google.com/ (accessed on 10 September 2021).
52. Arduino Nano. Available online: https://store.arduino.cc/arduino-nano (accessed on 10 September 2021).
53. Arduino Uno. Available online: https://store.arduino.cc/products/arduino-uno-rev3/ (accessed on 25 September 2021).
54. Arduino Mega. Available online: https://store.arduino.cc/products/arduino-mega-2560-rev3 (accessed on 25 September 2021).

55. Arduino Micro. Available online: https://store.arduino.cc/products/arduino-micro (accessed on 25 September 2021).
56. Semtech SX1308. Available online: https://www.semtech.com/products/wireless-rf/lora-gateways/sx1308 (accessed on 1 September 2021).
57. Semtech SX1301. Available online: https://www.semtech.com/products/wireless-rf/lora-gateways/sx1301 (accessed on 10 September 2021).
58. Semtech SX1276/SX1278 LoRa Chip. Available online: https://www.semtech.com/products/wirelessrf/loratransceivers/sx1276 (accessed on 10 September 2021).
59. Semtech SX1257. Available online: https://www.semtech.com/products/wireless-rf/lora-core/sx1257 (accessed on 10 September 2021).
60. HOPERF Chip RFM95/96/97/98. Available online: https://cdn.sparkfun.com/assets/learn_tutorials/8/0/4/RFM95_96_97_98W.pdf (accessed on 10 September 2021).
61. Murata Electronics. Available online: https://wireless.murata.com/pub/RFM/data/type_abz.pdf (accessed on 10 September 2021).
62. Arduino MKR WAN 1300. Available online: https://store.arduino.cc/mkr-wan-1300 (accessed on 10 September 2021).
63. Monteino. Available online: https://lowpowerlab.com/guide/moteino/ (accessed on 10 September 2021).
64. Libelium. Available online: http://www.libelium.com/extreme-range-wireless-sensors-connectivity-throughbuildings-incity-lora-868mhz-915mhz/ (accessed on 10 September 2021).
65. Lopy4. Available online: https://docs.pycom.io/gettingstarted/connection/lopy4/ (accessed on 10 September 2021).
66. Dragino LoRa Bee for Arduino. Available online: https://www.dragino.com/products/lora/item/109-lora-bee.html (accessed on 10 September 2021).
67. Dragino Concentrator PG1301. Available online: https://www.dragino.com/products/lora/item/149-lora-gps-hat.html (accessed on 10 September 2021).
68. Dragino OLG01. Available online: https://www.dragino.com/products/lora-lorawan-gateway/item/118-olg01.html (accessed on 25 September 2021).
69. Dragino OLG02. Available online: https://www.dragino.com/products/lora-lorawan-gateway/item/136-olg02.html (accessed on 25 September 2021).
70. Dragino LoRa Hat. Available online: https://www.dragino.com/products/lora/item/106-lora-gps-hat.html (accessed on 25 September 2021).
71. FZ0430 Voltage Sensor. Available online: http://www.ekt2.com/pdf/412_ARDUINO_SENSOR_VOLTAGE_DETECTOR.pdf (accessed on 10 September 2021).
72. ACS712. Available online: https://www.sparkfun.com/datasheets/BreakoutBoards/0712.pdf (accessed on 10 April 2021).
73. INA219. Available online: https://www.ti.com/lit/ds/symlink/ina219.pdf (accessed on 10 September 2021).
74. SHT1x. Available online: https://www.sensirion.com/environmental-sensors/humidity-sensors/digital-humidity-sensors-for-accurate-measurements/ (accessed on 10 September 2021).
75. SHT2x. Available online: https://www.sensirion.com/en/environmental-sensors/humidity-sensors/humidity-temperature-sensor-sht2x-digital-i2c-accurate/ (accessed on 10 September 2021).
76. SHT3x. Available online: https://www.sensirion.com/en/environmental-sensors/humidity-sensors/digital-humidity-sensors-for-various-applications/ (accessed on 10 September 2021).
77. DHT11. Available online: https://www.mouser.com/datasheet/2/758/DHT11-Technical-Data-Sheet-Translated-Version-1143054.pdf (accessed on 10 September 2021).
78. DHT22. Available online: https://www.sparkfun.com/datasheets/Sensors/Temperature/DHT22.pdf (accessed on 10 September 2021).
79. LiPo Rider Pro. Available online: https://wiki.seeedstudio.com/Lipo_Rider_Pro/ (accessed on 10 September 2021).
80. LiPo Rider Plus. Available online: https://wiki.seeedstudio.com/Lipo-Rider-Plus/ (accessed on 25 September 2021).
81. LiPo Rider v1.3. Available online: https://wiki.seeedstudio.com/Lipo_Rider_V1.3/ (accessed on 25 September 2021).
82. Solar Panel. Available online: https://wiki.seeedstudio.com/3W_Solar_Panel_138x160/ (accessed on 10 September 2021).
83. Battery. Available online: https://es.rs-online.com/web/p/baterias-recargables/1449408/ (accessed on 10 September 2021).
84. Wemos D1 Mini. Available online: https://www.wemos.cc/en/latest/d1/d1_mini.html (accessed on 10 September 2021).
85. NodeMCU. Available online: https://components101.com/development-boards/nodemcu-esp8266-pinout-features-anddatasheet (accessed on 10 September 2021).
86. Wemos D1 Mini Pro. Available online: https://www.wemos.cc/en/latest/d1/d1_mini_pro.html (accessed on 10 September 2021).
87. Wemos D1 R1. Available online: http://www.esp8266learning.com/wemos-d1-esp8266-based-board.php (accessed on 10 September 2021).
88. ETSI. *Electromagnetic Compatibility and Radio Spectrum Matters (ERM); Short Range Devices (SRD); Radio Equipment To Be Used in the 25 MHz to 1000 MHz Frequency Range with Power Levels Ranging Up To 500 mW. European Harmonized Standard. EN 2019, 300*; ETSI: Sophia Antipolis, France, 2019; Volume 3.

sensors

MDPI

Article

Towards Efficient Electricity Forecasting in Residential and Commercial Buildings: A Novel Hybrid CNN with a LSTM-AE based Framework

Zulfiqar Ahmad Khan, Tanveer Hussain, Amin Ullah, Seungmin Rho, Miyoung Lee and Sung Wook Baik *

Intelligent Media Laboratory, Digital Contents Research Institute, Sejong University, Seoul 143-747, Korea; mzulfiqar3797@gmail.com (Z.A.K.); tanveerkhattak3797@gmail.com (T.H.); qamin3797@gmail.com (A.U.); smrho@sejong.ac.kr (S.R.); miylee@sejong.ac.kr (M.L.)

* Correspondence: sbaik@sejong.ac.kr

Received: 30 January 2020; Accepted: 28 February 2020; Published: 4 March 2020

Abstract: Due to industrialization and the rising demand for energy, global energy consumption has been rapidly increasing. Recent studies show that the biggest portion of energy is consumed in residential buildings, i.e., in European Union countries up to 40% of the total energy is consumed by households. Most residential buildings and industrial zones are equipped with smart sensors such as metering electric sensors, that are inadequately utilized for better energy management. In this paper, we develop a hybrid convolutional neural network (CNN) with an long short-term memory autoencoder (LSTM-AE) model for future energy prediction in residential and commercial buildings. The central focus of this research work is to utilize the smart meters' data for energy forecasting in order to enable appropriate energy management in buildings. We performed extensive research using several deep learning-based forecasting models and proposed an optimal hybrid CNN with the LSTM-AE model. To the best of our knowledge, we are the first to incorporate the aforementioned models under the umbrella of a unified framework with some utility preprocessing. Initially, the CNN model extracts features from the input data, which are then fed to the LSTM-encoder to generate encoded sequences. The encoded sequences are decoded by another following LSTM-decoder to advance it to the final dense layer for energy prediction. The experimental results using different evaluation metrics show that the proposed hybrid model works well. Also, it records the smallest value for mean square error (MSE), mean absolute error (MAE), root mean square error (RMSE) and mean absolute percentage error (MAPE) when compared to other state-of-the-art forecasting methods over the UCI residential building dataset. Furthermore, we conducted experiments on Korean commercial building data and the results indicate that our proposed hybrid model is a worthy contribution to energy forecasting.

Keywords: buildings energy management; deep learning; energy consumption prediction; LSTM; autoencoder; load forecasting; smart sensors

---

1. Introduction

Electrical energy consumption has recently been accelerating due to rapid population and economic growth [1]. According to the World Energy Outlook (2017), global energy demand is predicted to increase by 1.0% compound annual growth rate (CAGR) over the period of 2016-40 [2]. Residential buildings play a vital role in this consumption, constituting 27% of total global energy usage, and have a substantial impact on overall energy consumption [3]. In the US, buildings make up 40% of their national overall energy usage [4]. Due to the high level of electricity consumption in commercial and residential buildings, efficient smart electrical energy prediction and its management are becoming more important because the load forecasting directly affects the control and planning of power systems' operation. A research study estimated that a 1% decrease in forecasting errors can save £10 million per year for the UK power system [5]. Therefore, appropriate energy planning plays a vital role in saving energy, as well as being an economical solution. Future energy planning is possible through computationally intelligent electricity forecasting methods [6,7].

Electricity consumption prediction is a multivariate time series problem where the sensors generate data that may contain uncertainty [8,9], redundancy, missing values, etc. Due to irregular trend components and seasonal patterns, it is difficult to accurately predict electricity consumption by employing traditional machine learning models [10]. On the other hand, deep learning models yield ultimately better results and are less error prone. Deep learning models are aggressively studied in several applications such as CNNs, which are superior at recognizing images, and recurrent neural networks (RNNs) [11], which perform well in natural language processing (NLP) [12] and speech recognition problems. In recent studies, many researchers integrated multiple models in the aforementioned domains to achieve convincing results that are applicable in real-world scenarios. Utilizing hybrid techniques, CNN with LSTM has achieved state-of-the-art results for various domains, such as convincing results for emotion recognition [13], speech processing [14], activity recognition [15] and also in the medical domain, where it shows superior performance in detecting arrhythmias [16]. Similar hybrid models are used in the energy forecasting domain to achieve state-of-the-art results.

Several techniques have been developed for energy consumption prediction, including ARIMA [17], SVM and SVR [18], time series [8], neuro fuzzy and linear regression (LR) models [19] and artificial neural networks [20]. These prediction models are grouped into four major groups: statistical, machine learning (ML), deep learning and hybrid models. Energy forecasting related studies are grouped based on this categorization and their descriptions along with the dataset used and strategy followed is given in Table 1.

Among the statistical-based models, Fumo and Biswas [21] used a linear regression model for residential energy prediction and observed time resolution effects on the model's performance. Daily energy consumption prediction is proposed in Reference [22] by using multiple-linear regression with genetic programming. They integrated five variables through genetic programming and then fed them into their proposed prediction model. The performance of this model is increased by removing unnecessary variables, but independent variables correlation leads to the problem of multicollinearity and it is also challenging to get explanatory variables via linear regression models. Therefore, such models are not recommended for electricity prediction.

In the machine learning approaches category, SVR was used to forecast electricity consumption in buildings [23] and improved the performance of the model by adding temperature variables. Another approach based on random forest was developed in Reference [24], in which the authors predicted the following week's energy by using human dynamics. In the machine learning approach, if the model does not have many features, then it generates complex decision boundaries. However, these models drain into an overfitting problem if the data is increased or the correlation between variables is complicated. If a model is overfitted, it greatly affects the prediction accuracy and hence is not recommended for use in residential or commercial buildings energy forecasting.

Deep learning models are widely used for electricity prediction, in which Reference [25] used a sequence-to-sequence model for electricity consumption prediction in buildings and achieved the highest possible performance. The authors of Reference [1] used stacked AE and reduced noise disturbance and randomness from the electricity consumption data via deep features. These models extracted important features in cases where they had complex attributes and a lot of redundant data. However, modeling the spatial and temporal features of electricity consumption data is difficult for deep learning models.

Among these approaches, some recent studies show combinations of models for electricity consumption prediction. The authors of Reference [26] integrated CNN with the LSTM model for electricity prediction, where the CNN layers were used to extract spatial features and LSTM was utilized for modeling temporal information. The combination of CNN with Bi-directional LSTM was presented in Reference [27] where the CNN layers were used to extract important information and the Bi-directional LSTM used these features in both the forward and backward direction to make a final prediction. These models achieved the best results but still the error rate was too high for them to be implemented for accurate electricity consumption prediction in real-world scenarios.

Table 1. The four types of prediction models for energy consumption.

Category	Paper	Learning Strategy	Dataset	Description
Statistical models	[21]	LR	Electricity consumption	Analysis of electricity prediction using LR according to time resolution.
	[22]	Multiple regression (MR)		Develops two models: ML and genetic algorithm (GA), where GA is used to select critical information from the dataset followed by optimal prediction via the ML model.
	[28]	MR		Uses backward elimination and a multicollinearity process for suitable variable selection and uses a MR model for medium-term electricity prediction.
Machine learning-based models	[23]	SVR	Electricity load	Adds a temperature variable to improve the performance of SVR for electricity prediction.
	[24]	Random forest regressor	Electricity consumption	Avoids overfitting by using an ensembled method and transforms the data from time to frequency domain to solve the input data computational complexity.
DL-based models	[25]	Seq2seq	Electricity load	Collects data from real smart meters and develops a sequence-to-sequence-based prediction model for short-term electricity prediction in buildings.
	[1]	Stacked AE (SAE)	Electricity consumption	Combines SAE with an extreme learning machine (ELM), where SAE is used to extract features and ELM is used as a prediction model.
	[29]	DRNN based on pooling	Electricity load	Uses pooling based DRNN, addresses the overfitting problem in a naïve deep learning network and tests the method in a real environment on smart meters in Ireland.
	[30]	Seq2seq	Electricity consumption	Uses a sequence-to-sequence model based on modified LSTM.
Hybrid models	[26]	CNN-LSTM		CNNs are used to extract spatial features and LSTM is used for modeling temporal features.
	[27]	CNN-bidirectional LSM		CNNs are used to extract spatial features and bidirectional LSTM is used for these features for final prediction.

We proposed a hybrid model of CNN LSTM-AEs' synergy for electricity prediction in residential and commercial buildings. CNN layers are used to extract spatial features and their output is fed into LSTM-AE, followed by a dense (fully connected) layer for final prediction. Finally, the time resolution is changed to observe if further improvement can be made using the CNN with a LSTM-AE model. For the first time, a hybrid model of CNN and LSTM-AE is developed and tested to predict residential and commercial power consumption. The following are the main contributions of this research work:

- The input dataset is passed through a preprocessing step where redundant, outlier or missing values are removed, and the data are normalized to achieve satisfactory prediction results.
- A novel hybrid model is developed in this work for accurate future energy prediction. The proposed model integrates CNN with LSTM_AE in which the CNN layers are used to extract spatial features from input data and then LSTM-AE are used to model these features.
- The experimental results demonstrate that the proposed CNN with LSTM-AE model has the best performance compared to other models. The evaluation metrics record the smallest value for MSE, MAE, RMSE and MAPE for energy consumption prediction.

2. Proposed Framework

Prediction of electrical power consumption in residential and commercial buildings is very important to provide better energy management services. Due to the impact of unpredictability or the noisy arrangement of data, accurate electricity consumption prediction is a challenging task. For these reasons, the forecasting model sometimes generates incorrect prediction results. Moreover, several methods have been developed based on traditional networks with high error rates. The traditional methods have the problems of needing to learn from scratch, overfitting or short-term memory challenges if the data increase or the correlation between variables is complicated. These issues can be easily solved using sequential learning models, through modeling the spatial and temporal features for electricity consumption is also challenging. Therefore, in this paper, we developed a CNN with LSTM-AE model and a data preprocessing step to efficiently predict electricity consumption in residential and commercial buildings. The overall architecture of the proposed framework for electricity consumption is shown in Figure 1. Further, each section of the proposed framework for electricity consumption is discussed in the next sections.

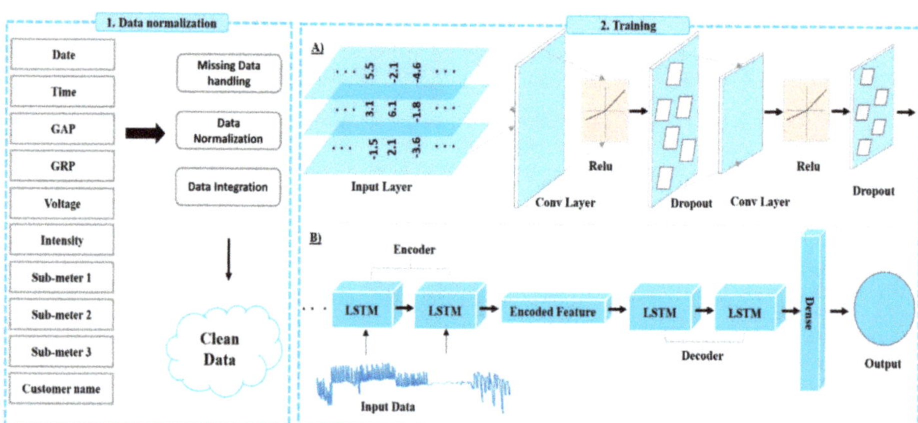

Figure 1. Proposed framework for electricity consumption prediction.

2.1. Data Preprocessing

This section offers detailed analysis about the collection and refinement of data. The data is collected from smart meters which are installed at the edge of the electricity network and connect all appliances to a main board. Normally, the data are gathered annually or monthly, which generates noise and abnormalities in the data due to measurement or human error, meter problems and climate change, if the meters are installed for a long time. Before training, the data need to be refined and normalized for good results.

The tested datasets include null, redundant and outlier values. Similarly, samples from the datasets are not all in the same range and need to be normalized before training for accurate prediction.

Null, redundant and outlier values are extracted from the datasets and are discussed in this section. Also, different normalization techniques were applied to get the odd range values within a specified limit. These techniques include Min-Max scalar, standard transform, Max-Abs scalar, quantile and power transform, as shown in Figure 2. After detailed analysis of each technique, finally, we selected standard transform for data normalization because it centers and scales each feature independently.

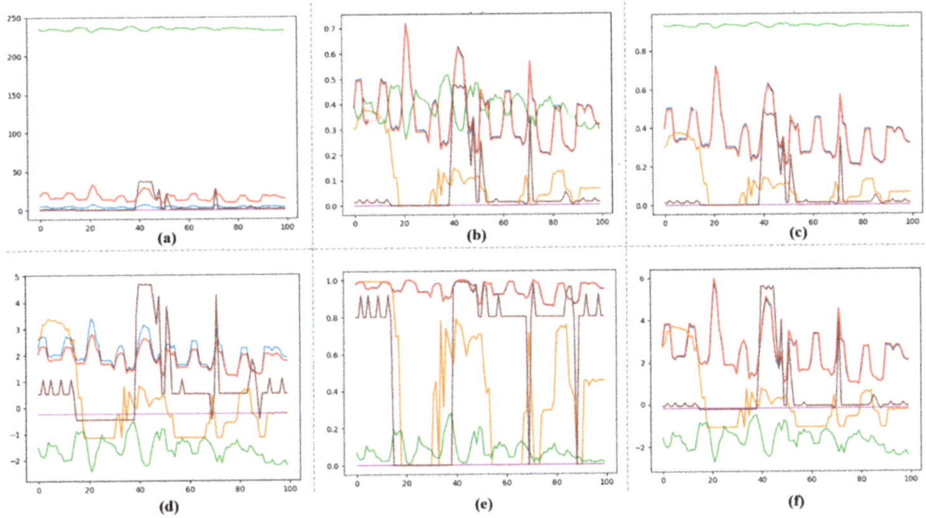

Figure 2. Data normalization techniques, where (**a**) original data in the dataset, (**b**) the range of data after applying Min-Max scalar, (**c**) the range of data after applying Max-Abs scalar, (**d**) the range of values after applying power transform, (**e**) the data plot after quantile transformation, and (**f**) the range of data after applying standard transformation.

The range of each feature is different in the original dataset, as shown in Figure 2a where the ranges of features are between 0–10, 0–50 and 200–250. After applying Max-Min normalization technique, the range of these features lies between 0 and 0.7, as visualized in Figure 2b. Similarly, after processing data with Max-Abs, the ranges are normalized between 0 to 0.8, as shown in Figure 2c. After normalizing data with quantile transformation, the features range is achieved between 0 to 1, as visualized in Figure 2e. However, we needed to transform the input data in a way such that the negative values also exist in the features to achieve good results. The range of power transformation is between −2 to 5 as visualized in Figure 2d, and standard transform is −2 to 6 as given in Figure 2f. However, the computational complexity of power law transformation is higher than standard transformation. Also, standard transform processes each feature independently. Due to these reasons, finally, we selected standard transform for data normalization.

2.2. ANN

ANN is a type of strong mathematical modeling tool inspired by the human nervous system. An early ANN model is MLP [31] which includes input, hidden and output layers. Each neuron relates to the next and previous layer neurons, which are similar in MLP with several input and output links. The value retrieved from the previous layer is summed up with some weight for each neuron individually, and a bias term. Finally, activation function "f" is used to transform the sum, which may be different for each neuron, as shown in Figure 3.

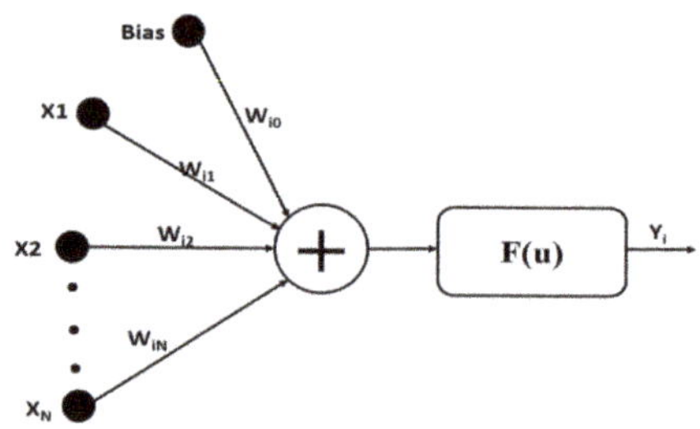

Figure 3. The simple neuron operation in ANN, where "X" represents the input data, "W" represents the weights, "F" is the activation function and "Yi" is the output.

2.3. CNN

CNN was specially developed for grid topology data processing [32]. For example, visual data, i.e., images and videos, are viewed as a two-dimensional grid and time series data are viewed as one-dimensional data. The CNN [33–35] uses a weight sharing concept that provides high accuracy in nonlinear problems, such as energy consumption prediction. Convolution-pooling layers of one dimension are shown in Figure 4. When the convolution is applied to the input data, I1, I2, I3, I4, I5 and I6 are converted to a features map C1, C2, C3, C4. Next, a pooling layer is applied to sample the feature-maps of the convolution layer. The pooling layer procedure is important for extracting high-level convolution features; after applying the pooling layer, the dimension of the features map is reduced to 2.

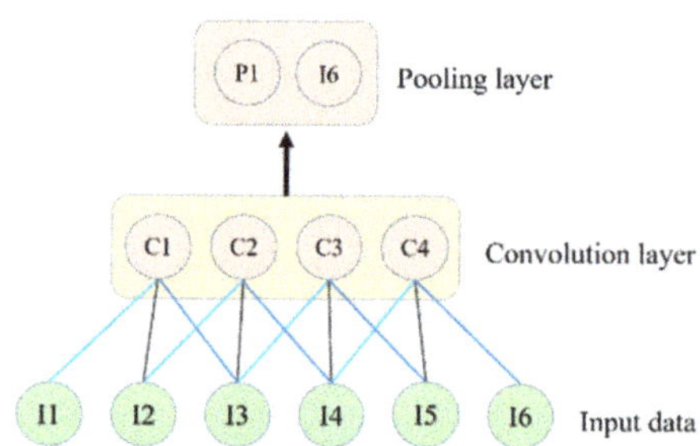

Figure 4. The operation of convolution layers and pooling layers over input data.

2.4. LSTM

The recurrent neural network (RNN) is another popular deep learning architecture, where connections between units form a directed graph along with the sequence information from the input, as depicted in Figure 5. The RNN processes a sequence of input data by using their internal state and turns into a vanishing gradient problem, which has a major negative effect on the model accuracy. An enhanced version of RNN is LSTM [36], which overcomes the vanishing gradient problem via the concept of gates (input, forget, and output) and memory cells. The LSTM operation is illustrated by the following equations and its structure is shown in Figure 5.

$$f_t = \Phi\left(\hat{W}_f \cdot [h_{t-1}, x_t] + B_f\right) \tag{1}$$

$$i_t = \Phi\left(\hat{W}_i \cdot [h_{t-1}, x_t] + B_i\right) \tag{2}$$

$$C_t = \tanh\left(\hat{W}_C \cdot [h_{t-1}, x_t] + B_C\right) \tag{3}$$

$$C_t = f_t \times C_{t-1} + i_t \times C_t \tag{4}$$

$$o_t = \Phi\left(\hat{W}_o \cdot [h_{t-1}, x_t] + B_o\right) \quad (5)$$

$$h_t = o_t \times \tanh(\Phi(C_t). \quad (6)$$

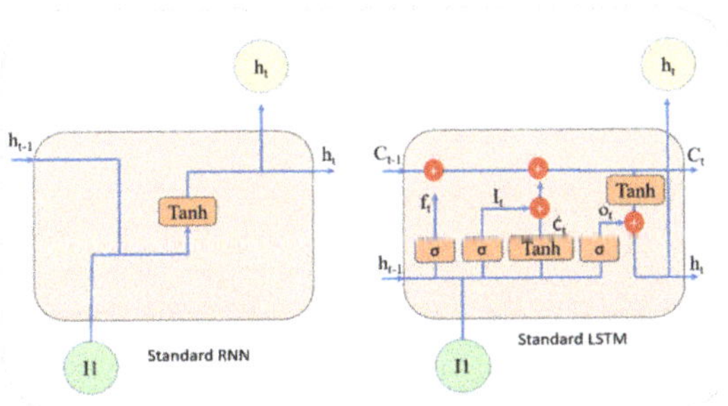

Figure 5. Standard architecture of RNN and LSTM.

In Equation (1), the network input is x_t, h_t is the output of the hidden layer, Φ represents the sigmoid function, the cell state is C_t and the state candidate values are represented through $\dot{C}_t$. $\hat{W}_i$, $\hat{W}_o$, $\hat{W}_f$ and $\hat{W}_C$ are the weights for the input, output, forget gate and memory cells, while B_i, B_o, B_f and B_C represent the bias for the input, output, forget gate and cell, respectively. The input gate decides whether input data will be reserved or not, the forget gate verifies if data will be lost or not, the cell records the processing state and the output is delivered through the output gate. This architecture is specially designed to address the vanishing gradient problem in RNNs.

2.5. LSTM-AE

Autoencoders (AE) are generally used in representation learning to understand unsupervised inputs in a feature vector. The conventional method utilizing an LSTM-AE is illustrated in Figure 6. We employed sequence-to-sequence AE for a time-series sequence dataset. The optimal goal is to predict the short-term electricity consumption of residential and commercial buildings. AE consists of an encoder and a decoder, where the input sequence is first encoded and then decoded. Let xt be the input features and F the feature space. The encoder function applied is: φ: xt →F that learns important features and encodes the features vector F. In the decoder, Ð= F → X, which intends to reconstruct the input by utilizing internal representations [37]

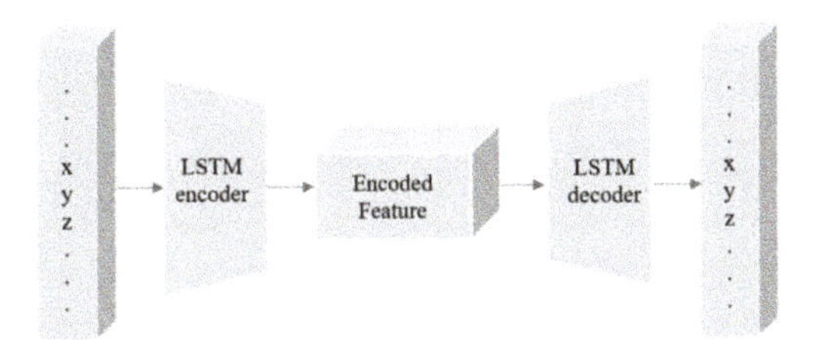

Figure 6. The internal structure of LSTM-AE where the first LSTM layer used as an encoder and the second is a decoder.

We employed LSTM cells for the execution of the encoder and decoder, which are capable of learning from temporal dependencies from one sequence and another. Formally, for input samples sequence $X(N)$, the AE function is applied ΦAE: φ Ð, which outputs samples $x(N)$.

$$\Phi_{AE}(X(N)) = x(N) \quad (7)$$

2.6. Training

In our proposed framework, the refined input data is passed to the training step. The training step includes two sub-sections where "A" demonstrates the CNN architecture and "B" shows the LSTM-AE architecture. The proposed hybrid model combines CNN with LSTM-AE to predict hourly and daily electricity consumption for residential and commercial buildings. The CNN layers include an input layer, hidden layers and an output layer, which extract features for LSTM-AE. The hidden layers include convolution, dropout, pooling and ReLU layers. Two convolution layers with the RELU activation function and dropout layer after each convolution are employed. Initially, the CNN extracts feature from the refined input data, then the output CNN features are fed into the LSTM encoder, which encodes the input sequences of four time-steps. The repeated vector layer replicates these encoded sequences twice from the model. These encoded sequences are inputted into another LSTM for decoding and finally a dense layer is used to produce the output prediction for the input sequence. The LSTM has problems modeling spatial features, so in this work we used CNN to extract spatial features and then fed them to the LSTM. Normally, the LSTM fails to learn temporal dependencies from one sequence to another, so in this work we developed a hybrid network to tackle these issues and developed a reliable solution for accurate electricity prediction. In this architecture, we used two 1D-convolutional layers, where two dropout layers are inserted after each convolutional layer, two encoder LSTM layers, one repeated vector layer, two decoder LSTM layers and finally one fully connected layer. As a result, the total number of layers are 10 in the proposed architecture and the model size is 445 KB with 33,811 parameters. The filter size for first convolution layer is 8 while for the second layer it is 16 and the kernel size is one for both convolution layers.

The proposed method works better than other state-of-the-art models because we integrated multiple architectures to develop a hybrid model (CNN-LSTM-AE), where CNN is used to extract spatial features from the input dataset and then feed these features to LSTM-AE. The simple LSTM model works well but is unable to learn temporal dependency between sequences, while LSTM-AE is capable of learning from temporal dependencies from one sequence and another. This is experimentally proven and the results are discussed in the Section 3. Therefore, we claim that our model works well and show convincing results when compared to other models.

3. Results

This section provides detailed discussion about the experimental setup, datasets, evaluation metrics, evaluation of the UCI dataset, evaluation of the Korean commercial building dataset and finally a comparative analysis of the proposed hybrid network with other baseline models.

3.1. Experimental Setup

We evaluated and validated the efficiency of the proposed hybrid CNN with LSTM-AE model using residential and commercial buildings datasets. We trained our hybrid model on TITAN X (Pascal)/PCLe/SSE2 GPU with an Intel Core i5-6600 processor, with 64 GB memory over the Ubuntu 16.4 LTS operating system. This model was implemented in Python (V3.5) in Keras (V2.2.4) with a TensorFlow (V1.12) backend and employed Adam as the optimizer. Several experiments were conducted to find the optimal selection of the hyper perimeter of each model. After extensive experiments, finally we decided to train the model over 50 epochs with 1000 as the batch size and a 0.2 validation split.

3.2. Datasets

In this paper, we used two datasets: the household electric power consumption dataset available on the UCI machine learning repository [38] and our own commercial data. A number of time-series variables were used in the proposed architecture to predict the global active power consumption. The UCI dataset contains actual power consumption data, with one-minute resolution, collected from a single residential building in France between 2006 and 2010. A total of 2,075,269 records are present in the dataset, with 25,979 missing values that are handled in the preprocessing step of the proposed framework. The dataset is then grouped into hourly and daily resolution to predict the electricity consumption for the short term. Table 2 shows the electricity consumption variables of the UCI dataset, which include date, time, global active power, global reactive power, voltage, intensity, submetering_1, submetering_2 and submetering_3 variables. The time variable includes months, days, years, hours and minutes. The submetering shows the electricity consumption in the home, where submetering_1-3 corresponds to the kitchen, laundry room and living room, respectively.

Table 2. Feature representation and detailed description of the residential dataset, namely the "individual household electricity consumption dataset".

Variable	Description
Date	Presented in dd/mm/yyyy format.
Time	Time variable given in hours, minutes and seconds (hh:mm:ss)
Global active power	Minutely given average active and reactive power for individual house.
Global active power	
Voltage	One-minute average voltage
Intensity	Current intensity for every minute.
Submetering (1, 2, 3)	Active electricity related to kitchen, laundry room and living room of residential home, while only one submetering_1 sensor in commercial dataset is related to office electricity.

Our new dataset is similar to the UCI dataset but with some differences which are mentioned below:

- The UCI dataset was derived from residential buildings while the proposed dataset was generated in commercial buildings.
- The UCI dataset has three consumption sensors: submeters 1, 2 and 3, while our dataset includes only one electricity consumption sensor.
- The UCI dataset includes 1-minute resolution, while the proposed dataset has 15-minute resolution.

3.3. Evaluation Metrics

The proposed method is evaluated on four standard metrics: MSE, MAE, RMSE and MAPE. The mathematical formulas of these metrics are given in Equations (8)–(11). RMSE is the percentage of difference between predicted and testing variables, MAE represents the percentage of difference between the predicted variables, MSE represents the average square value between the testing and predicted variables, while the last metric MAPE expresses the prediction accuracy in percentage. The training and validation loses for both UCI and Korean commercial building dataset are shown in Figure 7, where "A", "B", "C" and "D" represent the loses for residential building hourly data, residential building daily data, Korean commercial building hourly data and Korean commercial building daily data, respectively.

There are a total of 960,000 records in our dataset, with null and redundant values that are removed in the preprocessing step. Next, we normalized the input data to train the proposed model efficiently. For training purposes, 75% of the data are used from each dataset, while the remaining 25% are used for testing. This means that the first three years data of the UCI dataset are used for training, while the last year's data are used for testing. Furthermore, we performed several experiments on different deep models for comparison, such as CNN, LSTM, LSTM-AE and the CNN with LSTM-AE models.

$$MSE = \frac{1}{n}\sum_{1}^{n}(y - \hat{y})2 \tag{8}$$

$$MAE = \frac{1}{n}\sum_{1}^{n}|y - \hat{y}| \quad (9)$$

$$RMSE = \sqrt{\frac{1}{n}\sum_{1}^{n}(y - \hat{y})} \quad (10)$$

$$MAPE = \frac{100\%}{n}\sum_{t-1}^{n}\left|\frac{A_t - F_t}{A_t}\right| \quad (11)$$

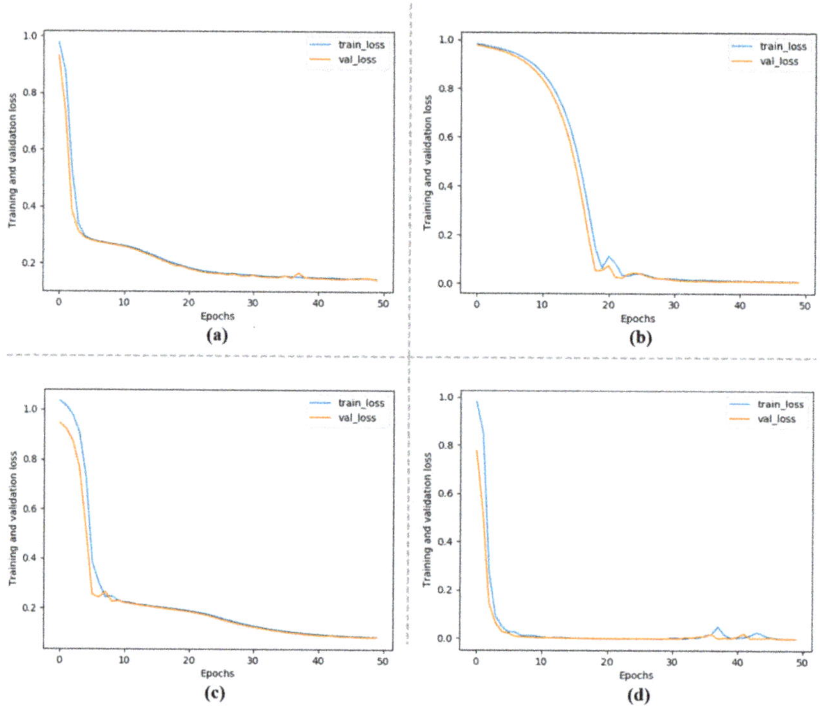

Figure 7. Training and validation loss during training.

3.4. Performance Evaluation over UCI Dataset

To validate the robustness of the proposed hybrid model, we performed experiments on several deep learning models with variable sets of resolutions. The results achieved for each model over hourly data are shown in Figure 8. First, we used CNN to check the performance of the model, and obtained values of 0.37, 0.47 and 0.67 for MSE, MAE and RMSE, respectively. On the other hand, when using LSTM, we observed 0.35, 0.45 and 0.61 for MAE, MSE and RMSE, correspondingly. Moreover, with the combined CNN-LSTM we obtained 0.31, 0.44, and 0.58 for MSE, MAE, RMSE, and with the LSTM-AE model values of 0.26, 0.38 and 0.56 for MSE, MAE, and RMSE, respectively. Inspired by the results of LSTM-AE, we combined CNN with LSTM-AE and recorded the smallest values: 0.19, 0.31 and 0.47 for MSE, MAE and RMSE, respectively.

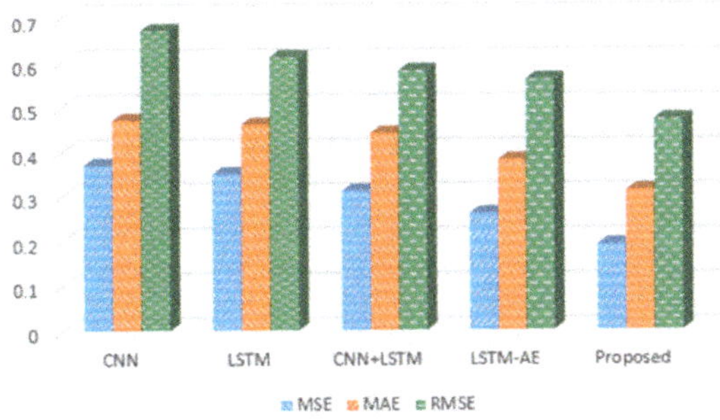

Figure 8. The MSE, MAE and RMSE error rates of different deep learning models for hourly electricity prediction.

Next, the performance of the aforementioned deep learning models for daily data was tested. For the MSE, MAE and RMSE evaluation metrics, our method performed best compared to the baseline models. In more detail, CNN achieved values of 0.006, 0.05 and 0.07 for MSE, MAE and RMSE, respectively, while LSTM reduced its error rate (compared with the hourly rate) to 0.05, 0.13 and 0.22 for MAE, MSE and RMSE. Furthermore, we combined the CNN with LSTM and achieved 0.007, 0.06, and 0.08 for MSE, MAE, and RMSE, whereas LSTM-AE showed values of 0.01, 0.07 and 0.11 for MSE, MAE, and RMSE, respectively. Finally, we tested the proposed CNN with LSTM-AE hybrid model and obtained the lowest values of all, at 0.0004, 0.01 and 0.02 for MSE, MAE and RMSE, respectively, as shown in Figure 9b.

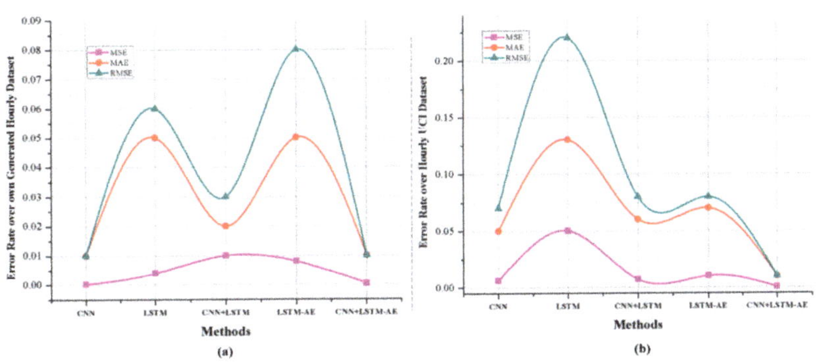

Figure 9. The detailed results of different deep learning-based models for one day resolution data where (**a**) demonstrates MSE, MAE and RMSE for the Korean commercial building dataset and (**b**) shows these error rates over UCI dataset.

3.5. Performance Evaluation over Newly Generated Dataset

The aforementioned models were also tested on our newly generated dataset, and the proposed model recorded convincing values for the tested evaluation metrics. The dataset was tested on both hourly and daily data resolution, as shown in Figure 10 where (a) shows electricity consumption prediction for hourly data, while (b) indicates electricity prediction for daily data. The difference between actual and predicted values is very narrow, but better performance is evident for the proposed model, especially for daily data future load prediction.

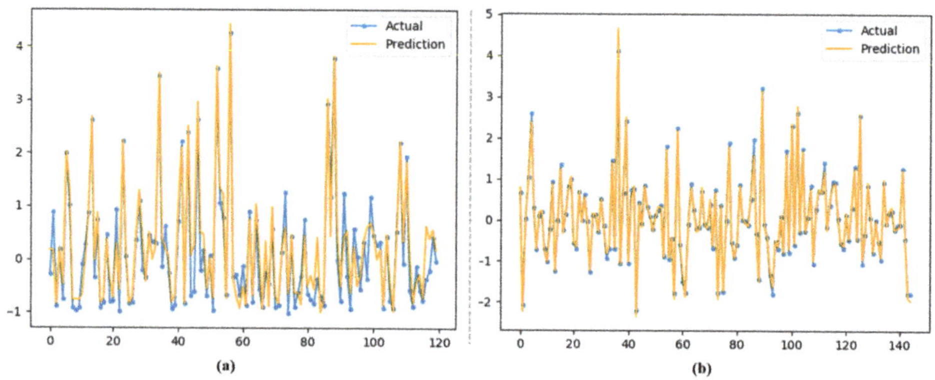

Figure 10. Visualization of performance of our proposed CNN with LSTM-AE over testing data for electricity prediction. (**a**) electricity consumption prediction for hourly data; (**b**) electricity prediction for daily data.

For hourly electricity prediction on the Korean commercial building dataset, the proposed model stands in third place, LSTM-AE is second and LSTM is first. For daily electricity prediction, the proposed model achieved the lowest error rates of 0.0003, 0.01 and 0.01 for MSE, MAE and RMSE, respectively. Figure 9a shows the prediction performance of the proposed hybrid model for hourly electricity consumption, while Figure 11 demonstrates the daily energy prediction error rate for each model.

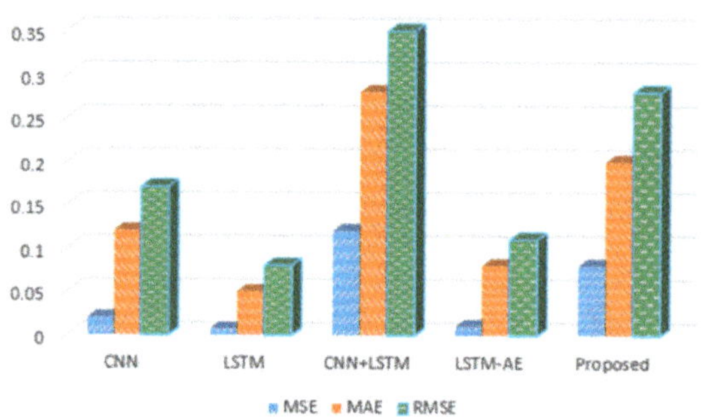

Figure 11. The results achieved by different deep learning-based models for daily resolution of data on our own dataset.

3.6. Comparison with other Baseline Models

The performance of the proposed hybrid model was evaluated and compared with other competitive baseline models, which were similarly used for the same dataset. The results were compared for both hourly and daily data. For hourly prediction, the proposed method was compared with References [26,27,30,39] and achieved the smallest error rate among these models, as shown in Table 3. For daily prediction, the proposed model performance was compared with References [26,27,30,40,41] and achieved better results, as demonstrated in Figure 12. For instance, the proposed hybrid model recorded the smallest error rates of 0.19, 0.31 and 0.47 for the hourly dataset, and recorded 0.01, 0.08, 0.11 and 0.69 for the daily dataset.

Table 3. The comparative analysis of the proposed method with other state-of-the-art Deep Learning and traditional techniques for hourly data resolution.

	Methods	MSE	MAE	RMSE	MAPE
Deep Learning Methods	Kim, T.-Y et al. [26]	0.35	0.33	0.59	-
	Kim, J, -Y et al. [39]	0.38	0.39	-	-
	Marino et al. [30]	-	-	0.74	-
	Le et al. [27]	0.29	0.39	0.54	-
Traditional Machine Learning models	ARMA [42]	-	-	0.30	-
	SVM [43]	-	1.12	1.25	-
	Linear Regression [41]	-	-	-	1.03
	SVR [41]	-	-	-	1.29
	Gaussian Process [41]	-	-	-	0.82
	Proposed	**0.19**	**0.31**	**0.47**	**0.76**

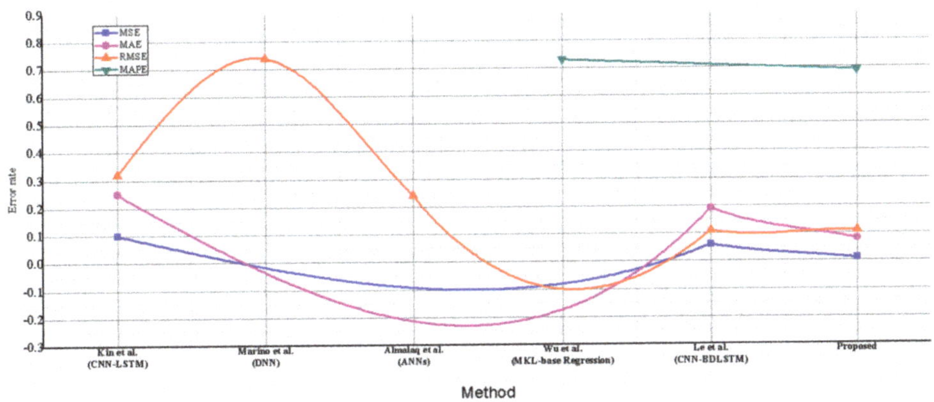

Figure 12. Comparative analysis of the proposed hybrid CNN with LSTM-AE model with the methods developed by Kim et al. [26], Marino et al. [30], Almalaq et al. [40], Wu et al. [41] and Le et al. [27]. In the figure, our model performance is compared with other state-of-the-art models in term of MSE, MAE, RMSE and MAPE. Our model attains the smallest values for each metric.

4. Conclusions

In this article, we developed a novel framework for the prediction of electricity consumption in residential and commercial buildings, and evaluated it using two datasets including the UCI household electricity consumption prediction and Korean commercial building data. Initially, the input data are preprocessed to remove missing, redundant and outlier values. Next, we apply different normalization techniques for better representation of the input data, which yields an effective model. Further, we developed a novel hybrid CNN with LSTM-AE model. The proposed model has three modules for predicting electricity consumption: CNN, LSTM-AE and FC. Primarily, two CNN layers are used to extract information from several variables in the dataset, which are then fed to LSTM-AE, which converts the sequence into an encoded features vector and then decodes it through another LSTM. The encoded feature vector layer duplicates these encoded sequences and finally a dense layer is used to produce the output prediction. The experimental results of the proposed hybrid model outperform other state-of-the-art models for electricity consumption prediction, in terms of different performance metrics such as MSE, MAE, RMSE and MAPE.

Author Contributions: Conceptualization, Z.A.K.; methodology, Z.A.K.; software, Z.A.K.; validation, T.H. and A.U.; formal analysis, M.L.; investigation, A.U.; resources, S.W.B.; data curation, Z.A.K.; writing—original draft preparation, Z.A.K.; writing—review and editing, T.H and A.U.; visualization, T.H.; supervision, S.W.B.; project administration, M.L.; funding acquisition, S.W.B. and S.R. All authors have read and agreed to the published version of the manuscript.

Funding: This work was supported by the National Research Foundation of Korea (NRF) grant funded by the Korea government (MSIT) (No. 2019M3F2A1073179).

Conflicts of Interest: The authors declare no conflict of interest.

References

1. Li, C.; Ding, Z.; Zhao, D.; Yi, J.; Zhang, G. Building energy consumption prediction: An extreme deep learning approach. *Energies* **2017**, *10*, 1525. [CrossRef]
2. Sieminski, A. International energy outlook. *Energy Inf. Adm. (EIA)* **2014**, *18*.
3. Nejat, P.; Jomehzadeh, F.; Taheri, M.M.; Gohari, M.; Majid, M.Z.A. A global review of energy consumption, CO_2 emissions and policy in the residential sector (with an overview of the top ten CO_2 emitting countries). *Renew. Sustain. Energy Rev.* **2015**, *43*, 843–862. [CrossRef]
4. Amarasinghe, K.; Wijayasekara, D.; Carey, H.; Manic, M.; He, D.; Chen, W.-P. Artificial neural networks based thermal energy storage control for buildings. In Proceedings of the IECON 2015-41st Annual Conference of the IEEE Industrial Electronics Society, Yokohama, Japan, 9–12 November 2015.
5. Bunn, D.; Farmer, E.D. *Comparative models for electrical load forecasting*; Wiley: New York, NY, USA, 1985.
6. Ullah, A.; Haydarov, K.; Ul Haq, I.; Muhammad, K.; Rho, S.; Lee, M.; Baik, S.W.J.S. Deep Learning Assisted Buildings Energy Consumption Profiling Using Smart Meter Data. *Sensors* **2020**, *20*, 873. [CrossRef] [PubMed]
7. Ullah, F.U.M.; Ullah, A.; Haq, I.U.; Rho, S.; Baik, S.W.J.I.A. *Short-Term Prediction of Residential Power Energy Consumption via CNN and Multilayer Bi-directional LSTM Networks*; IEEE: Piscataway, NJ, USA, 2019.
8. Deb, C.; Zhang, F.; Yang, J.; Lee, S.E.; Shah, K.W. A review on time series forecasting techniques for building energy consumption. *Renew. Sustain. Energy Rev.* **2017**, *74*, 902–924. [CrossRef]
9. Kim, K.-H.; Cho, S.-B. Modular Bayesian Networks with Low-Power Wearable Sensors for Recognizing Eating Activities. *Sensors* **2017**, *17*, 2877.
10. Ahmad, M. Seasonal Decomposition of Electricity Consumption Data. *Rev. Integr. Bus. Econ. Res.* **2017**, *6*, 271–275.
11. Hussain, T.; Muhammad, K.; Ullah, A.; Cao, Z.; Baik, S.W.; Albuquerque, V.H.C.d. Cloud-Assisted Multiview Video Summarization Using CNN and Bidirectional LSTM. *IEEE Trans. Ind. Inform.* **2020**, *16*, 77–86. [CrossRef]
12. Kwon, S.J.S. A CNN-Assisted Enhanced Audio Signal Processing for Speech Emotion Recognition. *Sensors* **2020**, *20*, 183.
13. Wang, J.; Yu, L.-C.; Lai, K.R.; Zhang, X. Dimensional sentiment analysis using a regional CNN-LSTM model. In Proceedings of the 54th Annual Meeting of the Association for Computational Linguistics, Berlin, Germany, 7–12 August 2016; Volume 2, Short Papers. pp. 225–230.
14. Sainath, T.N.; Vinyals, O.; Senior, A.; Sak, H. Convolutional, long short-term memory, fully connected deep neural networks. In Proceedings of the 2015 IEEE International Conference on Acoustics, Speech and Signal Processing (ICASSP), Brisbane, QLD, Australia, 19–24 April 2015; pp. 4580–4584.
15. Ullah, A.; Ahmad, J.; Muhammad, K.; Sajjad, M.; Baik, S.W. Action recognition in video sequences using deep bi-directional LSTM with CNN features. *IEEE Access* **2017**, *6*, 1155–1166. [CrossRef]
16. Oh, S.L.; Ng, E.Y.; San Tan, R.; Acharya, U.R. Automated diagnosis of arrhythmia using combination of CNN and LSTM techniques with variable length heart beats. *Comput. Biol. Med.* **2018**, *102*, 278–287. [CrossRef] [PubMed]
17. Kaur, H.; Ahuja, S. Time series analysis and prediction of electricity consumption of health care institution using ARIMA model. In Proceedings of the Sixth International Conference on Soft Computing for Problem Solving, Patiala, India, 23–24 December 2016.
18. Paudel, S.; Elmitri, M.; Couturier, S.; Nguyen, P.H.; Kamphuis, R.; Lacarrière, B.; Le Corre, O. A relevant data selection method for energy consumption prediction of low energy building based on support vector machine. *Energy Build.* **2017**, *138*, 240–256. [CrossRef]

19. Pombeiro, H.; Santos, R.; Carreira, P.; Silva, C.; Sousa, J.M. Comparative assessment of low-complexity models to predict electricity consumption in an institutional building: Linear regression vs. fuzzy modeling vs. neural networks. *Energy Build.* **2017**, *146*, 141–151. [CrossRef]
20. Ascione, F.; Bianco, N.; De Stasio, C.; Mauro, G.M.; Vanoli, G.P. Artificial neural networks to predict energy performance and retrofit scenarios for any member of a building category: A novel approach. *Energy* **2017**, *118*, 999–1017. [CrossRef]
21. Fumo, N.; Biswas, M.R. Regression analysis for prediction of residential energy consumption. *Renew. Sustain. Ernergy Rev.* **2015**, *47*, 332–343. [CrossRef]
22. Amber, K.; Aslam, M.; Hussain, S. Electricity consumption forecasting models for administration buildings of the UK higher education sector. *Energy Build.* **2015**, *90*, 127–136. [CrossRef]
23. Chen, Y.; Xu, P.; Chu, Y.; Li, W.; Wu, Y.; Ni, L.; Bao, Y.; Wang, K. Short-term electrical load forecasting using the Support Vector Regression (SVR) model to calculate the demand response baseline for office buildings. *Appl. Energy* **2017**, *195*, 659–670. [CrossRef]
24. Bogomolov, A.; Lepri, B.; Larcher, R.; Antonelli, F.; Pianesi, F.; Pentland, A. Energy consumption prediction using people dynamics derived from cellular network data. *EPJ Data Sci.* **2016**, *5*, 13. [CrossRef]
25. Kong, W.; Dong, Z.Y.; Jia, Y.; Hill, D.J.; Xu, Y.; Zhang, Y. Short-term residential load forecasting based on LSTM recurrent neural network. *IEEE Trans. Smart Grid* **2017**, *10*, 841–851. [CrossRef]
26. Kim, T.-Y.; Cho, S.-B. Predicting Residential Energy Consumption using CNN-LSTM Neural Networks. *Energy* **2019**. [CrossRef]
27. Le, T.; Vo, M.T.; Vo, B.; Hwang, E.; Rho, S.; Baik, S.W.J.A.S. Improving electric energy consumption prediction using CNN and Bi-LSTM. *Appl. Sci.* **2019**, *9*, 4237. [CrossRef]
28. Vu, D.H.; Muttaqi, K.M.; Agalgaonkar, A. A variance inflation factor and backward elimination based robust regression model for forecasting monthly electricity demand using climatic variables. *Appl. Energy* **2015**, *140*, 385–394. [CrossRef]
29. Shi, H.; Xu, M.; Li, R. Deep learning for household load forecasting—A novel pooling deep RNN. *IEEE Transact. Smart Grid* **2017**, *9*, 5271–5280. [CrossRef]
30. Marino, D.L.; Amarasinghe, K.; Manic, M. Building energy load forecasting using deep neural networks. In Proceedings of the IECON 2016-42nd Annual Conference of the IEEE Industrial Electronics Society, Florence, Italy, 23–26 October 2016; pp. 7046–7051.
31. Orbach, J. Principles of Neurodynamics. Perceptrons and the Theory of Brain Mechanisms. *Arch. Gen. Psychiatry* **1962**, *7*, 218–219. [CrossRef]
32. LeCun, Y.; Bengio, Y. *Convolutional networks for images, speech, and time series. The handbook of brain theory and neural networks*; MIT Press: Cambridge, MA, USA, 1995.
33. Qiu, Z.; Chen, J.; Zhao, Y.; Zhu, S.; He, Y.; Zhang, C. Variety identification of single rice seed using hyperspectral imaging combined with convolutional neural network. *Appl. Sci.* **2018**, *8*, 212. [CrossRef]
34. Li, C.; Zhou, H. Enhancing the efficiency of massive online learning by integrating intelligent analysis into MOOCs with an application to education of sustainability. *Sustainability* **2018**, *10*, 468. [CrossRef]
35. An, Q.; Pan, Z.; You, H. Ship detection in Gaofen-3 SAR images based on sea clutter distribution analysis and deep convolutional neural network. *Sensors* **2018**, *18*, 334. [CrossRef]
36. Hochreiter, S.; Schmidhuber, J. Long short-term memory. *Neural Comput.* **1997**, *9*, 1735–1780. [CrossRef]
37. Sutskever, I.; Vinyals, O.; Le, Q.V. Sequence to sequence learning with neural networks. In *Proceedings of the Advances in neural information processing systems*; NIPS: San Diego, CA, USA; pp. 3104–3112.
38. UCI. Individual household electric power consumption Data Set. Available online: https://archive.ics.uci.edu/ml/datasets/individual+household+electric+power+consumption (accessed on 4 March 2020).
39. Kim, J.-Y.; Cho, S.-B. Electric energy consumption prediction by deep learning with state explainable autoencoder. *Energies* **2019**, *12*, 739. [CrossRef]
40. Almalaq, A.; Edwards, G. Comparison of Recursive and Non-Recursive ANNs in Energy Consumption Forecasting in Buildings. In Proceedings of the 2019 IEEE Green Technologies Conference (GreenTech), Lafayette, LA, USA, 3–6 April 2019; pp. 1–5.
41. Wu, D.; Wang, B.; Precup, D.; Boulet, B.J.I.T.o.S.G. *Multiple Kernel Learning based Transfer Regression for Electric Load Forecasting*; IEEE: Piscataway, NJ, USA, 2019.

42. Chujai, P.; Kerdprasop, N.; Kerdprasop, K. Time series analysis of household electric consumption with ARIMA and ARMA models. In Proceedings of the International MultiConference of Engineers and Computer Scientist, Hong Kong, China, 13–15 March 2013; pp. 295–300.
43. Rajabi, R.; Estebsari, A. Deep Learning Based Forecasting of Individual Residential Loads Using Recurrence Plots. In Proceedings of the 2019 IEEE Milan PowerTech, Milan, Italy, 23–27 June 2019; pp. 1–5.

MDPI
St. Alban-Anlage 66
4052 Basel
Switzerland
www.mdpi.com

Sensors Editorial Office
E-mail: sensors@mdpi.com
www.mdpi.com/journal/sensors

www.ingramcontent.com/pod-product-compliance
Lightning Source LLC
Chambersburg PA
CBHW061826210926
51850CB00006B/814